HOMOGENEOUS CATALYSIS

HOMOGENEOUS CATALYSIS

THE APPLICATIONS AND CHEMISTRY OF CATALYSIS BY SOLUBLE TRANSITION METAL COMPLEXES

GEORGE W. PARSHALL

Central Research & Development Department
E. I. du Pont de Nemours and Company
Experimental Station
Wilmington, Delaware

A Wiley-Interscience Publication

JOHN WILEY & SONS, New York • Chichester • Brisbane • Toronto

Library of Congress Cataloging in Publication Data

Parshall, George William, 1929–
Homogeneous catalysis.

"A Wiley-Interscience publication."
Includes index.
1. Catalysis. 2. Transition metal compounds.
3. Chemistry, Organic—Synthesis. I. Title.

QD505.P37 541.3′95 79-27696
ISBN 0-471-04552-7

Printed in the United States of America

10 9 8 7 6 5 4 3 2

PREFACE

The catalysis of organic reactions by soluble metal complexes has become a major synthesis tool, both in the laboratory and in the chemical industry. The fundamental research to support this development has become fairly sophisticated in some areas but remains primitive in others. This specialization is reflected in the books and reviews that have been written on the subject. These publications, some of which are listed at the end of Chapter 1, provide a rather unbalanced view of the field. Much attention is devoted to some academically fashionable areas, but little or no coverage is given to other catalytic reactions of great practical importance.

The object of this book is to provide a balanced description of homogeneous catalytic reactions that are useful, either in the organic synthesis laboratory or in industry. To the best of my knowledge all major industrial processes catalyzed by soluble transition metal complexes are described. In addition the more useful laboratory procedures that have found their way into *Organic Syntheses, Macromolecular Syntheses,* and *Synthesis* are reviewed. For each reaction, the description of the practical application is followed by a short review of our knowledge of the mechanism of the reaction. The review of mechanism is often distressingly brief because so little is reported on the subject.

GEORGE W. PARSHALL

Wilmington, Delaware
March 1980

CONTENTS

1. Growth of Homogeneous Catalysis **1**

2. Transition Metal Chemistry Relevant to Catalysis **5**

2.1 Bonding Capacity of Transition Metal Ions, 5
2.2 Metal-Ligand Interactions, 8

Electronic and Steric Effects, 10

2.3 Fundamental Reactions, 12

Ligand Replacement, 12
Oxidation and Reduction, 13
Oxidative Addition and Reductive Elimination, 14
Insertion and Elimination Reactions, 16

2.4 Characteristics of Catalysts, 17

3. Reactions of Olefins and Dienes: Isomerization and Hydrogenation **22**

3.1 Olefin Coordination, 22

Coordination of Dienes, 24
Bonding Schemes, 24
Delocalized Bonding, 25

3.2 Reactions of Coordinated Olefins, 27

Nucleophilic Attack, 27
Olefin Insertion Reactions, 29

3.3 Isomerization, 31

Mechanisms of Double-Bond Migration, 33
Carbon Skeletal Isomerization, 35

3.4 Hydrogenation, 36

Hydrogenation of Simple Olefins, 37
Mechanism of Olefin Hydrogenation, 39

3.5 Selective Hydrogenation of Polyenes, 41
3.6 Asymmetric Hydrogenation, 43

4. Polymerization and Addition Reactions of Olefins and Dienes **48**

4.1 Olefin Polymerization, 48

Polyethylene, 49
Polypropylene, 51
Ethylene-Propylene Copolymers, 52

4.2 Polybutadiene and Polyisoprene, 53
4.3 Oligomerization of Olefins, 56

Aluminum-Catalyzed Oligomerization of Ethylene, 56
Shell Higher Olefins Process, 58
Olefin Dimerization, 59

4.4 1,4-Hexadiene, 63
4.5 Dimerization and Trimerization of Dienes, 65

Diene Dimerization, 66
Cyclododecatriene Synthesis, 68

4.6 Additions to Olefins and Dienes, 70

Hydrocyanation, 70
Hydrosilylation, 71

5. Reactions of Carbon Monoxide **77**

5.1 Coordination of Carbon Monoxide, 77
5.2 CO Insertion Processes, 79
5.3 Acetic Acid Synthesis, 80
5.4 Carboxylation of Olefins, 82
5.5 Hydroformylation, 85

Cobalt Catalysts, 86
Rhodium Catalysts, 89

5.6 Decarbonylation, 90
5.7 Isocyanates from Nitro Compounds, 93
5.8 CO Oxidation and the Shift Reaction, 94

5.9 CO Hydrogenation, 95

Ethylene Glycol, 95
Alcohol Synthesis, 96
Vinyl Acetate Synthesis, 97

6. Oxidation of Olefins and Dienes **101**

6.1 Wacker Acetaldehyde Synthesis, 102
6.2 Acetoxylation of Olefins and Dienes, 104

Vinyl Acetate and Glycol Acetates from Ethylene, 105
Allylic Oxidation, 108
Diene Oxidation, 109

6.3 Other Palladium-Catalyzed Oxidations, 111

Substitution, 111
Oxidative Coupling of Olefins, 112

6.4 Glycol Acetate Syntheses, 113
6.5 Olefin Epoxidation by Hydroperoxides, 115

7. Arene Reactions **120**

7.1 Benzene as a Ligand, 121
7.2 Palladium-Catalyzed Reactions, 123

Arene-Olefin Coupling, 124
Arene-Arene Coupling, 126
Oxidative Substitution, 127
Oxidative Carbonylation, 128

7.3 Copper-Catalyzed Oxidations, 129

Decarboxylation, 130
Phenol Coupling, 132

7.4 Coupling Reactions of Aryl Halides, 135

Stoichiometric Coupling, 135
Catalytic Coupling, 137

7.5 Arene Hydrogenation, 139

Cobalt Carbonyl Systems, 139
Ziegler Systems, 140
Allylcobalt Catalysts, 141
Ruthenium Complexes, 143

8. Acetylene Reactions **147**

8.1 Coordination Chemistry of Acetylenes, 148

8.2 Acetylide-Catalyzed Reactions, 150

Oxidative Coupling, 150
Addition to Aldehydes and Ketones, 151
Chloroprene Synthesis, 152

8.3 Additions to Acetylenes, 155

Acetaldehyde Synthesis, 155
Vinyl Acetate Synthesis, 156
Vinyl Chloride Synthesis, 157
Hydrocyanation, 157
Chlorination, 158

8.4 Acetylene-CO Reactions, 159

Acrylate Synthesis, 159
Bifurandione Synthesis, 160
Hydroquinone Synthesis, 161

8.5 Oligomerization, 163

Cyclooctatetraene Synthesis, 164
Trimerization, 165
Cotrimerization Reactions, 168

9. Carbene Complexes in Olefin Metathesis and Alkane Reactions **171**

9.1 Alpha-Hydrogen Elimination, 172
9.2 Olefin Metathesis Processes, 173
9.3 Olefin Metathesis Mechanisms, 176
9.4 Alkane Reactions, 179

H/D Exchange, 179
Hydrogenolysis and Isomerization, 180

10. Oxidation of Hydrocarbons by Oxygen **185**

10.1 Reactions of O_2 with Metal Complexes, 186
10.2 Reaction of O_2 with Hydrocarbons, 188
10.3 Adipic Acid Synthesis, 190

Cyclohexane Oxidation, 190
Oxidation of Cyclohexanol and Cyclohexanone, 193

10.4 Oxidation of Cyclododecane, 194
10.5 Acetic Acid Synthesis, 195

Butane Oxidation, 196
Acetaldehyde Oxidation, 198

10.6 Oxidation of Methylbenzenes, 200

Toluene Oxidation, 201
Xylene Oxidation, 201
Other Substrates, 202
Mechanism, 202

11. Condensation Polymerization and Related Processes **208**

11.1 Polyester Synthesis, 208

Transesterification, 209
Direct Esterification, 212

11.2 Polyurethane Synthesis, 214
11.3 Polyamides and Intermediates, 216

Dichlorobutene Isomerization, 218
Dicyanobutene Synthesis, 219

12. Trends in Homogeneous Catalysis **222**

12.1 New Feedstocks, 222
12.2 Enhanced Selectivity, 225
12.3 Hybrid Catalysts, 227
12.4 New Catalyst Chemistry, 229

Photoactivated Catalysts, 229
Metal Cluster Catalysts, 229

Index **233**

HOMOGENEOUS CATALYSIS

1 GROWTH OF HOMOGENEOUS CATALYSIS

Soluble metal complexes, especially those of the transition metals, are used extensively in industry to catalyze syntheses of organic compounds. In 1977 roughly 9 million tons of organic chemicals were produced in the United States by homogeneous catalytic processes (Table 1.1). Some of these processes involved more than one catalytic step. To keep perspective on the importance of these processes, however, one should recall the much greater volume of production with heterogeneous catalysts. Ammonia synthesis alone roughly equaled the combined output of all homogeneous catalytic processes.

The growth of homogeneous catalysis is even more impressive than the scale of production. Although soluble metal salts were used commercially as catalysts for acetylene reactions as early as 1910, broad use of soluble catalysts did not begin until the 1940s. Several factors combined to create a favorable climate for development of new processes. In Germany wartime restrictions on raw materials led to new CO-based processes for production of fuels and plastics. More broadly, the introduction of many new polyester, polyamide, and vinyl polymers between 1940 and 1960 called for worldwide development of new processes for production of the monomers.

The impetus for new process technology came during a period of rapid development in the organic chemistry of the transition metals. Fischer, Wilkinson, Ziegler, and Natta were doing the research that won them Nobel prizes in chemistry. Their research provided the basis for homogeneous catalysis as we know it today. Industry has built on their fundamental discoveries to create about two dozen major processes that are catalyzed by soluble transition metal complexes. This development has continued to the present. During the 1½-year period in which this book was written, three major new homogeneous or organometallic catalytic processes were announced.

The major virtue of homogeneous catalysis that has led to its widespread adoption by industry is selectivity, the ability to produce pure products in high yield.

Table 1.1 Homogeneous Catalysis in the U.S. Chemical Industry

Chapter	Reactions and Products	Approximate 1977 Production[a] (thousands of metric tons)
5	*Carbonylations*	
	Oxo alcohols (hydroformylation)	780[b]
	Acetic acid (from methanol)	180[b]
4	*Olefin polymerization*	
	Polybutadiene (coordination catalysis)	345
	Ethylene polymers (solution processes)	500[b]
4,11	*Olefin additions*	
	α-Olefins	177
	Adiponitrile	200[b]
	Diene dimers and trimers	40[b]
	Chloroprene	166
6	*Olefin oxidation*	
	Acetaldehyde	410[b]
	Propylene oxide (Oxirane process)	418[b]
10	*Alkane and arene oxidation*	
	Terephthalic acid and esters	2277
	Adipic acid	698
	Acetic acid	986[b]
	Benzoic, isophthalic acids	86[b]
11	*Condensation polymerization*	
	Polyester fiber	1660[c]
Total		8923

[a] From "Synthetic Organic Chemicals—1977," U.S. International Trade Commission, unless shown otherwise.
[b] Amount estimated to arise from homogeneous catalytic processes.
[c] *Chem. Eng. News,* December 4, 1978.

This characteristic is very important in the preparation of pharmaceuticals and polymer intermediates that must be extremely pure. Many processes that illustrate this selectivity are described in the succeeding chapters. Perhaps the most impressive is the hydrogenation of prochiral olefins to produce single optical isomers in more than 90% yield. This latter process is used commercially to produce L-dopa which is used in the treatment of Parkinson's disease (Section 3.4).

Much of the selectivity observed with soluble catalysts originates in the process control that is attainable in the liquid phase. Not only are temperature and mixing better controlled than in heterogeneous systems, but also the nature of the active catalytic species is regulated more effectively. Control of catalyst and ligand concentration is better than that attainable on the surface of a solid. As a result a soluble catalyst is apt to be homogeneous in the sense that only a few kinds of catalytic species are present, as well as in the traditional sense that it operates in a single phase.

The application of homogeneous catalysis has been accompanied by growth in our understanding of the chemistry involved. Mechanistic studies have usually followed commercial application, but the information from these studies has been valuable in the optimization of reaction conditions and in development of new catalysts. Studies of the mechanisms of homogeneous catalytic reactions have advanced rapidly because most of the techniques of physical organic chemistry can be used with little modification. Many of the mechanistic principles developed with soluble catalysts also apply to heterogeneous catalysts for which much less mechanistic information is available.

In the present volume an attempt is made to present a balanced picture of this rapidly growing field. The chapter on the organic chemistry of the transition metals is brief because ample treatment of this topic can be found elsewhere. More attention is devoted to applications, reaction chemistry, and mechanistic studies of technologically significant homogeneous catalytic reactions. The reader may note that some industrially important processes have received little mechanistic study in the past. One hopes that some fundamental research will be done on these significant but unfashionable reactions.

The attempt to maintain perspective on the entire field limits the amount of attention that can be devoted to each topic. The reader who wishes more information on a particular reaction will be helped by the extensive selection of edited volumes that have appeared in the last 5–10 years. These books contain chapters on selected topics in catalysis written by specialists. In general they provide excellent in-depth coverage of the chosen topics. The catalysis of hydrogenation, polymerization, and carbon monoxide reactions occupy complete volumes that reflect the amount of effort applied to these fields.

GENERAL REFERENCES

Wender, I., and P. Pino, *Organic Synthesis with Metal Carbonyls*, 1st ed., Wiley, 1968; 2nd ed., 1977, provide expert coverage of a broad variety of topics in homogeneous catalysis, not just metal carbonyl reactions.

Taqui Khan, M. M., and A. E. Martell, *Homogeneous Catalysis by Metal Complexes*, Academic Press, 1974, 2 vols., discuss molecular activation reactions and basic principles of catalysis in considerable detail.

Schrauzer, G. N., *Transition Metals in Homogeneous Catalysis*, Marcel Dekker, 1971, provides a collection of chapters on fundamental processes by several authors.

Ugo, R., *Aspects of Homogeneous Catalysis*, D. Reidel, offers a continuing series of volumes that contain chapters on many aspects of catalysis by workers in the field.

Stone, F. G. A., and R. West, *Advances in Organometallic Chemistry*, Vol. 17, *Catalysis and Organic Syntheses*, Academic Press, 1979, present chapters by authoritative authors on many of the homogeneous catalytic reactions of olefins and dienes.

Homogeneous Catalysis, Advances in Chemistry Series **70** and **132**, American Chemical Society, 1968, 1974, are collections of symposia papers with a strong industrial flavor.

Rylander, P. N., *Organic Syntheses with Noble Metal Catalysts,* Academic Press, 1973, provides good descriptions of reactions, both stoichiometric and catalytic, of interest in organic synthesis.

Several series publications contain chapters on homogeneous catalysis. Some prominent examples are:

Advances in Catalysis, Academic Press.
Advances in Organometalic Chemistry, Academic Press.
Organometallic Reactions, Wiley-Interscience.
Catalysis—Specialist Periodical Reports, Chemical Society.
Catalysis Reviews, Marcel Dekker.

Books on Special Topics

B. R. James, *Homogeneous Hydrogenation,* Wiley-Interscience, 1973.
F. J. McQuillin, *Homogeneous Hydrogenation in Organic Chemistry,* D. Reidel, 1976.
J. Falbe, *Carbon Monoxide in Organic Synthesis,* Springer-Verlag, 1970.
J. C. W. Chien, *Coordination Polymerization,* Academic Press, 1975.
C. E. Schildknecht and I. Skeist, *Polymerization Processes,* Wiley-Interscience, 1977.
J. Boor, *Ziegler-Natta Catalysts and Polymerization,* Academic Press, 1978.

2 TRANSITION METAL CHEMISTRY RELEVANT TO CATALYSIS

The individual steps of homogeneous catalytic reactions are usually reactions that are well known from organometallic and coordination chemistry. Typical steps include ligand complexation and dissociation, oxidative addition and reductive elimination, "insertion" of unsaturated ligands into M—C and M—H bonds, and abstraction of H atoms from alkyl groups [1,2]. Such reactions and the bonding principles on which they are based are described here in elementary fashion. Much more comprehensive and rigorous treatments are available in the standard texts cited at the end of this chapter.

2.1 BONDING CAPACITY OF TRANSITION METAL IONS

Homogeneous catalysis in the context of this volume concerns syntheses that are catalyzed by soluble complexes of the transition metals (members of the Ti, V, Cr, Mn, Fe, Co, Ni, and Cu triads). The transition metals possess energetically accessible *d* orbitals that are only partially filled with electrons (in at least one normal oxidation state). Both the number and the shape of these orbitals contribute to the extraordinary range of reaction pathways exhibited by transition metal compounds.

The typical transition metal atom has one *s*, three *p*, and five *d* orbitals that possess geometrical and energetic characteristics suitable for bonding. In selected cases these nine orbitals permit the formation of bonds to nine ligands. A striking example is the hydride complex $ReH_7(PEt_2Ph)_2$ [3], which contains seven covalent Re—H bonds and two P → Re coordinate bonds. The coordinate bonds are formed by donation of an electron pair from a phosphine ligand to a vacant orbital of the rhenium atom. In principle the Re—H bonds are similar to the C—H bonds in methane. The seven valence electrons of rhenium are deployed in seven orbitals in which they pair with the lone electrons of the seven hydrogen atoms. Although this compound is exceptional in its high coordination number, it illustrates the principle that a metal ion binds ligands in both covalent and coordinate modes. Much of the "magic" of catalysis lies in this versatility. Coordination to a metal ion induces changes in the electron distribution in a complex ligand such as carbon monoxide or an olefin. These changes modify the reactivity of the ligand molecule, sometimes dramatically.

Simple rules have been devised to predict the existence and stability of coordination compounds of the transition metals. The observation that the nine outer orbitals of a transition metal can accommodate 18 electrons gave rise to the "18-electron rule" or "inert gas rule" for prediction of the stability of complexes. As the rule was originally applied to metal carbonyl compounds [4], a compound would be stable when the sum of the valence electrons from the metal atom and those donated by the ligands (two electrons from each carbon monoxide) equaled that of an inert gas atom. For example, in $Cr(CO)_6$, the chromium provides 6 electrons, and the carbon monoxide ligands supply 12 for a total of 18. If one adds the 18 core electrons of the chromium atom, the metal is surrounded by 36 electrons, a stable configuration like that in krypton. For convenience one usually omits the core electrons from the calculation and simply counts the electrons that occupy the outer orbitals of the metal. A total of 18 electrons from the metal valence orbitals and from the donor ligands forecasts stability. This simple empirical rule successfully predicts the stability of the known mononuclear and dinuclear metal carbonyls except $V(CO)_6$, a paramagnetic monomer with an electron count of 17. Interestingly, this compound reduces easily to $[V(CO)_6]^-$, a stable ion with 18 electrons in its potentially bonding orbitals.

The 18-electron rule is broadly applicable to prediction of the stability of organometallic compounds. Rather elaborate electron counting rules have been developed [1], but for most complexes, the following system works well. Simple covalently bonded ligands such as chloro, methyl, and hydrido are considered to contribute one electron to a bond with the metal atom. Electron pair donors like amines, CO, phosphines, isonitriles, and olefins contribute two electrons. Other common ligands are assigned the following electron donor capabilities:

3—π-allyl, nitrosyl (usually)

4—dienes

5—π-cyclopentadienyl

6—π-arene

As with the simple metal carbonyls an electron count of 18 (sum of metal valence electrons and ligand electrons) leads one to expect a moderate degree of stability, at least toward thermal degradation and nucleophilic attack. The formal charge on an ionic complex such as $[V(CO)_6]^-$ is subtracted from the total electron count. Counterions (e.g., NH_4^+ or PF_6^-) that are not covalently or coordinately bonded to the metal atom are omitted from the calculation.

As an example of electron counting in a complex with several types of ligands, consider the versatile hydroformylation catalyst $Rh(CO)(H)(PPh_3)_3$. The rhodium atom possesses 9 valence electrons. The carbon monoxide molecule contributes 2 electrons, the hydrogen atom 1, and each of the three triphenylphosphine ligands supplies 2 electrons for a total of 18, the magic number for stability. The prediction of stability seems vindicated by the observation that the compound melts without change at 172–174°C under nitrogen. The predicted stability is limited to resistance to reactions at the metal center. The rhodium hydride, like many organometallic compounds, is vulnerable to electrophiles and to free radicals that can attack a ligand such as H directly.

Although the 18-electron rule is empirical in origin, its success can be explained qualitatively. If the nine potentially bonding metal orbitals are filled, the compound is said to be *coordinatively saturated.* Attack on the metal atom by a nucleophile would add electrons to a high-energy metal atomic orbital or to an antibonding molecular orbital and is unfavorable. Hence, ligand substitution reactions usually proceed by a mechanism like the S_N1 process in organic chemistry. Generally, one ligand dissociates to create a "coordination-deficient" complex that will readily bind other potential ligands.

The generality of the ligand dissociation-association process had led to formulation of a "16- and 18-electron rule" [1] for *reactivity* of transition metal complexes. This rule has been very useful in discernment of preferred reaction pathways in homogeneous catalysis. It has many exceptions and is largely limited to *non-radical* reactions of the Group VI—VIII metals, but it provides a useful conceptual framework for the chemistry of Chapters 3 to 6.

Basically, the 16- and 18-electron rule forecasts that a catalytic reaction will proceed by a series of ligand dissociation and association steps in which intermediates with 16 and 18 electrons alternate. A simple example in a stoichiometric reaction is the replacement of two carbonyl groups in $Cr(CO)_6$ by triphenylphosphine ligands:

$$\underset{\mathbf{18e}}{Cr(CO)_6} \xrightarrow[\text{slow}]{-CO} \underset{\mathbf{16e}}{Cr(CO)_5} \xrightarrow{PPh_3} \underset{\mathbf{18e}}{Cr(CO)_5(PPh_3)} \xrightarrow{-CO}$$

$$\underset{\mathbf{16e}}{Cr(CO)_4(PPh_3)} \xrightarrow{PPh_3} \underset{\mathbf{18e}}{Cr(CO)_4(PPh_3)_2}$$

The initial $Cr(CO)_6$ molecule, as forecast by the 18-electron rule, is fairly resistant to S_N2 nucleophilic attack. Here, the rate of substitution is only slightly dependent on triphenylphosphine concentration [5]. (The analogous hexacarbonyl complexes of Mo and W are more susceptible to nucleophilic attack by an S_N2-type mechanism.) The major pathway for substitution is dissociation of a CO ligand to give a reactive $Cr(CO)_5$ moiety. This 16-electron species rapidly binds a triphenylphosphine molecule to form a stable 18-electron compound. Further dissociation of CO to give a 16-electron, five-coordinate $Cr(CO)_4(PPh_3)$ species is even slower, but it is a major reaction path for formation of the final bis(triphenylphosphine) complex. The CO dissociation processes have high activation energies, but dissociation can be promoted very conveniently by photolysis [6]. The postulated $Cr(CO)_5$ intermediate has been characterized spectroscopically in noble gas matrices [7].

Similar alternation of 16- and 18-electron intermediates also occurs in many catalytic reactions, as illustrated in the discussion of the cyclic nature of catalysis in Section 2.4. As already mentioned however, there are many exceptions to the 16- and 18-electron rule. A conspicuous class of exceptions are the square planar complexes of rhodium(I), platinum(II) and gold(III). These 16-electron complexes show considerable kinetic stability and often react by dissociation to give labile 14-electron complexes. An outstanding example is "Wilkinson's catalyst," which is perhaps the most versatile soluble catalyst. Its structure [8] contains a planar array of four ligand donor atoms about the central rhodium atom:

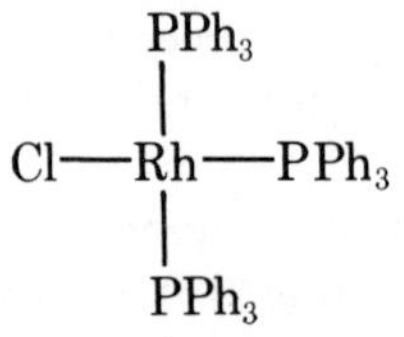

```
        PPh3
         |
   Cl—Rh—PPh3
         |
        PPh3
```

The Rh—Cl link can be viewed as a normal covalent bond, while the three Rh—P bonds are formed by donation of electron pairs from phosphorus to rhodium. A major pathway for reactions of this compound involves dissociation of a triphenylphosphine ligand to give a highly reactive 14-electron intermediate [9]. The role of this intermediate in hydrogenation of olefins is discussed in Section 3.4.

2.2 METAL-LIGAND INTERACTIONS

The utility of transition metal complexes in catalysis lies not only in the abundance of *d* orbitals but in their symmetry as well. As illustrated in Figure 2.1, simple *d* orbitals can lie either on or between atomic axes. Hybrids of the five *d* orbitals with the outer *p* and *s* orbitals can assume a great variety of shapes and orientations. Covalent bonds between the metal atom and ligands such as alkyl groups or hydrogen or halogen atoms are qualitatively like those in organic compounds. The most important differences lie in the coordinate bonds between ligand molecular orbitals and metal atomic orbitals. The symmetries of the orbitals are very important to the stability of the coordinate bond. This point is illustrated by the interaction between a transition metal atom and the dinitrogen molecule [10].

The N_2 molecule is inert to most reagents, but it forms surprisingly stable complexes with electron-rich transition metal compounds. In doing so, it is sometimes activated toward reducing agents. The ultimate example is the metalloprotein nitrogenase, which catalyzes the reduction of N_2 to NH_3 at soil temperature and atmospheric pressure. The strength of the coordinate bond to transition metals such as iron or molybdenum (the two metals in nitrogenase) results from a combination

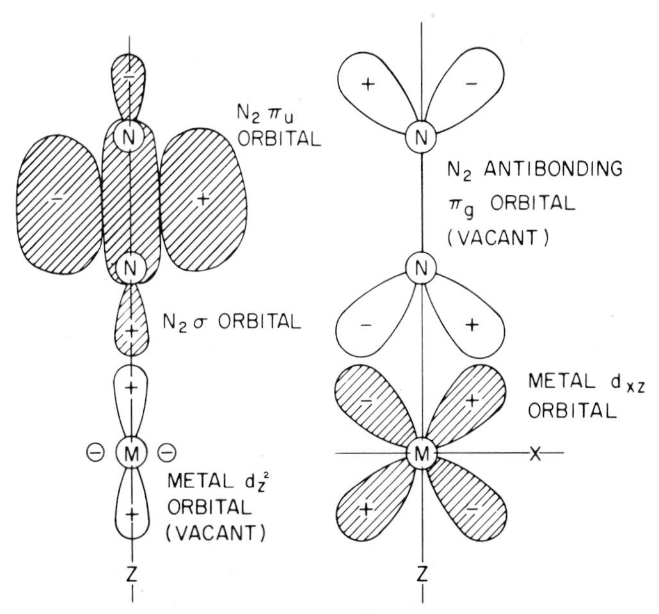

Figure 2.1 Cross sections in the xz plane of transition metal d_{z^2} and d_{xz} orbitals and of N_2 molecular orbitals. An identical set of N_2 π-orbitals occurs in the yz plane along with a metal d_{yz} orbital that has the same shape as the d_{xz} orbital.

of two bond types. Even though N_2 is a very weak base that does not interact with protonic acids or Lewis acids such as PF_5, it appears to act as an electron pair donor to a vacant metal orbital. As illustrated in the left portion of Fig. 2.1, a filled *sigma* orbital localized on one nitrogen atom can donate to a vacant metal orbital. (The d_{z^2} orbital is shown although a p_z orbital would function equally well.) This transfer of electron density to the metal, even though slight, enhances a second kind of interaction illustrated in the right side of the figure. A filled metal d orbital (xz) has the proper symmetry to donate electrons to the vacant π_g orbital of the N_2 molecule. This interaction is significant, even though N_2 does not react with powerful conventional nucleophiles like CH_3^- or RS^-. These two electron transfer interactions are synergistic and produce an M—N bond of moderate strength, even though each is very weak in the absence of the other.

The combination of σ-donor and π-acceptor bonds like that shown in Figure 2.1 stabilizes complexes of many ligands that are isoelectronic with the N_2 molecule. The most important from a practical viewpoint is carbon monoxide, which forms a large family of metal carbonyl complexes. The nature of the M—CO bond is discussed in Chapter 5, along with the applications of carbon monoxide in synthesis. Other ligands that can bond to metals as N_2 does in Figure 2.1 are the RN≡C and RC≡N molecules and the CN^-, N_3^-, and $RC{\equiv}C^-$ ions.

Interestingly, the transfer of electron density from the metal ion to the antibonding π_g orbital of the ligand seems to weaken the N≡N or C≡O bond only slightly. Although the infrared stretching frequencies of the ligand may shift

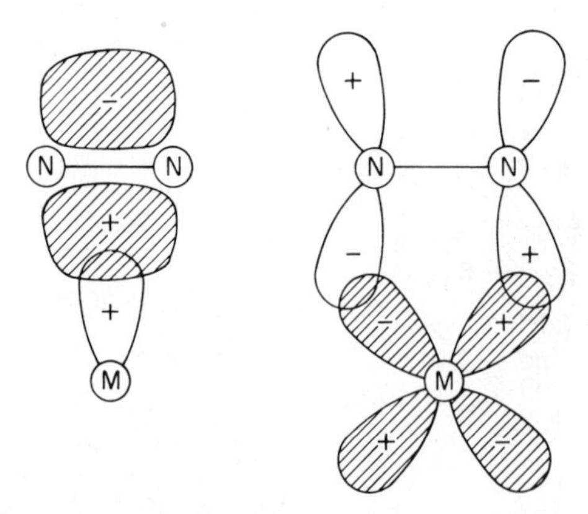

Figure 2.2 "Side-on" interaction of a dinitrogen molecule with *d* orbitals of a transition metal ion. *Left:* Donation from a π_u-bonding orbital of N_2 to a vacant metal orbital. *Right:* Back-donation from a filled *d* orbital to a π_g-antibonding orbital of the ligand.

substantially, the N≡N or C≡O bond length changes little on complexation. The most important effect of "end-on" coordination from a catalytic viewpoint may be to provide a facile pathway for electron transfer from a low valent metal ion to the ligand, as in the reduction of N_2 or CO by Ti^{2+} or Zr^{2+} complexes.

The dinitrogen molecule can also bond to a metal ion in a "side-on" fashion in contrast to the end-on approach of Figure 2.1. Although all well-characterized complexes of N_2 have the end-on structure [11], the side-on structure has been suggested repeatedly for the transition state in metal-catalyzed reactions of dinitrogen. The side-on bond, illustrated in Figure 2.2, is analogous to that observed in olefin and acetylene complexes. The bonding of these important catalytic substrates is discussed in Chapters 3 and 8.

In the side-on structure, the ligand-to-metal bond involves donation from one lobe of a filled π orbital to a vacant metal orbital. Again, as in the end-on mode, this donation increases the electron density on the metal ion and facilitates electron transfer to a π-antibonding orbital of the ligand. In both modes of bonding, the metal *d* orbitals have the right symmetry for donation to the antibonding π_g-ligand orbitals. Although no structural data are available for side-on N_2 complexes, olefinic and acetylenic C—C bonds may lengthen substantially when the ligand coordinates to a metal ion. This weakening of the C—C bond would be expected if the antibonding π_g orbitals are populated to a significant extent.

Electronic and Steric Effects

The effectiveness of a metal complex as a catalyst is strongly dependent on the interactions among the ligands arrayed about the central metal ion. A classic example is the "*trans*-effect" or "*trans*-influence" [12] by which one ligand labilizes another that is *trans* to it in a square planar or octahedral complex. For example, in Wilkinson's catalyst, $RhCl(PPh_3)_3$, the Rh—P distance *trans* to a triphenylphosphine ligand is 0.1 Å longer than that *trans* to the chloro ligand [8]. The interactions between ligands occur by two general mechanisms. Electronic effects are transmitted via metal-ligand bonds. Steric effects are largely interligand repulsions in sterically crowded complexes. The electronic effect is evident in the Rh—P bond lengthening in Wilkinson's catalyst. However, steric effects [13] are also evident in this compound, since the four ligands are twisted out of planarity by the bulk of the triphenylphosphine ligands.

The electronic characteristics of catalysts for olefin and diene reactions generally differ greatly from those of complexes that catalyze oxidation reactions. Although there are several exceptions, the catalysts for hydrogenation, dimerization, and carbonylation of olefins tend to be low valent complexes that are stabilized by "soft" or polarizable ligands. This class of ligands is typified by CO, N_2, phosphines, and the larger halide ions. Oxidation catalysts are usually high valent ions surrounded by "hard," nonpolarizable ligands such as water, alcohols, amines, hydroxide, or carboxylate ions. Ligands of this latter class attach to the metal ion by a simple σ-donor bond, often quite ionic. The "soft" ligands frequently bind by a combination of σ-donor and π-acceptor interactions, as illustrated by the dinitrogen example presented earlier.

Even within a series of ligands such as PX_3, great differences in bonding characteristics are observed. Trialkylphosphines are primarily σ-donor ligands that transfer a great deal of electron density to a metal ion. At the opposite extreme, PF_3 is a strong electron acceptor, much like CO. Intermediate characteristics are seen with triarylphosphine and $P(OR)_3$ ligands. These effects are cumulative when several similar ligands are attached to a metal ion as in $Ni(PX_3)_4$. The effect is illustrated in the reaction of protonic acids with this class of complexes:

$$Ni(PX_3)_4 + H^+ \rightleftharpoons [HNi(PX_3)_4]^+$$

(The protonation reaction is an important way to generate the M—H bond, as will be described in Section 2.4.) When the ligand is a trialkylphosphine, so much electron density accumulates on the nickel atom that it becomes a strong base. Even acids as weak as ethanol protonate $Ni(PEt_3)_4$ [14]. In contrast $Ni(PF_3)_4$ shows no significant basicity. Trialkyl and triaryl phosphite complexes are protonated by various mineral acids, consistent with an intermediate electron density on the metal [15].

Steric effects are less easy to quantify experimentally but are just as important to the function of a soluble catalyst as are the electronic effects. One useful approach to determination of the effective bulk of a ligand is to measure the "cone angle" subtended by the ligand in a space-filling molecular model of a complex [13] (Figure 2.3). The magnitude of the cone angle is an approximate measure of the space occupied by the ligand in the coordination sphere of the metal. Thus, four $P(OMe)_3$ ligands (cone angle 107°) fit comfortably in a tetrahedral array in $Ni[P(OMe)_3]_4$ while tri-*o*-tolyl phosphite, with a cone angle of 141°, must distort considerably to form an analogous complex. In fact, $Ni[P(O\text{-}o\text{-}Tol)_3]_4$ dissociates spontaneously at room temperature to form $Ni[P(O\text{-}o\text{-}Tol)_3]_3$. The catalytic effectiveness of this complex is largely attributable to the ease with which the coordinatively unsaturated

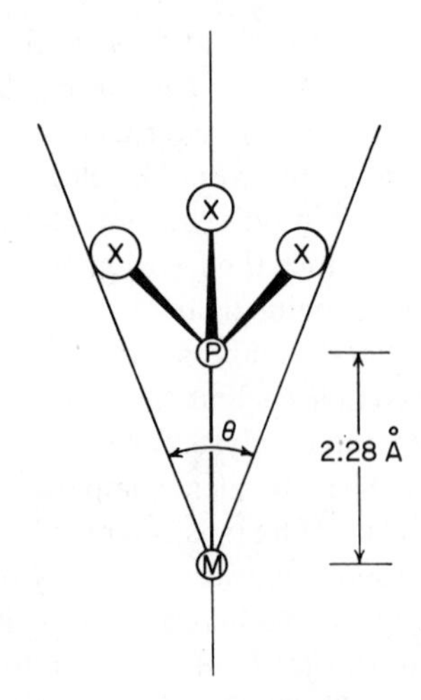

Figure 2.3 Method of measuring the cone angle θ of a ligand PX_3 coordinated to a metal atom. Reprinted with permission from C. A. Tolman, J. Am. Chem. Soc. *92,* 2957 (1970). Copyright by the American Chemical Society.

tris(phosphite) complex forms. Similarly, the major catalytic pathway in reactions of Wilkinson's catalyst, $RhCl(PPh_3)_3$, involves dissociation of a triphenylphosphine ligand, a reaction promoted by the bulk of the phosphine (cone angle 145°).

2.3 FUNDAMENTAL REACTIONS

The major reaction pathways in homogeneous catalysis are simply the fundamental reactions of coordination chemistry and organometallic chemistry. These reactions appear in many combinations and sequences both in the catalytic cycle and in the catalyst transformations that provide access to the cycle. The individual reactions are discussed in the following sections. Even these basic reactions are not thoroughly understood from a mechanistic viewpoint. It has become clear that processes such as ligand replacement, insertion, and oxidative addition are much more complex than was believed 10 years ago.

Ligand Replacement

Even the simplest of transition metal reactions, the replacement of one two-electron donor ligand by another, can occur by several different mechanisms. The simplest mechanism was illustrated in the replacement of one CO in $Cr(CO)_6$ by a triphenylphosphine molecule. In this instance the major pathway displays first-order kinetics like the S_N1 process in organic chemistry. A carbon monoxide dissociates from the metal to create a vacant coordination site to be filled by the phosphine ligand. When, however, one considers the analogous substitution reactions of $Mo(CO)_6$ and $W(CO)_6$, another pathway involving second-order kinetics becomes significant [5]. As in the S_N2 process of organic chemistry, it seems likely that the entering phosphine ligand actively displaces carbon monoxide from the metal. A nominally seven-coordinate transition state is probably involved. The contribution of the S_N2-like mechanism to the Mo and W reactions may reflect the slightly larger size of these atoms compared to chromium. The expanse of the outermost *d*-orbitals in the heavier transition metals may ease the formation of intermediates with high coordination numbers.

Another, more complex ligand substitution process has been recognized recently. This process, which involves paramagnetic (odd electron count) intermediates, appears to operate in the substitution of CO in binuclear carbonyl complexes such as $Mn_2(CO)_{10}$ and $Co_2(CO)_8$ [16]. It may also operate with mononuclear complexes such as $HCo(CO)_4$ and with polynuclear clusters such as $Co_4(CO)_{12}$. In the case of the important catalyst precursor $Co_2(CO)_8$, the following scheme is probably significant:

$$Co_2(CO)_8 \underset{\Delta}{\overset{h\nu \text{ or}}{\rightleftharpoons}} 2 \cdot Co(CO)_4$$

$$\cdot Co(CO)_4 + R_3P \rightarrow CO + \cdot Co(CO)_3(PR_3)$$

$$\cdot Co(CO)_4 + \cdot Co(CO)_3(PR_3) \rightarrow Co_2(CO)_7(PR_3)$$

The rate-limiting step is dissociation of the binuclear $Co_2(CO)_8$, which itself is sluggish in S_N1 or S_N2 substitution processes. The paramagnetic $\cdot Co(CO)_4$ molecule reacts rapidly with the incoming phosphine ligand. Radical recombination reactions of $\cdot Co(CO)_4$ and $\cdot Co(CO)_3(PR_3)$ produce the observed substitution products, $Co_2(CO)_7(PR_3)$ and $Co_2(CO)_6(PR_3)_2$. Evidence has been presented for a similar mechanism for substitution of the CO ligands of $HRe(CO)_5$ by triphenylphosphine [17].

An extremely important unanswered question is why the paramagnetic $\cdot Co(CO)_4$ is more labile than the analogous diamagnetic $Ni(CO)_4$ molecule. The significance of such odd-electron intermediates in ligand substitution, oxidative addition, and insertion reactions is just beginning to be appreciated.

Oxidation and Reduction

Many important homogeneous catalytic reactions are oxidations of organic substrates catalyzed by soluble metal complexes. In these oxidations the metal ion commonly cycles between two relatively stable oxidation states such as Co(II) and Co(III). Both one-electron and two-electron redox cycles are significant. Some especially important one-electron cycles are Co(II)/Co(III) and Mn(II)/Mn(III) and Cu(I)/Cu(II). The Pd(0)/Pd(II) two-electron couple is involved in many oxidations of olefins. Practical examples of such oxidation cycles appear in Chapters 6 and 10.

Although some one-electron oxidations involve simple transfer of an electron between a metal ion and an organic substrate, transfer of an electron *and* a ligand is prevalent in both one- and two-electron processes. An example selected from an important catalytic process is the reaction of an alkyl hydroperoxide with a cobalt(II) ion (solvation omitted):

$$ROOH + Co^{2+} \rightarrow RO\cdot + [Co^{III}\text{-}OH]^{2+}$$

This reaction effects transfer of an OH radical, complete with one electron from the O—O bond, to the cobalt ion. The product is a complex of Co(III) with an OH^- ion, at least in a formal sense.

The simple electron transfer reaction is illustrated by the process in which Cu(II) is reduced to Cu(I) by reaction with an acetyl radical:

$$CH_3CO\cdot + Cu^{2+} \rightarrow [CH_3CO]^+ + Cu^+$$

Both the ligand transfer and electron transfer processes are important in hydrocarbon oxidation reactions. Both processes are involved in the catalytic oxidation of acetaldehyde to acetic anhydride (Chapter 10).

Oxidative Addition and Reductive Elimination

In addition to the conventional oxidation and reduction reactions discussed earlier, the organometallic chemist also likes to speak of oxidative addition and its inverse process, reductive elimination. The oxidative addition concept [1,18] has been a powerful tool for understanding homogeneous catalytic processes. In recent years, however, it has become apparent that oxidative addition is a label for a family of reactions rather than a precise mechanistic description.

A convenient example of oxidative addition is the reaction of an alkyl halide with a zerovalent platinum complex [19,20]:

$$R\text{—}X + Pt(PEt_3)_3 \xrightarrow[\text{addition}]{\text{oxidative}} RPtX(PEt_3)_3 \xrightarrow[\text{dissociation}]{\text{ligand}} \text{trans-}R\text{—}Pt(PEt_3)_2\text{—}X + PEt_3$$

The initial platinum complex $Pt(PEt_3)_3$ can be regarded as a platinum atom solvated by the phosphine ligands. The addition of the alkyl halide oxidizes the platinum from a formal oxidation state of 0 to +2, if R and X are viewed as anionic ligands in the context of ligand field theory. This formal oxidation and the addition of the elements of RX is similar to the familiar quaternization of a tertiary amine or phosphine by an alkyl halide. The analogy to quaternization is quite appropriate if RX is benzyl chloride because, in this instance, the zerovalent platinum complex appears to react as a nucleophile that displaces Cl^- from the benzylic carbon in an S_N2 reaction [20]. In contrast to quaternization, however, the chloride ion subsequently bonds to the metal.

The complexity of organometallic reaction mechanisms is illustrated by the observation that benzyl bromide reacts with $Pt(PEt_3)_3$ by a different mechanism from that of the chloride. Several kinds of evidence suggest that oxidative addition of the bromide proceeds by a nonchain radical mechanism [21]. This radical process may be viewed as two sequential one-electron oxidations (phosphine ligands omitted):

$$RX + Pt^{0} \rightarrow R\cdot + Pt^{I}X \rightarrow RPt^{II}X$$

With other alkyl halides it seems likely that radical chain mechanisms are involved in the reaction with $Pt(PEt_3)_3$. Despite extensive study [20,22], the overall mechanistic picture of oxidative addition is unclear.

Many kinds of reactive X—Y bonds will "oxidatively add" to $Pt(PEt_3)_3$ and other electron-rich transition metal complexes. Some of these additions, such as the reaction of a Cl_2 molecule with Wilkinson's catalyst:

$$Rh^{I}Cl(PPh_3)_3 + Cl_2 \rightarrow Rh^{III}Cl_3(PPh_3)_3$$

are easy to view as oxidation processes. Other common additions, such as that of an HCl molecule or of a C—H bond of an arene, are less acceptable intuitively. Perhaps the most difficult to visualize as an oxidation is the addition of an H_2 molecule to form a dihydride, an important step in catalysis of olefin hydrogenation [9]:

$$Rh^{I}Cl(PPh_3)_3 + H_2 \rightleftharpoons Rh^{III}ClH_2(PPh_3)_3$$

As in the addition of an alkyl halide, the "oxidative" character of the addition of H_2 is a formalism. It is based on the definition of the hydrogen ligands in the product as hydride ions, even though the Rh—H bond is probably nonpolar in fact.

As indicated in the equation, the addition of hydrogen to Wilkinson's catalyst is reversible. On heating $RhClH_2(PPh_3)_3$ under vacuum, hydrogen is eliminated and the rhodium atom in the complex is "reduced" to its original oxidation state of +1.

The *reductive elimination* reaction has received much less study than oxidative addition, even though it is a key step in some homogeneous catalytic reactions. An important example is the reductive elimination of alkane from the alkylmetal hydrides, which are intermediates in olefin hydrogenation. In the hydrogenation of ethylene with Wilkinson's catalyst, an ethylrhodium hydride reductively eliminates ethane in the final step in the catalytic cycle:

$$(C_2H_5)(H)Rh^{III}Cl(PPh_3)_3 \longrightarrow C_2H_6 + RhCl(PPh_3)_3$$

It is often assumed that the elimination is a concerted process in which two *cis* ligands bump together and form a C—H bond. In view of the complexity of the oxidative addition process, however, it seems unlikely that reductive elimination is always so simple.

Insertion and Elimination Reactions

The so-called insertion reaction is ubiquitous in the catalytic chemistry of olefins. A prime example is *insertion* of an olefin into a metal-hydrogen bond:

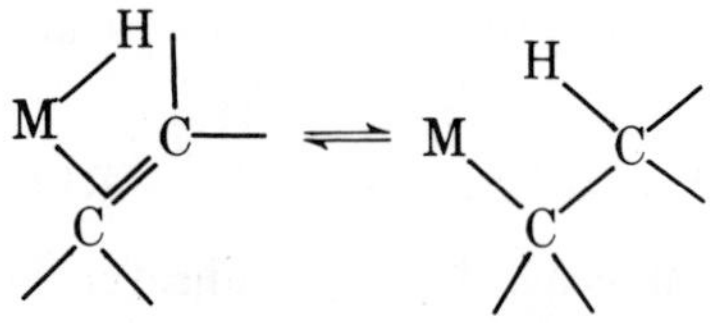

The term *insertion* is a useful description of the overall reaction, although it is not accurate in mechanistic detail. A better description would be migration of the hydrogen ligand from the metal atom to a carbon of the coordinated olefin. This distinction is useful in considering the reverse reaction *β-hydrogen elimination,* in which a β-hydrogen of an alkyl group migrates to the metal atom. The β-elimination reaction is very common in organometallic chemistry, but α-elimination processes that yield carbene complexes are also known (Chapter 9).

In all the well-characterized insertion reactions [23], the unsaturated substrate (olefin, diene, arene, CO, etc.) is coordinated to the metal ion prior to insertion. The migratory reagent (nominally H^-, R^-, or OH^-) is generally precoordinated, but this is not always true. Recent studies of Pd(II)-catalyzed olefin oxidation indicate that the attacking nucleophile approaches the coordinated olefin from solution rather than via the metal. The major function of the metal ion in such cases seems to be the activation of the unsaturated substrate toward nucleophilic attack. The combination of *sigma* and *pi* bonds between the unsaturated ligand and the metal (as described for N_2 earlier) provides a low-energy pathway for the rearrangement of the bonds involved in the insertion reaction. It appears that further advantages accrue when the nucleophilic reagent is in the coordination sphere of the metal. At the very least, precoordination of the nucleophile *cis* to the olefin provides a high local concentration of migratory reagent.

The bonding characteristics of olefins and of carbon monoxide that facilitate insertion are discussed in the introductions to Chapters 3 and 5, respectively. With olefins the best studied process is insertion into M—H bonds, although insertion into M—C bonds is obviously important in the polymerization of olefins. Carbon monoxide inserts readily into M—C bonds, but CO insertion into M—H is almost unknown. (Formylmetal complexes are thermodynamically unstable.) The reaction of CO with M—C bonds is the best studied insertion process and the one for which the migration mechanism is established most firmly. Stereochemical studies [23,24] of insertion in methylmanganese carbonyl compounds indicate that the acetyl group in the product occupies the site where the "inserted" CO was attached:

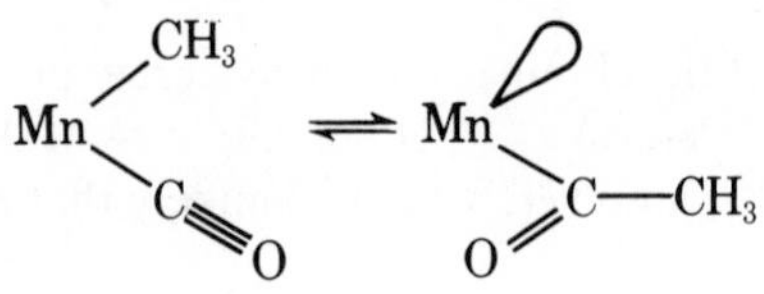

In other words the methyl group has migrated to the carbon of a coordinated carbon monoxide. In the reverse reaction, elimination of CO from a coordinated acetyl group, the methyl migrates from CO to a vacant coordination site on the metal.

The intimate mechanism of insertion and elimination is not known, even though the gross stereochemistry is known in a few instances. Molecular orbital calculations have been used to predict the movements of atoms and electrons in insertion processes [25,26], but these predictions have not been verified experimentally. It seems likely that several detailed mechanisms will be discovered for insertion just as for ligand replacement and for oxidative addition. By analogy with these other reactions it would not be surprising to discover that labile paramagnetic intermediates are sometimes involved in insertion. Indeed, both spectroscopic and chemical evidence point to a radical chain pathway for the insertion of an acetylene into a Pt—H bond in one instance [27]:

$$H{-}PtCl(PEt_3)_2 + RC{\equiv}CR \longrightarrow (R)C(H){=}C(R){-}PtCl(PEt_3)_2$$

2.4 CHARACTERISTICS OF CATALYSIS

The essential characteristics of a catalytic reaction are the same whether the catalyst is soluble or insoluble:

- The effect of a catalyst is purely kinetic. It does not make a thermodynamically forbidden reaction favorable, but it can dramatically accelerate an allowed reaction by providing a pathway for a low energy of activation.
- The microscopic catalyst site operates in a cyclic fashion through a series of reactions that are repeated each time a molecule of substrate is transformed.
- The active catalytic species is not necessarily the same compound that is put into the reaction mixture as a catalyst. Many transformations of the nominal catalyst may occur. The precatalytic reactions often give rise to an induction period before catalysis begins.

These points are exemplified in the isomerization of 1-butene to 2-butene catalyzed by a solution of $Ni(P(OEt)_3)_4$ and a protonic acid. The mechanism of the isomerization, which is discussed in detail in Chapter 3, is depicted in Figure 2.4. The transformation of the nominal catalyst NiL_4 to the "true catalyst" $[HNiL_3]^+$ takes place in two steps shown at the top of the figure. The cyclic nature of the catalytic process is indicated by writing the equations for the reaction with the organic substrate (1-butene) in the "clock" notation of Tolman [1]. A catalytic species enters the cycle at the 12-o'clock position. Through a series of four reactions

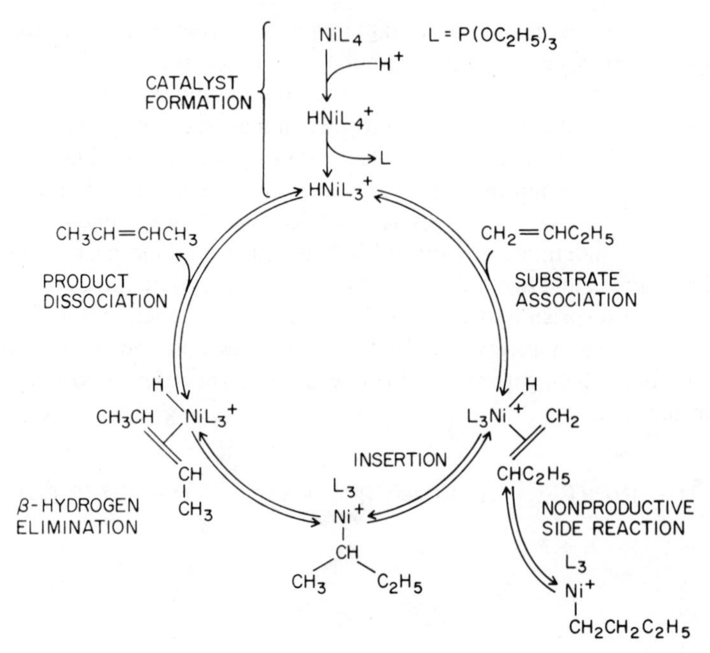

Figure 2.4 Mechanism for isomerization of 1-butene catalyzed by $Ni[P(OC_2H_5)_3]_4$ and a protonic acid.

and three intermediates displayed in a clockwise sequence, one molecule of 1-butene is isomerized and the catalytic species is regenerated. Two pairs of fundamental organometallic transformations—ligand association and dissociation, olefin insertion and β-hydrogen elimination—occur in the overall process. Each basic reaction appears repeatedly in subsequent chapters.

In the example of Figure 2.4 the active catalyst for the isomerization of 1-butene to 2-butene is formed by reaction of a protonic acid with tetrakis(triethyl phosphite) nickel(0) [28]. Neither catalyst component alone will isomerize the olefin under the standard mild reaction conditions but the *coordinatively unsaturated* nickel hydride $[HNiL_3]^+$ that is formed from them is highly active. The initial step in catalyst formation is protonation of the zerovalent nickel complex to form a cationic nickel "hydride," $[NHiL_4]^+$. Despite the nomenclature, which suggests negative charge on the H atom, it is likely that the Ni—H bond is relatively nonpolar. It may be regarded as a simple covalent bond in which both electrons happen to come from the metal.

The second catalyst formation step, *dissociation* of a phosphite ligand to give $[HiNL_3]^+$, is essential because the $[HNiL_4]^+$ complex is *coordinatively saturated* and cannot interact directly with 1-butene. As discussed in Section 2.1, a coordinately saturated complex has all its potentially bonding orbitals filled with electrons and must shed an electron donor ligand to bind the organic substrate. When a phosphite ligand has dissociated to give $[HNiL_3]^+$, the first step in the catalyst cycle, *substrate association,* occurs readily.

The olefin complex at the 3-o'clock position in the catalyst cycle contains the essential ingredients for isomerization of the olefin: an Ni—H bond and an

$$\text{Ni}\!-\!\!\!\begin{array}{c}\text{C}\\ \| \\ \text{C}\end{array}$$

bond. The latter bond is complex and is discussed in detail in Section 3.1. In the so-called insertion reaction the hydrogen atom migrates from the nickel to one of the carbon atoms of the olefinic double bond. The other carbon becomes σ-bonded to the nickel as an *n*-butyl or 2-butyl group, depending on which carbon atom has acquired the migratory hydrogen atom. As noted in the figure, formation of the *n*-butyl-nickel complex is unproductive and does not lead to olefin isomerization. Since, however, the insertion reaction is readily reversible, this side reaction does not seriously impair catalyst efficiency.

The productive step is addition of the hydrogen to the terminal carbon of the olefin to form a 2-butyl complex shown at the 6-o'clock position in the cycle. The next step is β-hydrogen elimination, a reaction that is the exact reverse of olefin insertion into the Ni—H bond. If a hydrogen from the CH_3 group adjacent to the Ni—C bond migrates to the nickel, the 1-butene complex is reformed, another unproductive reaction. If, on the other hand, a CH_2 hydrogen migrates, the 2-butene complex shown at 9 o'clock in the cycle is formed. Dissociation of the olefinic product completes the catalytic cycle by reforming the true catalyst $[HNiL_3]^+$.

This catalytic cycle for olefin isomerization illustrates the 16- and 18-electron rule (Section 2.1) very nicely. In Figure 2.4, 16 electron intermediates appear in the 6- and 12-o'clock positions. Eighteen electron intermediates are seen in the 3- and 9-o'clock positions in the cycle. This alternation of 16- and 18-electron intermediates is common for reactions catalyzed by Group VIII metal complexes.

Another important point about this example is that all the steps in the catalytic cycle for olefin isomerization are reversible. Hence, if one were to introduce 2-butene into the reaction mixture, it could be isomerized to 1-butene. In fact, if one starts with either 1-butene or 2-butene, the catalytic reaction ultimately produces an equilibrium mixture of about 69% *trans*-2-butene, 25% *cis*-2-butene, and 6% 1-butene [28]. This point is perhaps the most important of all concerning catalysis: *catalysis is a kinetic phenomenon.* It does not shift the equilibrium point in the mixture of products, but it dramatically enhances the rates of reactions by providing low-energy pathways. A catalyst can, however, affect product distribution by catalyzing competitive reactions at different rates. In the isomerization of 1-butene many catalysts give high, nonequilibrium concentration of *cis*-2-butene early in the reaction. Longer reaction periods yield the theoretical equilibrium mixture of the three olefins.

GENERAL REFERENCES

Cotton, F. A., and G. Wilkinson, *Advanced Inorganic Chemistry,* Wiley-Interscience, provide a good, readable description of the transition metal chemistry relevant to homogeneous catalysis. The principles of catalysis are presented in a lengthy chapter in the fourth edition (1980).

Heck, R. F., *Organotransition Metal Chemistry,* Academic Press, 1974, offers a broad treatment of the chemistry fundamental to metal-catalyzed transformations of organic compounds.

Basolo, F., and R. G. Pearson, *Mechanisms of Inorganic Reactions,* 2nd ed., Wiley, 1967, cover bonding and reaction mechanisms of coordination complexes.

Henrici-Olivé, G., and S. Olivé, *Coordination and Catalysis,* Verlag Chemie, 1976, provide extensive up-to-date coverage of coordination chemistry and relate it to popular topics in catalysis.

Shaw, B. L., and N. I. Tucker, *Organotransition Metal Compounds and Related Aspects of Homogeneous Catalysis,* Pergamon Press, 1975.

Muetterties, E. L., *Transition Metal Hydrides,* Marcel Dekker, 1971, provides chapters on synthesis, spectroscopy, and reactions of these compounds, which are prevalent in catalysis of olefin reactions.

Kochi, J. K., *Organometallic Mechanisms and Catalysis,* Academic Press, 1978, is especially good in the treatment of oxidation and electron transfer reactions.

Books on Specific Metals in Catalysis

P. W. Jolly and G. Wilke, *Organic Chemistry of Nickel,* Academic Press, 2 vols., 1974, 1975.

P. M. Maitlis, *The Organic Chemistry of Palladium,* Vol. 2, *Catalytic Reactions,* Academic Press, 1971.

F. R. Hartley, *The Chemistry of Platinum and Palladium,* Wiley, 1973.

E. A. Koerner von Gustorf, F-W. Grevels, and I. Fischler, *The Organic Chemistry of Iron,* Academic Press, 1978.

P. C. Wailes, R. S. P. Coutts, and H. Weigold, *Organometallic Chemistry of Titanium, Zirconium and Hafnium,* Academic Press, 1974.

SPECIFIC REFERENCES

1. C. A. Tolman, *Chem. Soc. Rev.,* **1,** 337 (1972).
2. J. K. Kochi, *Organometallic Mechanisms and Catalysis,* Academic Press, 1978.
3. J. Chatt and R. S. Coffey, *J. Chem. Soc., A,* 1963 (1969).
4. N. V. Sidgewick and R. W. Bailey, *Proc. Roy. Soc.,* **144,** 521 (1934); N. V. Sidgewick, *The Chemical Elements and Their Compounds,* Clarendon Press, 1950, p. 547.
5. J. R. Graham and R. J. Angelici, *Inorg. Chem.,* **6,** 2082 (1967).
6. J. A. Bowden and R. Colton, *Aust. J. Chem.,* **24,** 2471 (1971).
7. R. N. Perutz, J. J. Turner, *et al., J. Am. Chem. Soc.,* **97,** 4791, 4805 (1975).
8. M. J. Bennett and R. B. Donaldson, *Inorg. Chem.,* **16,** 655 (1977); J. Reed and P. Eisenberger, *J.C.S. Chem. Comm.,* 628 (1977).
9. J. Halpern and C. S. Wong, *J.C.S. Chem. Comm.,* 629 (1973). C. A. Tolman, P. Z. Meakin, D. L. Lindner, and J. P. Jesson, *J. Am. Chem. Soc.,* **96,** 2762 (1974); Y. Ohtani, M. Fujimoto, and A. Yamagishi, *Bull. Chem. Soc. Japan,* **50,** 1453 (1977).
10. K. G. Caulton, R. L. DeKock, and R. F. Fenske, *J. Am. Chem. Soc.,* **92,** 515 (1970); R. W. F. Hardy, R. C. Burns, and G. W. Parshall, "Bioinorganic Chemistry of Nitrogen Fixation," in G. L. Eichhorn, Ed., *Inorganic Biochemistry,* Elsevier, 1973, p. 745.
11. D. Thorn, T. H. Tulip, and J. A. Ibers, *J. Chem. Soc. Dalton,* 2022 (1979).
12. T. G. Appleton, H. C. Clark, and L. E. Manzer, *Coord. Chem. Rev.* **10,** 335 (1973).
13. C. A. Tolman, *Chem. Rev.,* **77,** 313 (1977).

14. R. A. Schunn, *Inorg. Chem.*, **15,** 208 (1976).
15. J. D. Druliner, A. D. English, J. P. Jesson, P. Meakin, and C. A. Tolman, *J. Am. Chem. Soc.*, **98,** 2156 (1976); C. A. Tolman, *Inorg. Chem.*, **11,** 3128 (1972).
16. M. Absi-Halabi and T. L. Brown, *J. Am. Chem. Soc.*, **99,** 2982 (1977).
17. B. H. Byers and T. L. Brown, *J. Am. Chem. Soc.*, **97,** 947 (1975).
18. J. Halpern, *Accounts Chem. Res.*, **3,** 386 (1970); J. P. Collman, *ibid,* **1,** 136 (1968).
19. D. H. Gerlach, A. R. Kane, G. W. Parshall, J. P. Jesson, and E. L. Muetterties, *J. Am. Chem. Soc.*, **93,** 3543 (1971).
20. J. A. Osborn, "Organotransition Metal Chemistry," in Y. Ishii and M. Tsutsui, Eds., Plenum Press, 1975, p. 65.
21. A. V. Kramer, J. A. Labinger, J. S. Bradley, and J. A. Osborn, *J. Am. Chem. Soc.*, **96,** 7145 (1974); A. V. Kramer and J. A. Osborn, ibid, **96,** 7832 (1974).
22. J. K. Stille and K. S. Y. Lau, *Accounts Chem. Res.*, **10,** 434 (1977).
23. F. Calderazzo, *Angew. Chem. Int. Ed.*, **16,** 299 (1977).
24. K. Noack and F. Calderazzo, *J. Organomet. Chem.*, **10,** 101 (1967).
25. H. Berke and R. Hoffmann, *J. Am. Chem. Soc.*, **100,** 7224 (1978).
26. D. L. Thorn and R. Hoffman, *J. Am. Chem. Soc.*, **100,** 2079 (1978).
27. H. C. Clark and C. S. Wong, *J. Am. Chem. Soc.*, **99,** 7073 (1977).
28. C. A. Tolman, *J. Am. Chem. Soc.*, **94,** 2994 (1972).

3 REACTIONS OF OLEFINS AND DIENES: ISOMERIZATION AND HYDROGENATION

The most diverse applications of homogeneous catalysis involve reactions of olefins and dienes. Some, like polymerization, hydroformylation, and oxidation, are carried out industrially on an immense scale. Others are primarily of interest to the organic chemist for laboratory syntheses. Nearly all these reactions, however, involve similar steps in which the olefin and the other reagents are activated by interaction with a soluble metal complex.

This chapter treats the basic principles involved in the reaction of an olefin with a transition metal complex. Bonding patterns and two elementary reactions are considered first. The application of these principles in simple catalytic reactions is illustrated by discussions of the isomerization and hydrogenation of olefins and dienes. These two reactions have limited practical application but are probably the best understood in terms of scope and mechanism. The more complex and useful C—C bond-forming reactions—polymerization, oligomerization, and hydrocyanation—appear in Chapter 4. Hydroformylation and carbonylation of olefins are considered in Chapter 5, along with other reactions of carbon monoxide. The oxidation of olefins to aldehydes, ketones, vinyl esters, and epoxides appears in Chapter 6. Olefin metathesis is discussed among carbene reactions in Chapter 9 because the mechanism appears to involve substantially different principles than those described here.

3.1 OLEFIN COORDINATION

Olefin complexes of transition metal ions are among the oldest known organometallic compounds. The first organic derivative of a transition metal to be characterized was the ethylene complex $K[Pt(C_2H_4)Cl_3]$. This compound is commonly known as Zeise's salt, in honor of the Danish pharmacist who isolated it in 1827 [1]. (In 1961 the author of this volume was elated when he discovered a platinum

Table 3.1 Stability of Olefin Complexes

Olefin (ol)	$Pd(ol)Cl_3^-$	$Cu(ol)^+$	$Ag(ol)^+$	
			H_2O	Glycol
Ethylene	17.4[a]	6360[b]	97.9[b]	22.3[c]
Propylene	14.5	—	87.2	9.1
1-Butene	11.2	—	—	7.7
cis-2-Butene	8.7	—	—	5.4
trans-2-Butene	4.5	—	—	1.4

[a] Equilibrium constant at 25° for $PdCl_4^{2-} + ol \rightleftharpoons Pd(ol)Cl_3^- + Cl^-$ (ref. 5).
[b] Association with M^+ in aqueous solution at 25° (ref. 6).
[c] Association with $AgNO_3$ in ethylene glycol at 40° (ref. 7).

complex of the olefinic ketone, mesityl oxide. Subsequent investigation showed that the complex had been prepared by Zeise in 1840.) The development of olefin coordination chemistry proceeded slowly until about 1950. At that time both the theoretical understanding and the preparative chemistry of olefin complexes began to blossom. Landmark papers by Dewar [2] and by Chatt [3] provided a model for the interaction between an olefin and a metal, which is used with little modification to this time. In the synthesis area literally hundreds of olefin complexes have been prepared since 1950 [4].

Simple olefin complexes are known for nearly all the transition metals. Commonly the metal ion in such complexes has a low formal oxidation state. With platinum, for example, olefin complexes of both Pt(O) and Pt(II) are well established, as in $Pt^0(C_2H_4)_3$ and $[Pt^{II}(C_2H_4)Cl_3]^-$, but stable olefin complexes of Pt(IV) are unknown. The formal oxidation state of the metal is probably less important, however, than the actual electron density at the metal center. In Zeise's salt, $[Pt(C_2H_4)Cl_3]^-$, which is used as an example for discussion of bonding principles, the metal ion is probably almost electroneutral as a result of metal-ligand interactions. In the case of olefin complexes of Ag^+, the equilibrium constants (Table 3.1) for complexation are strongly dependent on the solvent in which the measurement is made. Donation of electron density from a mild donor solvent such as water or an alcohol probably offsets the formal positive charge on the silver ion.

As shown in Table 3.1, the stability of an olefin complex depends on the nature of the olefin, as well as on the metal and its ligands. The general trend is that alkyl substituents on the C=C function decrease the stability of the complex that is formed. (The same qualitative trend of stability constants has been observed for many kinds of metal-olefin complexes.) It is usually assumed that the destabilizing effect of the substituents is due to steric interaction between the olefin and other ligands on the metal atom. The strength of the metal-olefin bond is greatly increased with strained cycloolefins such as cyclooctene. This effect is often assigned to relief of steric strain when the C═C bond order is reduced by complexation to a metal ion.

Coordination of Dienes

Conjugated dienes, typified by 1,3-butadiene, may coordinate to metals in several ways:

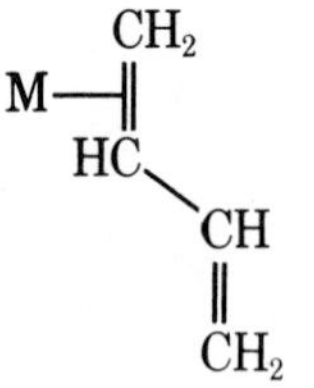

$$\underset{\mathbf{1}}{M\text{—}(CH_2{=}CH)\text{—}CH{=}CH_2} \qquad \underset{\mathbf{2}}{M(\eta^4\text{-}H_2C{=}CH\text{—}CH{=}CH_2)} \qquad \underset{\mathbf{3}}{\underset{|\atop M}{H_2C{=}CH}\text{—}\underset{|\atop M}{CH{=}CH_2}}$$

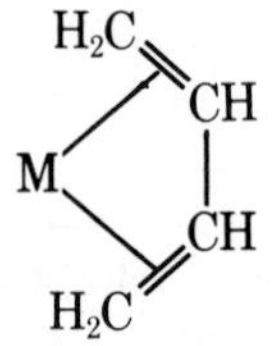

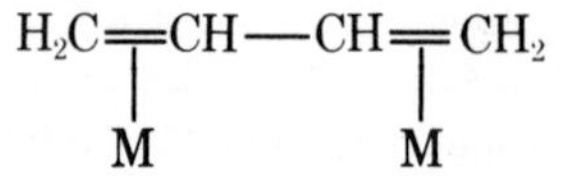

The coordination of a diene to two metal ions as in **3** is observed in some binuclear complexes [4]. Binuclear coordination is believed to occur on the surface of metallic catalysts but is unusual in soluble catalysts. On the other hand both monodentate and chelate coordination of a diene to a metal ion, as in **1** and **2**, are well known. The distinction between **1** and **2** is important because compounds that constrain a diene to bind in monodentate fashion give different products than those that have chelate bonding (**2**). This difference is apparent in the discussion of diene polymerization and the synthesis of 1,4-hexadiene in Chapter 4.

Bonding Schemes

The formal simplicity of Zeise's anion, $[Pt(C_2H_4)Cl_3]^-$, has made it a favorite subject for theoretical studies of olefin complexation [8]. It is also attractive because there is chemical and structural similarity with the anion $[Pd(C_2H_4)Cl_3]^-$, implicated as an intermediate in the Wacker catalytic synthesis of acetaldehyde (Chapter 6).

The crystal structure of Zeise's anion shows the olefin and the three chloride ligands in a square planar array about the platinum atom (Figure 3.1). The ethylene ligand is oriented perpendicular to the coordination plane [9]. In solution the olefin can rotate about the $Pt\text{—}\overset{C}{\underset{C}{\|}}$ axis, as demonstrated for other olefin complexes of the noble metals. The C—C bond distance (1.37(3)Å) is slightly longer than that (1.33 Å) in ethylene itself. The Pt—Cl bond *trans* to the olefin is a little longer than the *cis* Pt—Cl bonds, consistent with the *trans*-labilizing effect often observed for π-bonding ligands.

The bonding scheme most commonly used to account for the structure and reactivity of olefin complexes is illustrated in Figure 3.1. This scheme, propounded

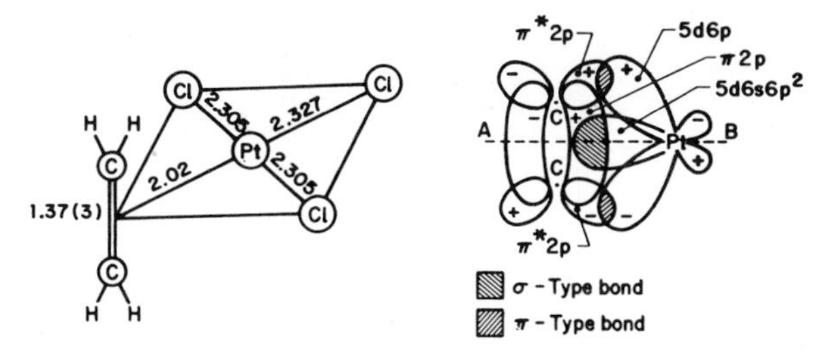

Figure 3.1. *Left:* The structure of Zeise's anion in the crystalline state (ref. 9). *Right:* A qualitatively useful ethyleneplatinum bonding scheme (ref. 2,3).

by Dewar [2] and by Chatt and Duncanson [3], is satisfactory in a qualitative sense. The cloud of electron density in one lobe of the π orbital of ethylene overlaps with a vacant *sigma* orbital of platinum. A second bond of *pi* symmetry is formed by overlap of a filled *dp* hybrid orbital of platinum with the π^* orbital of ethylene. The two bonding modes are complementary and synergistic in the transfer of electron density from ethylene to platinum and back again. The overall effect produces a Pt—C_2 bond of moderate stability. The slight lengthening of the C—C bond may be due to population of the π^*-antibonding orbital of the olefin.

Attempts to assess the quantitative contributions of the various atomic orbitals to the bonding molecular orbitals in Zeise's anion have had modest success. A recent SCF Xα scattered-wave calculation [10] indicates that the donation from the ethylene π orbital to platinum *d* orbitals is more complex than assumed in the qualitative scheme of Figure 3.1 but does not greatly alter the qualitative picture. The calculation suggests that the σ bond contributes much more to the metal-ligand attraction than the π bond does. It also provides a useful rationale for the electronic spectrum of Zeise's salt.

More recent *ab initio* calculations of olefin complexation with nickel [11] have attempted to diagnose the reactivity of the coordinated olefin ligand. A major conclusion is that the electron density on the olefin varies dramatically with changes in the oxidation state of the metal and with variations in the other ligands. With zerovalent metal atoms and strong donor ligands, substantial negative charge accumulates on the olefin and thus renders it susceptible to electrophilic attack. Higher oxidation states such as Pd(II) in the Wacker process catalyst (Chapter 6) lead to withdrawal of electron density and vulnerability to nucleophilic attack. These predictions are consistent with the observed reactivity of olefin complexes.

Delocalized Bonding

One important characteristic of coordinated dienes is that they react readily with electrophiles or nucleophiles to give substituted π-allyl ligands [12]. These nominally ionic ligands are intermediates in most catalytic reactions of conjugated dienes and also take part in some reactions of monoolefins. The π-crotyl (1-methylallyl) ligand can be formed by several addition or elimination reactions, as shown in Figure 3.2.

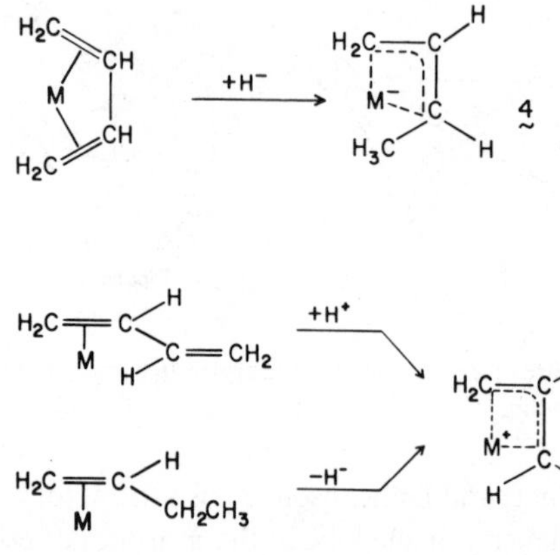

Figure 3.2. Formation of π-allylic complexes from butadiene complexes and a 1-butene complex.

The addition of a nucleophile to a chelate-bonded butadiene(**2**) is illustrated by hydride addition to give an *anti*-π-crotyl complex(**4**). The *anti* nomenclature indicates that the C_4 ligand has kept its original geometry and the methyl group points away from (anti) the hydrogen substituent on C-2 of the allyl group. Similarly addition of a proton to a chelated butadiene can give an analogous *anti*-π-crotyl cationic complex (not shown). If the butadiene is coordinated through only one double bond (**1**) or if the π-crotyl group arises by hydride abstraction from 1-butene (**6**), the product is usually the *syn*-π-crotyl complex (**5**) (methyl *syn* to the H on C-2). The *syn-anti* distinction is important because the conformation of the π-crotyl intermediate determines the geometry of products derived from it in catalytic reactions of dienes.

The four carbon atoms in a π-crotyl ligand are generally almost coplanar. The three allylic carbons are about the same distance from the metal atom and the two allyl C—C bonds are roughly the same length. This latter point is consistent with the formulation of the π-allyl ligand as a delocalized hybrid of two resonance structures:

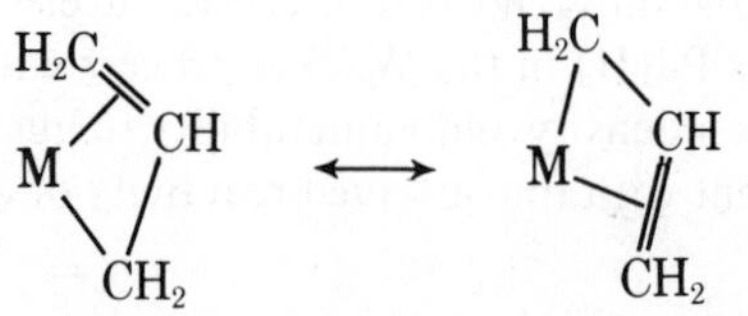

Because the allyl ligand can be regarded as a hybrid of a σ-alkyl and a π-olefin bond, it is counted as a 3-electron donor in the usual electron-counting schemes. It is commonly considered to occupy two coordination sites in an octahedral complex. These formalisms are sometimes useful in deciding whether or not a complex is coordinatively saturated.

The $\sigma + \pi$ nature of the allyl ligand is reflected in many of its reactions. For example, consider the carbonylation of a π-allyl complex:

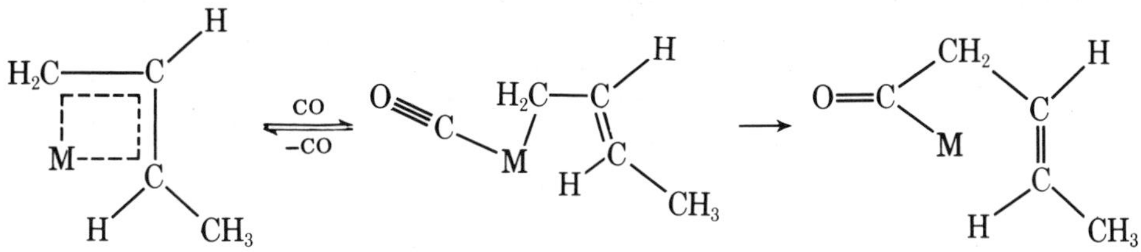

The π-olefin end of the *syn*-π-crotyl ligand swings free to create a vacant coordination site, which is subsequently occupied by carbon monoxide. In the second step in the carbonylation sequence the CO inserts in the C—M bond of the σ-crotyl ligand. Similar insertions also occur with π-crotyl ligands if there is an "insertable" ligand such as CO in an adjacent position in the coordination sphere of the metal. Whether the crotyl ligand is bonded in σ or π fashion, reaction tends to occur preferentially at the less substituted end of the allyl unit. As shown for the *syn* ligand, which yields a *trans* olefin, the configuration of the allyl group is preserved in the product. Several mechanisms for *syn-anti* isomerization are, however, available.

More extended delocalized bonding systems in hydrocarbon ligands are well known, although the π-allyl system is most prevalent in practical catalytic reactions [13].

3.2 REACTIONS OF COORDINATED OLEFINS

The homogeneous catalytic reactions of olefins usually comprise a series of simple steps, as illustrated for olefin isomerization in Chapter 2. Most of these steps are the general "building block" reactions described in that chapter—oxidative addition and reductive elimination, insertion and elimination, ligand association and dissociation. Two simple reactions specifically associated with olefin complexes are discussed in more detail here. These are nucleophilic attack on the coordinated olefin and insertion of an olefin into an M—C or M—H bond. These two processes are closely related and may be regarded as intermolecular and intramolecular versions of the same process.

Nucleophilic Attack

Molecular orbital calculations forecast that coordination of an olefin to a metal ion strongly affects the reactivity of the C=C bond. In a qualitative sense the mixing of metal and olefin orbitals imparts some of the metal character to the olefin. The most important practical effect is that complexation to a high valent or cationic metal ion activates an olefin toward attack by a nucleophile [12].

Some major applications of homogeneous catalysis such as the Wacker acetaldehyde synthesis depend on metal-promoted nucleophilic attack on an olefin. Palladium(II) salts are most commonly involved in catalytic oxidations (Chapter 6), but stoichiometric additions of nucleophiles to cationic olefin complexes of other metals have been studied extensively. The nucleophilic reagent may be an ion such as OH^-, X^-, $RCOO^-$, RS^-, or $CH(CN)_2^-$, or it may be a neutral molecule, for example, NH_3, NR_3, PR_3, SR_2. Generally, the reaction occurs by attack of an external nucleophile (Nu) on one end of the coordinated olefin:

$$M^+\!-\!\|(CH_2{=}CH_2) + \text{Nu} \longrightarrow M\!-\!CH_2\!-\!CH_2\!-\!Nu^+$$

This process occurs with rearrangement of carbon-metal bonds to produce a β-substituted ethyl derivative. Overall, electron density is transferred from the nucleophile to the metal ion. In many respects the addition resembles the final step in electrophilic bromination of an olefin:

$$Br^+ + \;>C{=}C< \;\longrightarrow \left[Br<\!\!(C\!-\!C) \right]^+ \xrightarrow{Br^-} Br\!-\!C\!-\!C\!-\!Br$$

In this analogy Br^+ behaves like M^+ in activation of the C=C bond to nucleophilic attack.

In the examples just given, the nucleophile attacks the olefin from the side opposite the metal. An equally logical situation is precoordination of the nucleophile to the metal followed by *cis*-attack:

$$M(\text{Nu})(CH_2{=}CH_2) \longrightarrow M\!-\!CH_2\!-\!CH_2\!-\!Nu$$

This reaction is the familiar "insertion" process, which is discussed in the following section. The distinction between insertion and external attack can often be made on the basis of the stereochemistry of the product. This stereochemical criterion has been used in the study of the palladium-catalyzed oxidations of olefins (Chapter 6).

The stereochemical distinction between *cis*-addition (insertion) and *trans*-addition has also been demonstrated for stoichiometric addition of an amine to an

α-olefin. Diethylamine was added to 1-butene, which was coordinated in the S-configuration in a platinum complex of known geometry [14]:

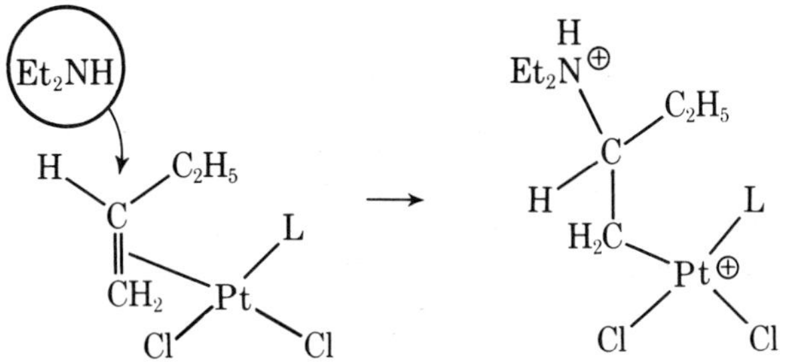

$$\text{Et}_2\text{NH} + \text{H(C}_2\text{H}_5\text{)C=CH}_2\text{–Pt(L)Cl}_2 \longrightarrow \text{Et}_2\overset{\oplus}{\text{N}}\text{H–C(H)(C}_2\text{H}_5\text{)–H}_2\text{C–}\overset{\oplus}{\text{Pt}}\text{(L)Cl}_2$$

L = S-α-methylbenzylamine

The product was the S-derivative in high optical purity, consistent with *trans* addition of the amine to the double bond.

Similar additions of amines and ammonia to ethylene complexes of iron, molybdenum, and tungsten have been reported [15]. The product is usually a β-metalloethylammonium salt:

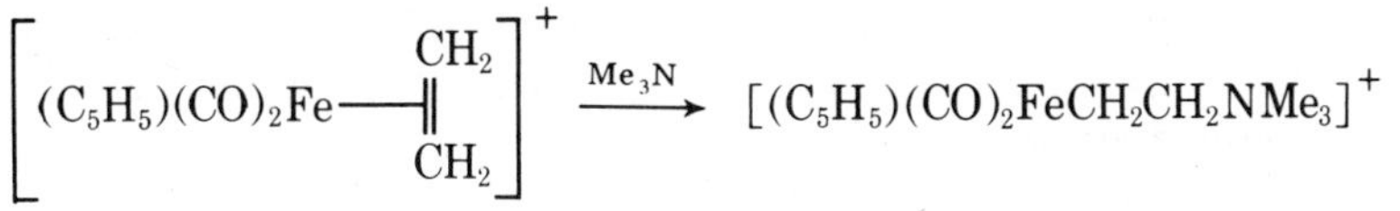

$$\left[(C_5H_5)(CO)_2Fe\text{—}\overset{CH_2}{\underset{CH_2}{\|}}\right]^+ \xrightarrow{Me_3N} [(C_5H_5)(CO)_2FeCH_2CH_2NMe_3]^+$$

Other nucleophiles, including phosphines, cyanide ion [15], enamines, and enolate ions [16], add in much the same way to give β-substituted ethyliron species. These reactions have not yet been incorporated in a catalytic cycle. When substituted olefins such as propylene or styrene are coordinated to the metal, a combination of steric and electronic effects determines the direction of addition. Typically the less substituted carbon atom remains bound to the metal [17]:

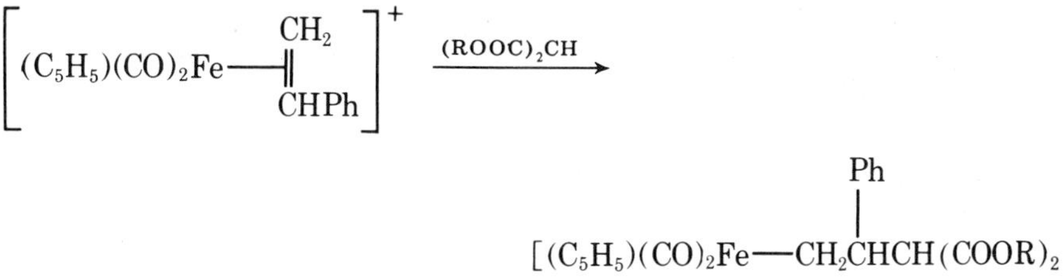

$$\left[(C_5H_5)(CO)_2Fe\text{—}\overset{CH_2}{\underset{CHPh}{\|}}\right]^+ \xrightarrow{(ROOC)_2CH} [(C_5H_5)(CO)_2Fe\text{—}CH_2\underset{}{\overset{Ph}{|}}{CH}CH(COOR)_2$$

Olefin Insertion Reactions

The olefin reactions described in this chapter, as well as those in Chapters 4 and 5, involve "insertion" into an M—C or an M—H bond as an essential step. In fact the two steps sometimes occur in sequence in a catalytic cycle:

$$\begin{array}{c} C{=}C \\ | \\ M{-}H \end{array} \rightleftharpoons M{-}C{-}C{-}H \xrightarrow{C=C}$$

$$\begin{array}{c} C{=}C \\ | \\ M{-}C{-}C{-}H \end{array} \longrightarrow M{-}C{-}C{-}C{-}C{-}H$$

The insertion of an olefin or conjugated diene into an M—H bond is well known in both stoichiometric and catalytic reactions. Curiously, insertion into an M—C bond is rare in stoichiometric reactions, even though it is invoked in most mechanisms for dimerization or polymerization of olefins.

Despite the importance of the insertion reaction in catalysis, there has been little study of the mechanism of olefin insertion or its reverse process, β-elimination. The points listed here seem to be supported by experience and by recent molecular orbital calculations [18]:

1. Insertion probably occurs by migration of the metal-bound *H*- or *C*-substituent to a terminal carbon of the C═C double bond (Section 2.4). Hence the process may be termed migratory insertion. One distinguishing feature of this mechanism is that the alkyl group arising from the insertion should keep the coordination site originally occupied by the olefin. This stereochemical criterion of mechanism is, however, difficult to apply and has not been demonstrated for olefin insertions.

2. The olefin must be precoordinated to the metal atom, preferably in a *cis* site in the coordination sphere. Even though some proposed mechanisms for olefin polymerization have involved direct reaction of an olefin molecule with an M—C bond, the stereochemistry of polypropylene synthesis seems to be explained more easily by mechanisms that involve precoordination.

3. The insertion process is readily reversible by transfer of a β-hydrogen of an alkyl group to the metal atom. However, elimination of β-alkyl groups is uncommon. The ease of β-hydrogen elimination is a major source of instability of alkyl derivatives of the transition metals. Alkyl derivatives such as methyl, benzyl, and neopentyl, which contain no β-hydrogen, are more stable than ethyl derivatives because the lowest energy pathway for decomposition is not available [19].

4. A corollary of point **2** is that a vacant site for olefin coordination may be necessary for the β-hydrogen elimination process:

$$M{-}CH_2CH_3 \longrightarrow M\begin{array}{l} {\diagup}H \\ \quad CH_2 \\ {\diagdown}\!\!/\!\!/ \\ \;\; CH_2 \end{array}$$

   Thermal decomposition of $(n\text{-}Bu)_2Pt(PPh_3)_2$ is inhibited by the presence of excess triphenylphosphine [20]. Presumably a phosphine must dissociate to create a vacant coordination site before β-hydrogen elimination to form 1-butene can occur.

The scarcity of olefin insertions into M—C bonds and of the reverse β-alkyl elimination reaction suggests that this process differs from the M—H insertion process in some important but unknown way. The activation energy for insertion into the M—C bond seems higher than for the M—H bond. The latter reaction is so facile that simple olefin metal hydride complexes are rare. One of the few examples is an ethylene niobium hydride, which undergoes insertion only on warming to 60° [21].

$$(\pi\text{-}C_5H_5)_2Nb(H)(CH_2{=}CH_2) \underset{}{\overset{C_2H_4}{\rightleftharpoons}} (\pi\text{-}C_5H_5)_2Nb(C_2H_5)(CH_2{=}CH_2)$$

Further insertion of ethylene into the Nb—C_2H_5 bond has not been observed, despite many attempts to accomplish the reaction. The β-hydrogen elimination occurs when the ethylniobium complex is heated under reduced pressure.

The activation energy for olefin insertion into M—C bonds, though higher than for M—H bonds, need not be very high. Ethylene polymerization with Ziegler-Natta catalysts, which presumably involves repetitive insertion into Ti—C bonds, occurs readily at room temperature. A significant peculiarity of this system is that it involves paramagnetic titanium(III) catalyst sites. Indeed, the most facile olefin insertions into M—C bonds seem to occur at paramagnetic centers. It is interesting to speculate how the presence of an unpaired electron might facilitate insertion. As indicated in Section 2.3, odd electron mechanisms are more prevalent in homogeneous catalysis than one might have guessed 10 years ago.

3.3 ISOMERIZATION

Double-bond migration in olefins is one of the simplest and most thoroughly studied catalytic reactions [22–25]. Industrially, soluble catalysts are used to isomerize olefins that are involved as intermediates in other homogeneous catalytic processes. For example, the synthesis of adiponitrile from butadiene and HCN (Section 4.6) includes two olefin isomerization steps:

$$CH_2{=}CHCH(CH_3)(CN) \rightleftharpoons CH_3CH{=}CHCH_2CN$$

$$CH_3CH{=}CHCH_2CN \rightleftharpoons CH_2{=}CHCH_2CH_2CN$$

The first step, conversion of a branched chain to a linear chain, involves cleavage of a C—C bond, a relatively uncommon mode of isomerization. The second step is a more common type in which an internal olefin equilibrates with a terminal olefin by hydrogen migration without disruption of the carbon skeleton of the olefin.

Another potentially useful olefin isomerization process also involves equilibration of internal and terminal olefins. Linear aldehydes can be prepared from internal olefins such as 2-pentene by using a catalyst that is active for both isomerization and hydroformylation of olefins. If the catalyst generates terminal olefin rapidly and hydroformylates the terminal olefin preferentially, respectable yields of linear aldehyde form by the sequence:

$$C_2H_5CH{=}CHCH_3 \rightleftharpoons C_3H_7CH{=}CH_2 \xrightarrow[CO]{H_2} C_3H_7CH_2CH_2CHO$$

A standard hydroformylation catalyst, $HCo(CO)_4$, is moderately effective in this regard (Chapter 5). This catalyst also isomerizes olefins without hydroformylation when the partial pressures of CO and H_2 are reduced to the minimum amount necessary to stabilize the complex [23].

Double-bond migration is coupled with hydrogenation in several potentially useful processes for selective hydrogenation of dienes and trienes to monoolefins (Section 3.5). To achieve selective hydrogenation of one of the two double bonds in 1,5-cyclooctadiene, it appears to be necessary to bring the double bonds into conjugation.

Double-bond migration catalyzed by soluble metal complexes is useful in laboratory-scale organic synthesis. Some difficulty accessible unsaturated steroids have been prepared by $RhCl_3$-catalyzed isomerization reactions [26]. This salt forms a soluble hydrate, $RhCl_3{\cdot}3H_2O$, which is probably the most convenient olefin isomarization catalyst for laboratory use. It is air-stable, commercially available, and easy to use. Typically, the olefin is heated with an ethanol solution of $RhCl_3{\cdot}3H_2O$. After several hours the mixture is cooled and diluted with water, and the olefin is isolated by conventional means [26]. Another commerically available isomerization catalyst is Wilkinson's compound, $RhCl(PPh_3)_3$. It has been used in several syntheses of natural products [27] and is faster than the simple $RhCl_3$ system [28].

Two other commercially available materials, $Fe_3(CO)_{12}$ and $PdCl_2$, are also useful isomerization catalysts [29] for the synthesis laboratory. As described here, these compounds catalyze double-bond migration by a different mechanism than the rhodium or nickel catalysts do. Hence different products may be isolated in kinetically controlled experiments.

Nickel hydrides may be the most significant isomerization catalysts from an industrial viewpoint. These complexes catalyze a wide range of double-bond migrations, in addition to carbon skeletal rearrangements of dienes (*vide infra*), that are not observed with other catalysts. The nickel hydrides are prepared by protonation of Ni(O) compounds such as $Ni(P(OR)_3)_4$ or by reduction of $NiX_2(PR_3)_2$ with alkylaluminum compounds [30,31].

Finally, catalysis of olefin isomerization is a kinetic phenomenon. If reactions are allowed to proceed to completion, equilibrium mixtures of olefins form. For instance, the ultimate product of 1-butene isomerization is an equilibrium mixture of 69% *trans*-2-butene, 25% *cis*-2-butene, and 6% 1-butene [25,30]. With many catalysts, however, the *cis*-2-butene is formed more rapidly than the *trans* isomer and can be isolated as a major product early in the reaction. Some of the more active catalysts facilitate equilibration of *cis* and *trans* isomers of olefins such as stilbene, for which double-bond migration is impossible.

Mechanisms of Double-Bond Migration

Two major mechanisms have been demonstrated for the double-bond migration process. The most common is a reaction in which a metal hydride reacts reversibly with an olefin to give a metal alkyl derivative. The alkyl group then undergoes β-hydrogen elimination to give either the original olefin or a new isomer. This hydride addition-elimination mechanism has been demonstrated for $[NiHL_4]^+$ catalysts [Section 2.4] and for many other metal hydride complexes [22,23]. The $RhCl_3$ and $RhCl(PPh_3)_3$ catalysts operate by addition and elimination of Rh—H functions, which form in the reaction mixtures [25,32].

The second major mechanism is a 1,3-hydrogen shift, which may involve π-allyl intermediates or transition states. The 1,3-shift mechanism prevails in isomerizations catalyzed by Pd(II) compounds, which are prone to form π-allyl derivatives. The distinction between these two major mechanisms is usually made on the basis of deuterium-labeling studies. Other mechanisms such as radical abstraction of hydrogen from allylic sites may be significant but are less well characterized.

The metal hydride addition-elimination mechanism is nicely illustrated by $RuHCl(PPh_3)_3$, a hydrogenation catalyst that also catalyzes isomerization of simple olefins. The mechanism of isomerization of 1-pentene by this catalyst [22] is shown in Figure 3.3. A coordinatively unsaturated complex labeled H—Ru in the figure reacts with 1-pentene to form the olefin complex shown at the 3-o'clock position in the cycle. Migratory insertion can occur in either of two ways. Addition of Ru—H to the olefin to form the 1-pentyl derivative (**7**) is a nonproductive side reaction. Addition in the opposite sense to give the 2-pentyl derivative (**8**) opens pathways to isomerization (although β-hydrogen elimination from the methyl group leads back to 1-pentene). β-Hydrogen elimination from C-3 can occur in two ways to give either *cis*- or *trans*-2-pentene as the complexes (**9**). Dissociation of the 2-pentene ligands completes the catalytic cycle. Readdition of Ru—H to 2-pentene

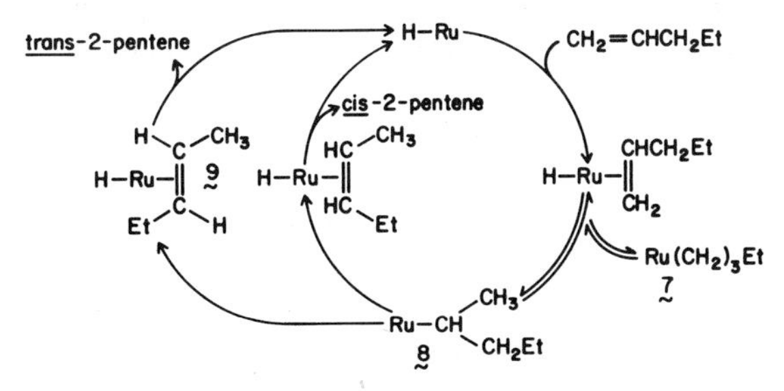

Figure 3.3. Catalytic cycle for isomerization of 1-pentene to *cis*- and *trans*-2-pentene by $RuHCl(PPh_3)_3$. The phosphine and chloride ligands are omitted for clarity.

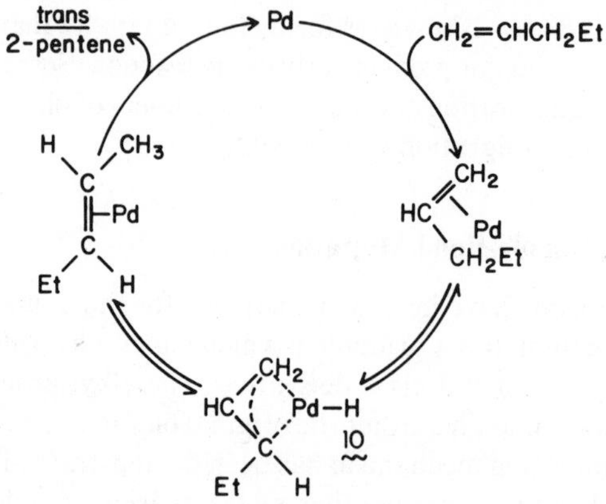

Figure 3.4. Isomerization of 1-pentene by $PdCl_2$ complexes. The chloride ligands are omitted for clarity.

can occur to give a 3-pentyl complex. This step is unproductive for pentenes because the 3-pentyl group can yield only 2-pentene, but, for long-chain olefins, repeated addition and elimination steps can move the double bond along the chain at random.

The β-hydrogen elimination to give *cis*-2-pentene is faster than the one that gives the *trans* isomer. Early in the reaction (50° in benzene), a 60:40 ratio of *cis* and *trans* 2-pentenes is observed. The factor that determines which isomer is formed in a single catalytic cycle is almost certainly the conformation of the 2-pentyl group at the time that β-hydrogen elimination occurs. The view along the C-2/C-3 bond axis may be represented as (C-3 in front):

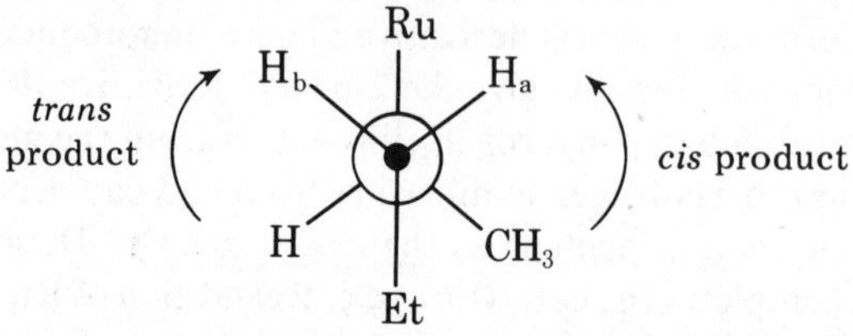

If the C-3 substituents in front rotate counterclockwise to place the Ru on C-2 and H_a on C-3 adjacent in eclipsed positions, *cis* elimination of Ru—H should yield *cis*-2-pentene. Rotation in the opposite sense would give the *trans* isomer.

1,3-Hydrogen shift, the other major mechanism for double-bond migration, is observed in the isomerization of 1-pentene by $Fe_3(CO)_{12}$ and $PdCl_2(NCPh)_2$ [29]. A number of detailed mechanisms have been proposed for the 1,3-shift. The simplest is a 1,3-suprafacial shift in which a hydrogen ion migrates from C-3 to C-1 in a coordinated terminal olefin without any direct metal-hydrogen interaction. Another conceptually simple proposal, metal-assisted proton migration, is shown in Figure 3.4. This mechanism probably operates in Pd^{2+}-catalyzed isomerization in nonpolar media such as benzene, the solvent chosen for mechanistic study [29]. Coordination of the olefin to the metal brings the allylic C-3 hydrogens of 1-pentene close to the metal atom. Transfer of hydrogen to the metal gives a π-allyl palladium hydride, **10.** The metal-bound hydrogen may return to C-3 to reform 1-pentene, or it may migrate to C-1 to form 2-pentene. With this catalyst the *trans* isomer shown in Figure 3.4 is favored kinetically as well as thermodynamically.

The oxidative addition of an allylic C—H bond to a $PdCl_2$ complex as shown in the cycle gives a species (**10**) that is formally Pd(IV), an uncommon oxidation state. However, the C—H cleavage in polar media usually occurs with proton dissociation: $C{-}H + Pd^{2+} \rightleftharpoons C{-}Pd^{+} + H^{+}$. It is tempting to suggest that a similar process occurs in olefin isomerization, but labeling studies with 1-pentene-1,2-d_2 indicate that only the deuterium on C-1 moves and that it moves by an intramolecular process [29]. This result suggests the intermediacy of a π-allylpalladium hydride as shown in the cycle.

The 1,3-hydrogen shift mechanism has also been suggested for double-bond migrations catalyzed by $Fe(CO)_5$, $Fe_2(CO)_9$, and $Fe_3(CO)_{12}$ [23]. A recent study indicates that $HCo(CO)_4$ catalyzes the isomerization of 3-phenylpropene to the 1-phenyl isomer by a Co—H addition/elimination mechanism [33].

Carbon Skeletal Isomerization

Olefin rearrangements that involve breaking of C—C bonds are rare but may be significant commercially. The conversion of a branched allyl cyanide to a linear isomer (Chapter 4) is an economically desirable process. On the other hand, isomerization of 1,4-hexadienes to methylpentadienes during hexadiene synthesis with nickel catalysts (Section 4.4) is an unwanted side reaction.

The species responsible for skeletal isomerization of dienes are nickel hydride complexes prepared by protonation of Ni(0) compounds [34] or by reaction of Ni(II) compounds with aluminum alkyls [31]:

$$NiL_3 + H^+ \rightleftharpoons HNiL_3 \qquad L{=}P(OC_7H_7)_3$$

$$\begin{matrix} Ni{-}Cl & & Ni{-}CH_2CHMe_2 & & \\ + & \rightarrow & + & \rightarrow & Ni{-}H + H_2C{=}CMe_2 \\ Al{-}CH_2CHMe_2 & & Al{-}Cl & & \end{matrix}$$

The nickel hydride complexes prepared in this way catalyze carbon skeleton rearrangements of several unconjugated dienes. The aluminum-containing catalysts seem to be more versatile than the cationic hydrides, but both systems transform *cis*-1,4-hexadiene to *trans*-2-methyl-1,3-pentadiene (70%) and 2,4-hexadienes in seconds at room temperature. Two different mechanisms are evident [31], but both involve cleavage of a β C—C bond in an alkylnickel chain. This β-carbon elimination is analogous to the β-hydrogen elimination in the double-bond migration process.

Figure 3.5 illustrates the rearrangement of 2,3-dimethyl-1,4-pentadiene by a C—C fragmentation process. This β-carbon elimination process involves some of the same intermediates as the synthesis of the diene from ethylene and isoprene

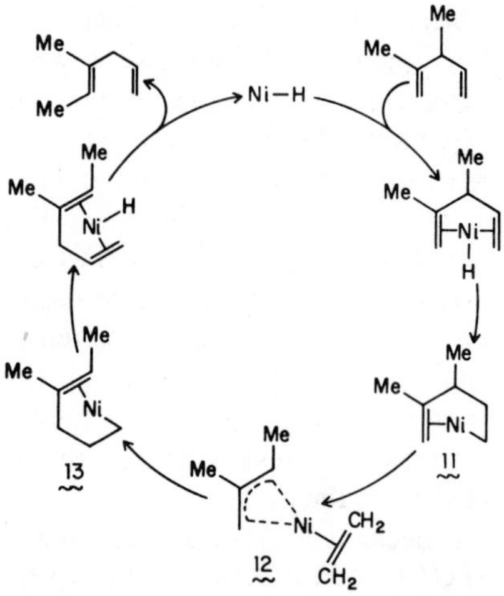

Figure 3.5 Isomerization of 2,3-dimethyl-1,4-pentadiene by a nickel hydride complex such as $NiHCl(PBu_3)_2$.

[35]. Initially an Ni—H complex coordinates to the diene. Insertion of a C=C into the Ni—H bond gives a dimethyl-4-pentenylnickel complex (**11**). Fragmentation of the β C—C bond forms a dimethyl-π-allyl complex (**12**) with an ethylene ligand. If the ethylene reinserts as shown, a 4-methyl-4-hexenyl complex (**13**) results and gives rise to 4-methyl-1,4-hexadiene by β-hydrogen elimination. Obviously, the cycle can also operate in reverse fashion to convert hexadienes to methylpentadienes.

A second rearrangement pathway via cyclopropylmethylnickel compounds has been proposed [31] for the isomerization of *cis*-1,4-hexadiene. This mechanism leads to conjugated diene products that are always more branched than the starting material. *trans*-1,4-Hexadiene does not undergo skeletal rearrangement but rapidly isomerizes to conjugated 2,4-hexadienes. This double-bond migration presumably involves conventional hydrogen migration processes [31,32].

3.4 HYDROGENATION

The hydrogenation of olefins has been studied extensively, perhaps more so than any other reaction catalyzed by soluble metal complexes [36,37]. This intensive study seems anomalous because soluble catalysts are seldom used for olefin hydrogenation in industry or in organic synthesis. Heterogeneous catalysts are usually more active and more convenient for practical applications such as the hydrogenation of cyclododecatriene to cyclododecane or of dicyanobutene to adiponitrile. The sole commercial use of a soluble catalyst for olefin hydrogenation is the reduction of an unsaturated amino acid to a precursor of the drug L-dopa. Even though this operation is conducted on a small scale, it is interesting because it involves asymmetric induction through use of an optically active catalyst. This selectivity is the major advantage of soluble catalysts.

Selectivity is the basis for another potential use of homogeneous hydrogenation catalysts. As discussed later, several metal complexes catalyze the hydrogenation of dienes and trienes to monoolefins with great specificity. Although these processes are not used commercially, they could be very useful for conversion of easily available butadiene dimers and trimers to polymer intermediates.

Another potential application, the hydrogenation of unsaturated polymers, is based on the mobility of a soluble catalyst in its reaction medium [38]. For example, in the hydrogenation of polyisoprene to the equivalent of an ordered copolymer of ethylene and propylene:

$$\left[CH_2\overset{CH_3}{\overset{|}{C}}{=}CHCH_2 \right]_n \xrightarrow{H_2} \left[CH_2\overset{CH_3}{\overset{|}{C}H}CH_2CH_2 \right]_n$$

a soluble catalyst brings its active site to the C=C bonds in the polymer chain. In contrast a heterogeneous catalyst requires that the polymer chain unfold to gain access to a catalyst site on the surface of a metal crystallite.

Hydrogenation of Simple Olefins

The reduction of a sterically accessible C=C bond is usually simple experimentally. One mixes hydrogen, olefin, and catalyst in an organic solvent at 25–100° and 1–3 atmospheres pressure. The reaction is usually clean, and the products are separated from catalysts by conventional techniques such as distillation or washing with water.

Dozens of soluble transition metal complexes catalyze hydrogenation of olefins, but four classes are preferred for practical hydrogenations:

- Wilkinson's catalyst, $RhCl(PPh_3)_3$ and the closely related $[Rh(diene)(PR_3)_2]^+$ complexes.
- Mixtures of platinum and tin chlorides.
- Anionic cyanocobalt complexes.
- Ziegler catalysts prepared from a transition metal salt and an alkylaluminum compound.

Each of these classes has particular advantages in synthesis. The first three classes are mechanistically distinct from one another, while the mechanism of the Ziegler catalysts is not well defined.

The best studied soluble catalyst for olefin hydrogenation is Wilkinson's catalyst, $RhCl(PPh_3)_3$ [39]. This moderately stable, commercially available compound catalyzes the hydrogenation of many sorts of olefins under mild conditions [40]. Terminal olefins such as 1-hexene are rapidly hydrogenated at room temperature and atmospheric pressure. The hydrogenations of internal olefins proceed slowly with this catalyst, but excellent results are often attained. This catalyst selectively reduces C═C bonds in the presence of other easily reduced functions such as nitro and —CH═O. It adds H_2 or D_2 cleanly *cis* to the C═C bond and usually produces little HD scrambling when it is used to introduce deuterium. Its application in synthesis is illustrated by the Organic Syntheses procedure for the hydrogenation of carvone [40]:

$$CH_2{=}C(CH_3)\text{-carvone} \xrightarrow{H_2} (CH_3)_2CH\text{-carvotanacetone}$$

Other uses, including steroid hydrogenation [41], are described in a recent book on homogeneous hydrogenation in organic chemistry [37]. One potential industrial application is the hydrogenation of dicyanobutenes to adiponitrile [42]:

$$NCCH_2CH{=}CHCH_2CN \xrightarrow[\text{base}]{H_2,RhCl(PPh_3)_3} NC(CH_2)_4CN$$

Even though Wilkinson's catalyst is slow in hydrogenation of internal olefins, the closely related catalysts $[Rh(diene)(PR_3)_2]^+$ [43] are useful with highly substituted olefins. These cationic catalysts are employed in the asymmetric hydrogenation of olefins.

Another family of catalysts for hydrogenation of C═C in the presence of C═O is generated by mixing platinum and tin chlorides [44]. The commercially available H_2PtCl_6 and $SnCl_2{\cdot}2H_2O$ react in methanol to form deep red solutions that contain species such as $[Pt(SnCl_3)_5]^{3-}$ [45]. These solutions catalyze hydrogenation of simple linear olefins and have been extensively studied for hydrogenation of vegetable oils to remove excessive unsaturation, which is responsible for flavor instability [46]. Other ligands such as phosphines are often added to modify the catalytic activity.

For hydrogenations in water with an inexpensive catalyst, solutions containing cobalt salts and excess cyanide ion are useful [47,48]. These solutions contain complex anions such as $[Co(CN)_5]^{3-}$ and $[HCo(CN)_5]^{3-}$. The catalysts are selective for hydrogenation of C═C bonds, which are conjugated with one another or with C═O, C≡N, or phenyl groups. In contrast to other diene hydrogenation catalysts, the cobalt cyanides are relatively unreactive with unconjugated dienes such as 1,5-cyclooctadiene.

For industrial applications such as hydrogenation of unsaturated polymers, Ziegler-type systems are useful [49]. These systems are prepared by mixing a hydrocarbon-soluble complex of a first-row transition metal with an alkane solution of an alkylaluminum compound. Typically, cobalt acetylacetonate or 2-ethylhexanoate is used with triethyl- or triisobutylaluminum [50]. Alkylithium reagents often replace the pyrophoric alkylaluminum compounds successfully [51]. The mixtures are dark, air-sensitive solutions that may contain some colloidal metal. Because of the presence of highly reactive alkyl-metal bonds, these catalysts react with functional groups such as OH, C═O, and C≡C—H.

In addition to these major classes, $Co_2(CO)_8$ and $[Co(CO)_3(PBu_3)]_2$ are useful hydrogenation catalysts. The latter is particularly useful because it is relatively stable and shows excellent selectivity in the hydrogenation of polyenes to mono-olefins. Other carbonyls such as $Cr(CO)_6$ and $Fe(CO)_5$ are also useful. They are less active but become effective when activated by heating or radiation to expel a carbonyl ligand and create a vacant coordination site. Polymer-attached metal complexes are potentially useful for large-scale hydrogenations because they facilitate separation of the catalyst from the product. The rapidly evolving areas of photoactivation and catalyst immobilization are discussed in Chapter 12.

Mechanism of Olefin Hydrogenation

Most soluble catalysts add hydrogen to a C═C bond very simply. The olefin and H_2 are brought together as ligands in the coordination sphere of the metal. A rearrangement of the H—M—(C‖C) complex to a metal alkyl is followed by some sort of M—C bond cleavage process. Catalysts differ in the mode of cleaving H_2 to form the metal hydride ligand and in the mechanism of cleavage of the metal-alkyl bond to form alkane. Three different H_2 cleavage mechanisms are observed for rhodium(I), platinum-tin, and cobalt cyanide catalysts. The third system may differ quite fundamentally in its hydrogenation mechanism.

At least three mechanistic pathways have been demonstrated for Wilkinson's catalyst [52–54]. The kinetically dominant mechanism is shown in Figure 3.6. The Rh^I species shown at the top of the catalyst cycle is probably a solvated three-coordinate complex formed by dissociation of a triphenylphosphine ligand from the parent compound:

$$RhCl(PPh_3)_3 \rightleftharpoons RhCl(PPh_3)_2 + Ph_3P$$

This Rh^I species is very coordination-deficient, with a formal electron count of 14. It readily undergoes "oxidative addition" of an H_2 molecule to form a dihydride, which is formally Rh(III) if the H ligands are regarded as H^-. In the presence of

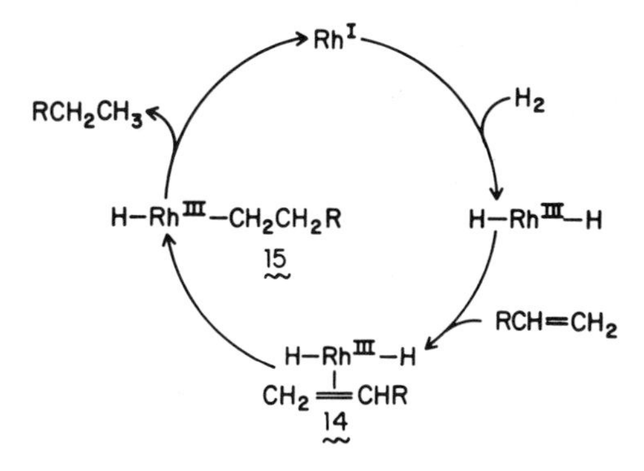

Figure 3.6. A major mechanistic pathway for olefin hydrogenation by Wilkinson's catalyst.

triphenylphosphine, $RhH_2Cl(PPh_3)_3$ is detected in solution by nmr [52]. When the concentration of triphenylphosphine is limited, the dihydride coordinates an olefin to form the complex (**14**). This complex, in turn, undergoes migratory insertion to produce the alkylrhodium hydride complex (**15**). The alkylrhodium hydride complex rapidly eliminates alkane and regenerates the catalytically active Rh^{I} species to complete the cycle. The same cycle probably operates for the $[Rh(diene)(PR_3)_2]^+$ catalysts. In this instance the Rh^{I} species is solvated $[Rh(PR_3)_2]^+$, which is formed by hydrogenation of the diene ligand. The sequence of reaction with H_2 and olefin is reversed with this catalyst [55]. The olefin coordinates *before* the oxidative addition of hydrogen, which is the rate-limiting step. In the hydrogenation of α-acetamidocinnamic acid, the alkyl hydride complex has been detected by nmr at low temperature.

The hydrogenation catalyst prepared by mixing H_2PtCl_6 or K_2PtCl_4 with stannous chloride follows a different reaction pathway. The catalyst solutions contain several anionic complexes, of which the best characterized is $[Pt(SnCl_3)_5]^{3-}$. The $SnCl_3^-$ ligands in this anion are very labile and dissociate to give vacant coordination sites for reaction with H_2 and with olefins. The ligands also inhibit reduction of the Pt(II) ion to metallic platinum. The $SnCl_3^-$ ligand appears to be a weak *sigma* donor and a good *pi* acceptor like carbon monoxide. These ligand-metal bond characteristics seem to favor stability of low valent metal centers.

A major difference from Wilkinson's catalyst is the mechanism of hydrogen activation. With the anionic platinum catalyst it occurs by heterolytic cleavage of H_2:

$$H_2 + [Pt(SnCl_3)_5]^{3-} \rightleftharpoons H^+ + [HPt(SnCl_3)_4]^{3-} + SnCl_3^-$$

The anionic platinum hydride, which can be isolated as a tetraalkylammonium salt [56], reacts with an olefin to give an alkyl complex, $[RPt(SnCl_3)_4]^{3-}$. Presumably the coordination of the olefin and the insertion reaction with the Pt—H bond occur much as indicated for Wilkinson's catalyst. The second major difference between the two systems lies in the cleavage of the metal-alkyl bond. In the platinum system, protonolysis by the acid formed in H_2 cleavage is the most likely reaction:

$$[RPt(SnCl_3)_4]^{3-} + H^+ \rightarrow RH + [Pt(SnCl_3)_4]^{2-}$$

The third major mechanism is based on homolytic cleavage of the dihydrogen molecule by metal-metal bonded species or by a paramagnetic complex. Two important examples are found in cobalt chemistry:

$$Co_2(CO)_8 + H_2 \rightleftharpoons 2\ HCo(CO)_4$$

$$2[Co(CN)_5]^{3-} \rightleftharpoons [Co_2(CN)_{10}]^{6-} \xrightarrow{H_2} 2[HCo(CN)_5]^{3-}$$

The kinetics of the cobalt cyanide system [57] are those expected if H_2 cleavage is the slow step. It is interesting that the two hydrogen cleavage reactions just shown occur with different oxidation states of the metal. The carbonyl reaction involves a formal transition form Co(0) to Co(I), whereas the cyanide complex changes from Co(II) to Co(III).

The reaction of the cobalt hydride with an olefin such as styrene is proposed to occur without precoordination [58]:

$$\begin{matrix} >C{=}C< \\ \vdots \quad \vdots \\ H{-}Co(CN)_5^{3-} \end{matrix} \longrightarrow \left[H{-}\overset{|}{\underset{|}{C}}{-}\overset{|}{\underset{|}{C}}{-}Co(CN)_5 \right]^{3-}$$

If this proposal is correct, it is an exception to the general rule that olefin insertion reactions involve a coordinated C=C bond. The rate and course of styrene hydrogenation change little with changes in cyanide ion concentration, whereas major effects would be expected if CN^- dissociation were required to free a coordination site for the olefin. Cyanide concentration does change the stereochemistry of diene hydrogenation, presumably by regulating formation of a π-allyl intermediate [48]. With either alkyl or allyl intermediates the reduced organic product is released from the metal by reaction with a second cobalt hydride anion:

$$H{-}\overset{|}{\underset{|}{C}}{-}\overset{|}{\underset{|}{C}}{-}Co(CN)_5^{3-} + HCo(CN)_5^{3-} \longrightarrow H{-}\overset{|}{\underset{|}{C}}{-}\overset{|}{\underset{|}{C}}{-}H + 2\,Co(CN)_5^{3-}$$

3.5 SELECTIVE HYDROGENATION OF POLYENES

The hydrogenation of dienes and trienes to monoolefins has several potentially useful applications in industry. Heterogeneous catalysts are used commercially for selective hydrogenation of polyunsaturated vegetable oils to give shortenings with better physical properties [59]. Typically, a linolenate ester with three C=C bonds per C_{18} chain may be reduced to a linoleate with two double bonds or an oleate with only one double bond per chain. In applications of this type, in which only moderate selectivity is required, heterogeneous catalysts excel because they are easy to separate from reaction products by decantation or filtration. Soluble catalysts are very difficult to separate from high boiling materials such as vegetable oils except when the catalyst can be extracted into a polar solvent.

Homogeneous catalysts are preferred over their heterogeneous analogs when high selectivity is required. One practical example is the hydrogenation of 1,5,9-cyclododecatriene to cyclododecene. This selective hydrogenation converts a readily available butadiene trimer (Section 4.5) to a precursor of dodecanedioic acid and laurolactam, two commercial polyamide intermediates. This hydrogenation is not operated industrially but may have considerable potential. Although heterogeneous catalysts such as palladium on alumina can be used for this hydrogenation, higher selectivity is available with soluble catalysts (Table 3.2). The two soluble catalysts prepared from nonprecious metals, $[Co(CO)_3(PBu_3)]_2$ and $NiI_2(PPh_3)_2$, are especially interesting because they are inexpensive and give very high yields of the

Table 3.2 Hydrogenation of 1,5,9-Cyclododecatriene (CDT) to Cyclododecene (CDE)[a]

Catalyst	Conditions: Temperature	Conditions: Pressure (atm)	Conditions: Time (hr)	Solvent	Conversion of CDT	Yield of CDE	Reference
Pd/Al_2O_3	150°	1.5	—	CDT	99%	91%	60
$[Co(CO)_3(PBu_3)]_2$	140°	30	1.7	C_6H_6	100%	99	61
$NiI_2(PPh_3)_2$	160°	80	6	CDT	100	99	62
$[Pt(SnCl_3)_5]^{3-}$	160°	100	10	Et_4NSnCl_3	99	87	56
$RuCl_2(PPh_3)_3$	25°	10	30	$C_6H_6/EtOH$	100	87	63
$RuCl_2(CO)_2(PPh_3)_2$	140°	5	1	C_6H_6	100	98	64
$RhCl_3(py)_3/NaBH_4$	75°	1	0.1	DMF	100	91	65

[a] Most of the hydrogenations employed commercially available *cis, trans, trans*-1,5,9-cyclododecatriene and yielded a mixture of *cis*- and *trans*-cyclododecene.

desired cyclododecene. Similar results are obtained in the hydrogenation of 1,5-cyclooctadiene to cyclooctene with some of the same catalysts [63,66].

The catalysts listed in Table 3.2 effect partial hydrogenation of either conjugated or unconjugated dienes and trienes. In contrast a second group of catalysts hydrogenate only conjugated dienes. Catalysts such as $[Co(CN)_5]^{3-}$ and $[Cr(CO)_3(\text{methyl benzoate})]$ reduce 1,3,7-octatriene, a linear butadiene dimer (Section 4.5), to mixtures of octadienes [47,67] by selective H_2 addition to the 1,3-diene function:

The unconjugated diene products are hydrogenated slowly, if at all, under standard reaction conditions. Other catalysts specific for conjugated dienes include $Cr(CO)_6$ [67–69] and $[Cr(CO)_2(C_5H_5)]_2$ [70].

A major distinction between the two classes of catalysts is that those in Table 3.2 are olefin isomerization catalysts, while the cobalt cyanide and chromium catalysts are not. The more versatile catalysts can convert unconjugated dienes or trienes to conjugated systems through double-bond migration (Section 3.3). Since most of the catalysts in this class are hydride complexes or form hydrides when treated with H_2, it is likely that the isomerization occurs by an M—H addition-elimination process.

The selectivity for hydrogenation of dienes in the presence of monoolefins arises from the exceptional stability of π-allyl complexes. Regardless of the H_2 cleavage mechanism, M—H addition to a conjugated diene is believed to generate a π-allyl intermediate. In a hydrogenation mixture that contains diene, monoolefin, and a platinum-tin chloride catalyst, the following reactions are believed to be in competition [56]:

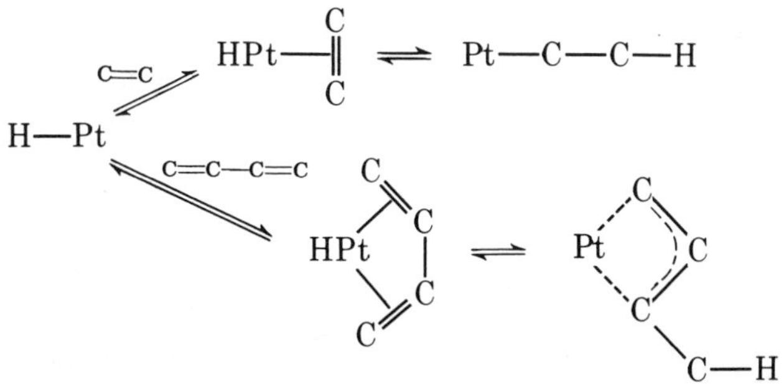

H—Pt + C═C ⇌ HPt—(C═C) ⇌ Pt—C—C—H

H—Pt + C═C—C═C ⇌ HPt(C═C—C═C) ⇌ Pt(π-C—C—C)—C—H

The reaction pathway involving the π-allyl intermediate is favored, especially when the olefin or diene must compete with excess ligand such as R_3P, CO, or $SnCl_3^-$ for a coordination site. Consequently, the diene in the reaction mixture is almost completely hydrogenated before the concentration of olefin increases to the point that olefin gains access to the catalyst. Similar competition for catalyst sites is believed to be responsible for selectivity in hydrogenation of dienes by heterogeneous catalysts.

The $Cr(CO)_6$ and $Cr(CO)_3$(arene) catalysts are said to catalyze hydrogenation of conjugated dienes specifically by 1,4-addition of hydrogen [68]. The diene coordinates in a *cisoid* configuration, and hydrogenation occurs according to the sequence:

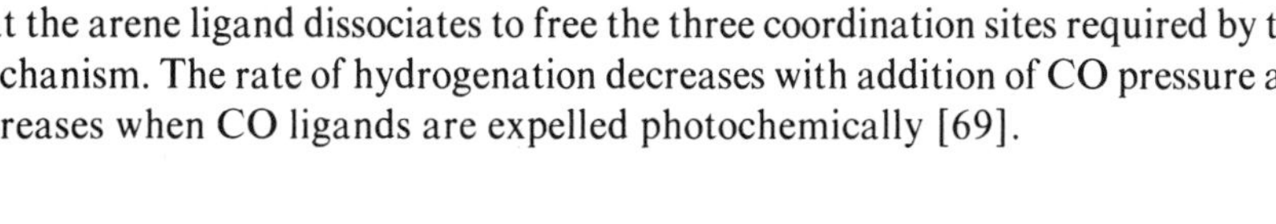

$$Cr(CO)_6 \xrightarrow[-2\,CO]{\text{diene}} (\text{diene})Cr(CO)_4 \xrightarrow[-CO]{H_2} H—Cr(\text{diene})(CO)_3—H \longrightarrow (\text{olefin})Cr(CO)_3$$

The high selectivity for 1,4-addition and the *cis* conformation of the olefinic product are consistent with this proposal. With the $Cr(CO)_3$(arene) catalysts it is proposed that the arene ligand dissociates to free the three coordination sites required by this mechanism. The rate of hydrogenation decreases with addition of CO pressure and increases when CO ligands are expelled photochemically [69].

3.6 ASYMMETRIC HYDROGENATION

Synthesis of optically active organic compounds from nonchiral starting materials is perhaps the most elegant application of homogeneous catalysis. This asymmetric induction can occur in many reactions catalyzed by transition metal complexes (Chapter 12), but the first commercial application [71] is the Monsanto synthesis of L-dopa, a drug used in treatment of Parkinson's disease.

The essential feature for selective synthesis of one optical isomer of a chiral substance is an asymmetric catalyst site that will bind a prochiral olefin preferentially in one conformation. This "recognition" of the preferred conformation is achieved by use of a chiral ligand on the metal. The ligand creates a "chiral hole" in the coordination sphere of the metal. In the L-dopa synthesis, a substituted cinnamic acid (**16**) is the prochiral olefin and a rhodium(I) complex bearing an optically active phosphine ligand is the catalyst [72]. The original process used a monodentate phosphine (**17**), but a chelating phosphine (**18**) gives higher optical yields and greater catalyst stability [73].

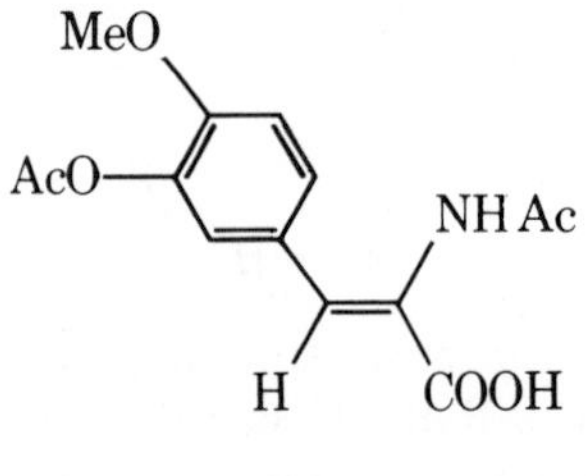

16 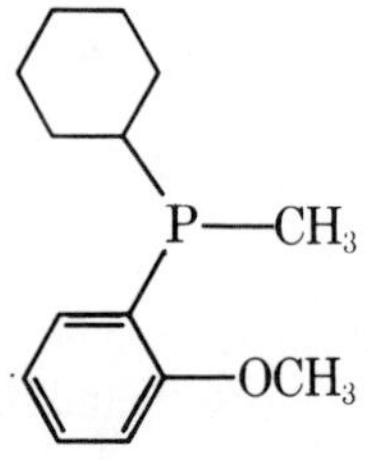 **17** 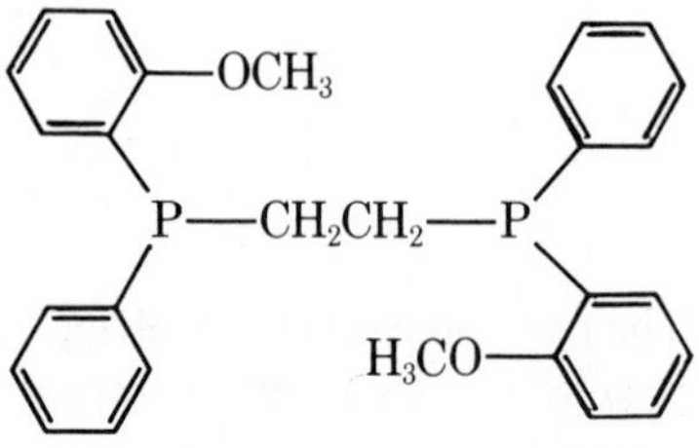 **18**

The phosphine is added to a methanol solution of a diene rhodium complex [43] to generate a cationic catalyst, $[RhH_2(PR_3)_2(MeOH)_n]^+$, after the diene ligand has been hydrogenated away. The cinnamic acid derivative coordinates to the rhodium through its C=C bond and the CO function of the acetamide group [74,75]. This rigid arrangement constrains the substrate to present one face specifically to the rhodium atom. Transfer of hydrogen ligands from rhodium to the bound face of the olefin generates one optical isomer specifically. In practice optical yields, expressed as enantiomeric excess of the desired conformation, approach 90%. The yields are very sensitive to solution pH and to the purity of the hydrogen used.

Rigidity and stability of the Rh—P bonds obviously contribute to the "recognition" ability of the complex. Diphosphine ligands often give superior optical yields because they form a more rigid complex. One of the first diphosphine ligands to be employed was DIOP [76], also known as the Kagan ligand. This ligand, used in conjunction with cationic rhodium catalysts, gives reasonable optical yields of hydrogenation products from many prochiral substrates. Because of its versatility and commercial availability, it is often the ligand of choice for exploratory studies of asymmetric hydrogenation. It is made from a natural product (tartaric acid) and contains its fundamental asymmetry at carbon centers rather than at phosphorus, as in the Monsanto catalysts. It suffers the disadvantage that the phosphorus atoms are separated by four carbon atoms, too far for stable chelate ring formation.

H

CH_3 O—C*—CH_2PPh_2

C

CH_3 O—C*—CH_2PPh_2

H

DIOP

It is usually assumed that the cationic rhodium(I) catalysts function by a mechanism much like that shown for Wilkinson's catalyst in Figure 3.6. This type of mechanism has been supported by studies on $[Rh(diene)(PR_3)_2]^+$ catalysts with nonchiral ligands [43]. Recently, it has been suggested that, with chelating diphosphine ligands such as **18,** the olefin coordinates to the catalyst before reaction with H_2 occurs [55,77]. Some *cis-trans* isomerization of the olefin occurs prior to hydrogenation in the L-dopa synthesis [78].

Many different chiral ligands have been explored in search of high optical yields in hydrogenation of olefins and ketones [79]. To date, success has been a matter of chance. Proper matching of ligand and substrate has been sought empirically because the detailed steric and bonding interactions in the active catalysts are not well known. Stereochemical and structural studies of these interactions seem a fruitful field of research.

REFERENCES

1. W. C. Zeise, *Pogg. Ann. Phys. Chem.*, **9,** 632 (1827).
2. M. J. S. Dewar, *Bull. Soc. Chim. France,* **18,** C79 (1951); *J. Am. Chem. Soc.*, **101,** 783 (1979).
3. J. Chatt and L. A. Duncanson, *J. Chem. Soc.*, 2939 (1953).
4. M. Herberhold, *Metal π-Complexes,* Elsevier, Vol. 1, *Complexes with Di- and Oligo-Olefinic Ligands,* 1966; Vol. 2, *Complexes with Monoolefinic Ligands,* 1974.
5. P. M. Henry, *J. Am. Chem. Soc.*, **88,** 1595 (1966).
6. O. N. Temkin, A. G. Ginzburg, and R. M. Flid, *Kinet. Katal.*, **5,** 221 (1964).
7. M. A. Muhs and F. T. Weiss, *J. Am. Chem. Soc.*, **84,** 4697 (1962).
8. F. R. Hartley, *Angew Chem. Int. Ed.*, **11,** 596 (1972).
9. J. A. J. Jarvis, B. T. Kilbourn, and P. G. Owston, *Acta Cryst.*, **B27,** 366 (1971).
10. N. Rösch, R. P. Messmer, and K. H. Johnson, *J. Am. Chem. Soc.*, **96,** 3855 (1974).
11. B. Akermark, M. Almemark, J. Almlöf, J-e. Backvall, and A. Stogard, *J. Am. Chem. Soc.*, **99,** 4617 (1977).
12. A. J. Birch and I. D. Jenkins, "Transition Metal Complexes of Olefinic Compounds," in H. Alper, Ed., *Transition Metal Organometallics in Organic Synthesis,* Vol. 1, Academic Press, 1976, pp. 2–75.
13. W. Keim, "The π-Allyl System in Catalysis," in G. N. Schrauzer, Ed., *Transition Metals in Homogeneous Catalysis,* Marcel Dekker, 1971, p. 59.

14. A. Panunzi, A. DeRenzi, and G. Paiaro, *J. Am. Chem. Soc.*, **92,** 3488 (1970).
15. W. H. Knoth, *Inorg. Chem.*, **14,** 1566 (1975).
16. M. Rosenblum, *Accounts Chem. Res.*, **7,** 122 (1974).
17. P. Lennon, A. M. Rosan, and M. Rosenblum, *J. Am. Chem. Soc.*, **99,** 8426 (1977).
18. D. L. Thorn and R. Hoffmann, *J. Am. Chem. Soc.*, **100,** 2079 (1978).
19. R. R. Schrock and G. W. Parshall, *Chem. Rev.*, **76,** 243 (1976).
20. G. M. Whitesides, J. F. Gaasch, and E. R. Stedronsky, *J. Am. Chem. Soc.*, **94,** 5258 (1972).
21. F. N. Tebbe and G. W. Parshall, *J. Am. Chem. Soc.*, **93,** 3793 (1971).
22. D. Bingham, D. E. Webster, and P. B. Wells, *J. Chem. Soc. Dalton,* 1514, 1519 (1974).
23. M. Orchin, *Adv. Catal.*, **16,** 1 (1966).
24. R. Cramer, *Accounts Chem. Res.*, **1,** 186 (1968).
25. M. Tuner, J. v. Jouanne, H-D. Brauer, and H. Kelm, *J. Mol. Catal.*, **5,** 425, 433, 447 (1979).
26. J. Andrieux, D. H. R. Barton, and H. Patin, *J. Chem. Soc. Perkin,* **I,** 359 (1977).
27. A. J. Birch and G. S. R. Subba Rao, *Tetahedron Letts,* 3797 (1968).
28. E. J. Corey and J. W. Suggs, *J. Org. Chem.*, **38,** 3224 (1973).
29. D. Bingham, B. Hudson, D. Webster, and P. B. Wells, *J. Chem. Soc. Dalton,* 1521 (1974).
30. C. A. Tolman, *J. Am. Chem. Soc.*, **94,** 2994 (1972).
31. R. G. Miller et al., *J. Am. Chem. Soc.*, **96,** 4211, 4221, 4229 (1974).
32. R. Cramer, *J. Am. Chem. Soc.*, **88,** 2272 (1966).
33. W. T. Hendrix and J. L. von Rosenberg, *J. Am. Chem. Soc.*, **98,** 4850 (1976).
34. L. W. Gosser and G. W. Parshall, *Tetrahedron Lett.*, 2555 (1971).
35. H. J. Golden, D. J. Baker, and R. G. Miller, *J. Am. Chem. Soc.*, **96,** 4235 (1974).
36. B. R. James, *Homogeneous Hydrogenation,* Wiley-Interscience, 1973.
37. F. J. McQuillin, *Homogeneous Hydrogenation in Organic Chemistry,* D. Reidel, 1976.
38. A. J. Birch and K. A. M. Walker, *Aust. J. Chem.*, **24,** 513 (1971); J. W. Kang, U.S. Patent 3,993, 855 (1977).
39. J. A. Osborn, F. H. Jardine, J. F. Young, and G. Wilkinson, *J. Chem. Soc. A,* 1711 (1966).
40. R. E. Ireland and P. Bey, *Org. Syn.*, **53,** 63 (1973).
41. W. Voelter and C. Djerassi, *Chem. Ber.*, **101,** 58 (1968).
42. F. N. Jones, U.S. Patent 3,459,785 (1969).
43. R. R. Schrock and J. A. Osborn, *J. Am. Chem. Soc.*, **98,** 2134, 2143, 4450 (1976).
44. R. D. Cramer, E. L. Jenner, R. V. Lindsey, and U. G. Stolberg, *J. Am. Chem. Soc.*, **85,** 1691 (1963); I. Yasumori and K. Hirabayashi, *Trans. Faraday Soc.*, **67,** 3283 (1971); W. Strohmeier and L. Weigelt, *Z. Naturforsch.*, **31b,** 387 (1976).
45. R. D. Cramer, R. V. Lindsey, C. T. Prewitt, and U. G. Stolberg, *J. Am. Chem. Soc.*, **87,** 658 (1965).
46. J. C. Bailar, *Plat. Met. Rev.*, **15,** 2 (1971); *J. Am. Oil Chem. Soc.*, **47,** 475 (1970).
47. J. Kwiatek, I. L. Mador, and J. K. Seyler, *Advances in Chemistry Series,* American Chemical Society, **37,** 201 (1963); J. Kwiatek, *Catal. Rev.*, **1,** 37 (1967).
48. M. S. Spencer and D. A. Dowden, U.S. Patent 3,009,969 (1961).
49. A. F. Halasa, U.S. Patent 3,872,072 (1975).
50. M. F. Sloan, A. S. Matlack, and D. S. Breslow, *J. Am. Chem. Soc.*, **85,** 4014 (1963); F. K. Shmidt, S. M. Krasnopolskaya, V. G. Lipovich, and B. A. Bazhenov, *Kinet. Katal.*, **15,** 86 (1974).
51. J. C. Falk and J. Van Fleet, *Org. Syn.*, **53,** unchecked procedure 1888 (1973).
52. C. A. Tolman, P. Z. Meakin, D. L. Lindner, and J. P. Jesson, *J. Am. Chem. Soc.*, **96,** 2762 (1974).

53. J. Halpern and C. S. Wong, *J.C.S. Chem. Comm.*, 629 (1973); J. Halpern, T. Okamoto, and A. Zakhariev, *J. Mol. Catal.*, **2,** 65 (1977).
54. Y. Ohtani, M. Fujimoto, and A. Yamagishi, *Bull. Chem. Soc. Japan,* **50,** 1453 (1977); **51,** 2562 (1978).
55. J. Halpern, private communication.
56. G. W. Parshall, *J. Am. Chem. Soc.*, **94,** 8716 (1972).
57. J. Halpern and M. Pribanic, *Inorg. Chem.*, **9,** 2616 (1970).
58. J. Halpern and L. Y. Wong, *J. Am. Chem. Soc.*, **90,** 6665 (1968).
59. "Bailey's Industrial Oil and Fat Products," D. Swern, Ed., Wiley-Interscience, 1964.
60. K. Katsuragawa and K. Yoshimitsu, *Jap. Kokai* **7,** 408, 481 (1974); *Chem. Abst.*, **81,** 17216 (1974).
61. A. Misono and I. Ogata, *Bull. Chem. Soc., Japan,* **40,** 2718 (1967); U.S. Patent 3,715,405 (1973).
62. L. W. Gosser, U.S. Patent 3,499,050 (1970).
63. J. Tsuji and H. Suzuki, *Chem. Lett.*, 1083 (1977).
64. D. R. Fahey, *J. Org. Chem.*, **38,** 80 (1973).
65. A. D. Shebaldova, V. I. Bystrenina, V. N. Kravtsova, and M. L. Khidekel, *Izvest. Akad, Nauk SSSR, Ser. Khim.*, 2101 (1975).
66. D. R. Fahey, U.S. Patent 3,717,585 (1973).
67. L. W. Gosser, U.S. Patent 3,673,270 (1972).
68. A. Rejoan and M. Cais, *Progress in Coordination Chemistry,* Elsevier, 1968, p. 32.
69. M. S. Wrighton and M. A. Schroeder, *J. Am. Chem. Soc.*, **95,** 5764 (1973).
70. A. Miyake and H. Kondo, *Angew Chem. Int. Ed.*, **7,** 631 (1968).
71. *Chem. Week.*, 71 (Nov. 20, 1974); *Ind. Res.*, 23 (Oct. 1974).
72. W. S. Knowles, M. J. Sabacky, and B. D. Vineyard, *Ann. N.Y. Acad. Sci.*, **295,** 274 (1977); W. S. Knowles and M. J. Sabacky, U.S. Patent 3,849,480 (1974).
73. B. D. Vineyard, W. S. Knowles, M. J. Sabacky, G. L. Bachman, and D. J. Weinkauff, *J. Am. Chem. Soc.*, **99,** 5946 (1977).
74. W. S. Knowles, *Proceedings, 1st International Symposium on Homogeneous Catalysis,* Corpus Christi, TX (1978).
75. J. M. Brown and P. A. Chaloner, *J.C.S. Chem. Comm.*, 321 (1978).
76. H. B. Kagan and T-P. Dang, *J. Am. Chem. Soc.*, **94,** 6429 (1972).
77. J. Halpern, D. P. Riley, A. S. C. Chan, and J. J. Pluth, *J. Am. Chem. Soc.*, **99,** 8055 (1977).
78. C. Detellier, G. Gelbard, and H. B. Kagan, *J. Am. Chem. Soc.*, **100,** 7556 (1978); K. F. Koenig and W. S. Knowles, ibid., **100,** 7561 (1978).
79. J. D. Morrison, W. F. Masler, and M. K. Neuberg, *Adv. Catal.*, **25,** 81 (1976).

4 POLYMERIZATION AND ADDITION REACTIONS OF OLEFINS AND DIENES

Organometallic catalysts, both soluble and insoluble, find wide practical application in C—C bond-forming reactions of olefins and dienes. The largest of these are the polymerizations and copolymerizations of ethylene, propylene, butadiene, and isoprene. Polymerizations of these hydrocarbons with catalysts based on transition metal complexes yield ordered polymers with physical properties different from those of free radical polymers. Similarly, oligomerization of olefins and dienes with soluble metal catalysts is used extensively to produce dimers, trimers, and other "low polymers." Additions of HCN and of Si—H compounds to olefins or dienes are also used extensively in industry. The mechanisms of these reactions, as discussed here, are based on the same elementary steps as those of Chapter 3. The major difference is that polymerization and oligomerization involve olefin insertion into an M—C bond in addition to the M—H insertion involved in isomerization and hydrogenation.

4.1 OLEFIN POLYMERIZATION

The major applications of organometallic catalysts in the polymerization of olefins and dienes are listed in Table 4.1. Both homogeneous and heterogeneous catalysts are used commercially, but the solid catalysts are increasingly preferred because they have technical advantages in many processes. Even the nominally homogeneous catalysts may not be soluble under reaction conditions. For example, catalytic solutions of $VOCl_3$ and i-Bu_2AlCl are clear to the naked eye, but light-scattering experiments suggest the presence of aggregates.

Both heterogeneous and homogeneous catalysts are discussed here because they are closely related mechanistically. The treatment is necessarily brief, but recent books and reviews that cover most aspects of olefin polymerization are listed at the end of the chapter. The references cited in these reviews represent many hundreds

Table 4.1 Polyolefin Production by Coordination Catalysis

Polymer	Major Catalysts	1977 U.S. Production (thousand metric tons)
High-density polyethylene	$TiCl_x/AlR_3$ Cr/silica	1640
Polypropylene	$TiCl_3/AlR_2Cl$ (insoluble)	1230
Ethylene/propylene/diene rubbers	$VOCl_3/AlR_2Cl$ (soluble)	160
cis-1,4-Polybutadiene	TiI_4/AlR_3 $Co(OOCR)_2/Al_2R_3Cl_3$	350
cis-1,4-Polyisoprene	$TiCl_3/AlR_3$	63

of man-years of research in the past quarter century. Curiously, despite this massive effort, little is known with certainty about the mechanism of coordination polymerization. Fundamental facts such as the oxidation state of the catalytic metal site and the nature of the insertion mechanism are not well established in many cases.

Polyethylene

The largest volume plastic in the United States is polyethylene. Two major varieties are produced. High-density polyethylene is a linear polymer with a density approaching 0.96 and a melting point of about 136°C. It is made by coordination polymerization as discussed here. Low-density polyethylene has a density near 0.92 and a wide melting range. It is a highly branched polymer prepared by free radical-initiated polymerization of ethylene at high pressure (about 2000 atmospheres). Many intermediate-density polyethylenes are prepared for specific applications by modification of the conditions used for manufacture of the high-density product. Controlled amounts of branching of the polymer chain are generated by introduction of terminal olefin comonomers such as 1-butene. Union Carbide has recently introduced a series of low-density polyethylene products synthesized with a modified coordination catalyst [1]. This catalytic process is economically advantageous because it operates at a much lower pressure than the radical-initiated process.

Two families of ethylene polymerization catalysts were developed in the early 1950s. The Ziegler catalysts are prepared by reaction of an alkylaluminum compound with $TiCl_4$ or $TiCl_3$ to give compositions that sometimes appear to be soluble in hydrocarbon solvents. The Phillips catalysts [2] are clearly insoluble materials prepared by deposition of chromium oxides on silica. Despite early acceptance of the Ziegler systems, chromium-based catalysts now dominate the production of polyethylene in the United States. The chromium systems now include chromocene-on-silica catalysts developed by Union Carbide [3].

Colloidal Ziegler catalysts are prepared by reaction of $TiCl_4$ with trialkylaluminum compounds in cyclohexane or heptane. The titanium is alkylated and reduced to Ti(III) in the form of a brown floc. In laboratory experiments this material polymerizes ethylene vigorously at room temperature and atmospheric pressure [4]. In commercial practice the catalyst solution is fed to a reactor along with ethylene at pressures varying from 10 to 160 atmospheres. Rapid polymerization occurs at 130–270°C to give a viscous solution of a highly linear polymer with a relatively narrow molecular weight range. This combination of properties is the major advantage of this process and compensates to some extent for the need to remove the corrosive chloride-containing catalyst from the product. Other Ziegler catalysts that are clearly insoluble are prepared from crystalline $TiCl_3$.

The heterogeneous chromium-containing catalysts give polyethylene with properties somewhat different from the Ziegler systems. The heterogeneous catalysts are generally noncorrosive and are left in the product, a substantial advantage because removal of catalyst from a viscous polymer solution is difficult. In addition the catalyst can also be used in a "gas phase" process in which polyethylene is grown directly on the surface of the catalyst in the absence of solvent. Polymerization pressures are fairly low (10–30 atmospheres).

The chromium-on-silica catalysts are obtained in two very different ways. A silica gel may be impregnated with aqueous chromate solution or with a hydrocarbon solution of $(Ph_3SiO)_2CrO_2$ to give a dispersion of Cr(VI) oxide sites on the surface of the support. These catalyst sites appear to be inactive before contact with a reducing agent such as CO, H_2, or ethylene. Reduction yields Cr sites with an oxidation state in the range of +2 to +5 (a subject of controversy) [5]. Alternatively, in a Union Carbide process, reaction of silica with $Cr(C_5H_5)_2$ gives low valent sites that are formally oxidized by the surface OH groups [3]:

$$(\text{Si—OH})_2 + \text{Cr}^{II}(C_5H_5)_2 \xrightarrow{-C_5H_6} (\text{Si—O})(\text{SiOH})\text{Cr}^{II}(C_5H_5) \longrightarrow (\text{Si—O})_2\text{Cr}^{IV}(\text{H})(C_5H_5)$$

It seems likely that chromium-on-silica catalysts prepared by both methods contain Cr—H sites that react with ethylene by a repetitive insertion process to give high molecular weight polymer:

$$\gt\text{Cr—H} \xrightarrow{C_2H_4} \text{—Cr—CH}_2\text{CH}_3 \xrightarrow{nC_2H_4} \gt\text{Cr(CH}_2\text{CH}_2)_n\text{C}_2\text{H}_5$$

Polymer grown in this manner is chemically bonded to the catalyst surface. In practice the polymer molecules are released from the surface by a molecular weight control agent such as hydrogen or by thermal cleavage. In either situation a metal hydride site is regenerated and can initiate growth of a new polymer chain.

$$\rangle Cr{-}CH_2CH_2R \xrightarrow{H_2} \rangle Cr{-}H + CH_3CH_2R$$

$$\rangle Cr{-}CH_2CH_2R \xrightarrow{\Delta} \rangle Cr{-}H + CH_2{=}CHR$$

R = polymer chain

The elimination of β-hydrogen in the thermal process generates a high molecular weight olefin. This olefin can copolymerize with ethylene to form a branch in a polyethylene chain. If olefin formation by β-elimination is frequent, highly branched, low-density polyethylene can be produced.

The fundamental processes for chain growth and termination for the titanium-based catalysts appear to be similar to those sketched here. The titanium catalysts are discussed in more detail in connection with their use in polypropylene production.

Polypropylene

Commercial polypropylene is a very regular polymer with properties that vary with the amount of crystallization that occurs during processing. Densities range from about 0.85 for amorphous material to 0.93 for crystalline isotactic polymer. Other properties change correspondingly. The basic polymer chain is ordered with respect to placement of the methyl groups at the chiral centers in each polymer unit. In the common isotactic polymer, half the polymer chains have methyl groups placed to the left of the extended chain, and the other chains have methyls to the right:

$$R{-}CH_2{-}C(CH_3)(H){-}CH_2{-}C(CH_3)(H){-}CH_2{-}C(CH_3)(H){-}CH_2{-}R'$$

This selectivity for ordering all the methyls on one side of the chain rather than randomly (atactic) is related to the symmetry of the catalyst site. It appears that a surface is required to generate a catalyst site with suitable geometric constraints. All commercial propylene polymerization catalysts are heterogeneous. Some homogeneous catalysts [6] produce syndiotactic polypropylene in which methyl groups are placed alternately to the left and right of the extended polymer chain.

The commercial polypropylene catalysts are modifications of the $TiCl_3/AlR_3$ systems developed by Ziegler and Natta. In one catalyst preparation used in laboratory polymerizations [7] a slurry of violet crystalline $TiCl_3 \cdot \frac{1}{3} AlCl_3$ in heptane is treated with diethylaluminum chloride in the presence of propylene. Rapid polymerization occurs at room temperature and 3–4 atmospheres' pressure to give highly isotactic polypropylene. The polymer slurry is treated with ethanol to kill the catalyst, and the polymer is purified by washing. Similar procedures and catalysts are used in laboratory preparation of isotactic poly-1-butene [8] and crystalline polystyrene [9].

The common commercial production of polypropylene resembles the laboratory

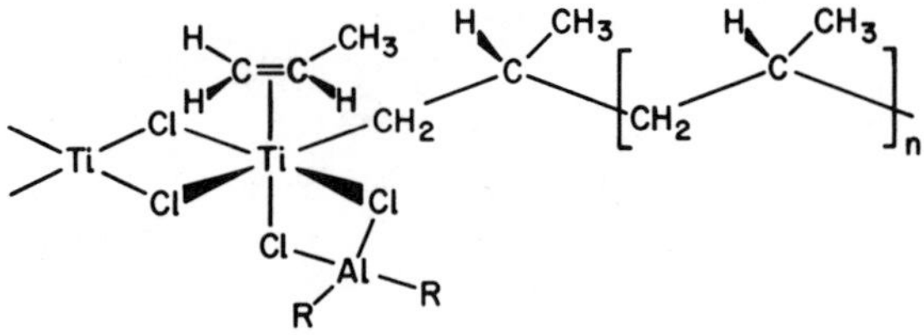

Figure 4.1 Proposed site for isotactic polymerization of propylene on the surface of a $TiCl_3$ crystal. The growing polymer chain extends into solution to the right.

preparation to some extent. Propylene is reacted with a hydrocarbon slurry of the alkylated $TiCl_3$ catalyst at 50–85° and 20–40 atmospheres pressure. In some recently developed systems small $TiCl_3$ aggregates are anchored to the surface of support materials such as $MgCl_2$. This procedure enhances the specific activity of the surface Ti sites. A major goal is to reduce the chloride concentration to the point that the catalyst can be left in the polymer without excessive corrosion to processing equipment. Inactivation and removal of catalyst residues contribute substantially to the cost of polypropylene manufacture.

The chemistry at the catalyst surface is not well known despite intensive study [10]. The treatment with AlR_2Cl is thought to alkylate surface titanium sites. The active site is assumed to be a monoalkylated titanium (3+) ion attached to the crystal by Ti—Cl—Ti bridges (Figure 4.1). It seems likely that alkylaluminum groups are attached at or near the site by Al—Cl—Ti coordination. The growing polymer chain is attached to the crystal by a Ti—C *sigma* bond. As shown in the figure, propylene coordinates to the titanium adjacent to the Ti—C bond [11]. The polymer grows by insertion of the coordinated olefin into the Ti—C bond. The propylene inserts with great regularity with respect to both head-to-tail orientation and to placement of all the methyl groups on the same side of the polymer chain. The regularity must arise from the stereochemistry of the "hole" in the coordination sphere in which the entering olefin molecule coordinates. To generate purely isotactic polypropylene, this coordination site must emerge unchanged from each insertion sequence. The movements of atoms and repopulation of orbitals have been calculated by both CNDO and *ab initio* methods [12,13].

Ethylene-Propylene Copolymers

Whereas the homopolymers of ethylene and propylene are usually plastics, a family of ethylene-propylene copolymers are elastic and are used in place of rubber in many industrial applications. The simple copolymers are often blended with more crystalline polymers as toughening agents. More commonly, though, a small amount of an unsymmetrical diene is copolymerized with ethylene and propylene to give an elastomer that can be cured by conventional rubber technology [14,15]. Common diene comonomers are *trans*-1,4-hexadiene (Section 4.4), dicyclopentadiene, and ethylidenenorbornene. Each of these dienes has a highly reactive double bond that readily copolymerizes to affix the diene to the polymer chain. The less reactive C—C bond in the diene remains intact to provide a site for cross-linking polymer chains in the "cure" of the final product. The cured products are essentially saturated and, in consequence, are more ozone-resistant than natural rubber or the synthetic polydienes discussed here.

Most of the ethylene-propylene copolymers and terpolymers are prepared with Ziegler catalysts based on a soluble vanadium compound such as $VOCl_3$ or VCl_4. Empirically, vanadium seems unique in its ability to incorporate the comonomers into the polymer chain in random sequence, an important characteristic to produce an elastomeric product. In laboratory experiments that simulate commercial practice [16] a reaction solvent such as hexane or chlorobenzene is chilled to 15° and saturated with an ethylene-propylene mixture at 1 atmosphere pressure. The diene (e.g., dicyclopentadiene) is added followed by the catalyst components, in this instance, $Al_2Et_3Cl_3$ and $VOCl_3$. After a short induction period of catalyst formation vigorous polymerization begins, and the three comonomers are added at rates to maintain the desired proportion in the polymer. The relative monomer reactivity is usually ethylene > propylene > diene. When the polymer solution becomes too viscous for effective stirring, the catalyst is killed by addition of alcohol. The catalyst residues are extracted with water, and the polymer is isolated by precipitation or evaporation.

The chemistry of catalyst formation has been studied extensively [10]. At the ratios of Al to V used commercially, VCl_4 and $VOCl_3$ are reduced to V(III) complexes that bear alkyl substituents [17,18]. These V(III) compounds are relatively unstable and decompose to catalytically inactive V(II) species in less than an hour at 15–40°. Fortunately, polymerizations are usually complete in minutes. The presence of chloride ion seems essential for the stability of the active catalyst. It is believed that the catalytically active vanadium complexes (either soluble or insoluble) are very similar to the Ti-based Ziegler catalyst depicted in Figure 4.1. Chloride bridges to aluminum and to other vanadium ions are believed to occupy four coordination sites. Olefin and alkyl ligands are situated in adjacent positions to facilitate the repetitive insertion process by which the polymer chain grows. In contrast to the well-defined sites on the surface of a $TiCl_3$ crystal, which constrain olefin molecules to insert in a very regular fashion, the vanadium catalyst sites must be quite flexible. Propylene, for example, enters the chain randomly with respect to both end-to-end placement and to stereoregularity under most process conditions used commercially.

4.2 POLYBUTADIENE AND POLYISOPRENE

Many of the desirable physical properties of natural rubber are due to the structure of the polymer chain:

$$\left[-CH_2-C(H)=C(CH_3)-CH_2- \right]_n$$

which arises from a *cis*-1,4 polymerization of isoprene. Synthetic elastomers with similar structures and physical properties are produced by coordination polymerization of butadiene and isoprene. Four major catalyst systems give this desired *cis* conformation in a 1,4-polymerization of the diene [19]:

1. alkyllithiums, usually BuLi.
2. iodide-modified Ziegler Ti catalysts.
3. cobalt-centered Ziegler catalysts.
4. allylnickel complexes.

The first three are used commercially for production of polybutadiene or polyisoprene in the United States. The butyllithium is sometimes referred to as an initiator rather than a catalyst, but, like the transition metal catalysts, it is selective for 1,4-polymerization of dienes [20].

Titanium-based Ziegler catalysts like those used to polymerize ethylene and propylene are also very effective for the polymerization of butadiene and isoprene. In a preparative experiment [21] addition of $TiCl_4$ and triisobutylaluminum to an isoprene solution in pentane produces 1,4-polyisoprene. The polymer has better than 94% *cis* conformation. The selectivity for *cis*-1,4-polybutadiene is enhanced by iodide modification of the standard Ziegler catalyst. These catalysts are prepared by reaction of TiI_4 with an alkylaluminum compound or by a three-component reaction such as $TiCl_4 + AlI_3 + AlR_3$. Such catalysts are suspensions of TiI_3 crystals that have been surface-alkylated like the polypropylene catalysts discussed earlier. They give polybutadiene, which is 90–93% *cis*-1,4 conformer [19]. Polymerization of isoprene with a clay-supported $VCl_3/Al(i\text{-}Bu)_3$ catalyst gives the thermodynamically preferred *trans*-1,4-polyisoprene [22]. This modified selectivity with a vanadium catalyst is consistent with a flexible catalyst site, as suggested to explain its randomness in copolymerization of ethylene and propylene.

Organocobalt catalysts based on hydrocarbon-soluble cobalt salts seem to be true homogeneous catalysts. These catalysts have achieved considerable commercial importance because they produce 96–98% *cis*-1,4-polybutadiene when used with a halide-containing aluminum compound. Interestingly, halogen-free cobalt catalysts can yield syndiotactic 1,2-polybutadiene with high specificity. The soluble commercial catalysts are prepared by reacting a cobalt(II) carboxylate with $Al_2R_3Cl_3$ [23]. These conditions are nicely illustrated by a laboratory preparation of *cis*-1,4-polybutadiene [24]. Diethylaluminum chloride and cobalt(II) octoate react in a benzene solution of butadiene in the presence of a small amount of water to produce a very active catalyst. Polymerization at 5° gives a high molecular weight 1,4-polybutadiene with about 98% *cis*-1,4 units.

The nickel-based catalysts are used less extensively than the cobalt systems, but much more is known about their chemistry as a result of extensive study [25–27]. It seems likely that the intermediates in the cobalt and nickel systems are similar. The nickel catalysts can be prepared in several different ways, but all methods appear to give relatively stable π-allyl derivatives in the presence of butadiene. The commercial process, alkylation of a nickel salt, probably proceeds through several steps:

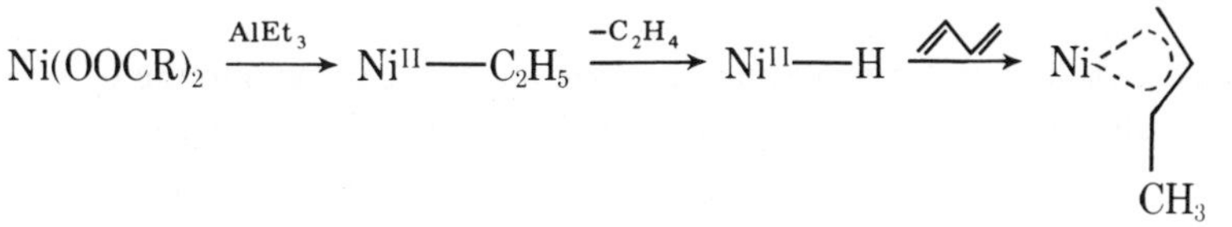

$$Ni(OOCR)_2 \xrightarrow{AlEt_3} Ni^{II}—C_2H_5 \xrightarrow{-C_2H_4} Ni^{II}—H \longrightarrow Ni \; (\pi\text{-crotyl}, CH_3)$$

Similar π-crotyl derivatives are obtained by protonation of zerovalent nickel complexes in the presence of butadiene.

Like cobalt, organonickel systems generally give high percentages of *cis*-1,4-polybutadiene. In the presence of iodide ion, though, the nickel catalysts give predominantly *trans*-1,4 product, the more stable isomer. This variation in product stereochemistry in butadiene polymerization closely parallels that in the codimerization of butadiene and ethylene to give *cis*- or *trans*-1,4-hexadiene (Section 4.5). The stereochemistry of the polybutadiene can be controlled by the concentration of potential ligands in the system. This effect can be used to prepare a "block" polymer that is half *cis*- and half *trans*-1,4-polybutadiene [28]. The polymerization is begun with π-allylnickel trifluoroacetate as the catalyst, which gives rise to a segment of *cis* polymer. However, before polymerization is complete, a trialkyl phosphite ligand is added. The ligand changes the course of the polymerization so that the second half of the growing polymer chain has the *trans* configuration.

The stereoselectivity of the catalyst can be explained on the basis of chelate versus monodentate coordination of the butadiene. Although this explanation is controversial, it accounts for most of the experimental observations [27]. The proposal is illustrated in Figure 4.2, which shows the chemistry involved in preparation of a *cis-trans* block polymer as just described.

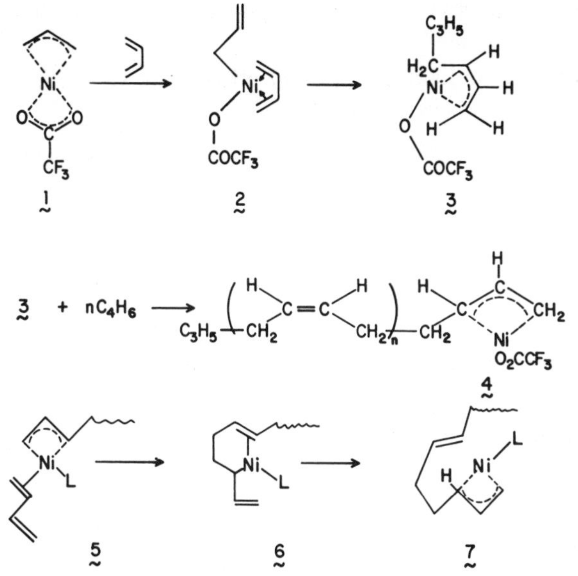

Figure 4.2 Nickel-catalyzed polymerization of butadiene.

The π-allylnickel catalyst (**1**) coordinates a butadiene through both double bonds in chelate fashion to form complex **2**. Insertion of the diene into the C—Ni bond of the σ-allyl ligand in **2** forms a new π-allylic ligand in **3**. The conformation of this allylic ligand ensures that the double bonds in the growing polymer chain in **4** have the *cis* configuration. If, however, a strongly bonding ligand L is added to the system, it occupies a coordination site on nickel and forces the butadiene to complex through only one C=C bond as in **5** (L = $P(OPh)_3$, anion omitted for clarity). Insertion of this single C=C bond gives **6**, which rearranges to a π-allylic structure (**7**). However, the configuration of the allylic ligand formed in this way leads to a *trans* C=C bond in the growing polymer chain. Thus, the presence or absence of ligand L determines the stereochemistry of the growing polymer chain.

4.3 OLIGOMERIZATION OF OLEFINS

The self-addition of olefins to form dimers, trimers, and low polymers is called oligomerization. This process can be identical in mechanism to the ethylene polymerization described earlier except that chain termination occurs much more frequently. In ethylene dimerization to 1-butene, chain transfer by β-hydrogen abstraction follows every insertion into an M—C bond:

$$\mathrm{M{-}H} \xrightarrow{C_2H_4} \mathrm{M{-}C_2H_5} \xrightarrow{C_2H_4} \mathrm{M{-}C_4H_9} \xrightarrow{-\mathrm{M{-}H}} \mathrm{H_2C{=}CHC_2H_5}$$

Reactions of this sort have several practical applications:

- Oligomerization of ethylene by organoaluminum compounds to give linear α-olefins or α-alcohols.
- Nickel-catalyzed oligomerization of ethylene to C_{10}–C_{20} α-olefins (Shell Higher Olefins Process).
- Dimerization of propylene to branched C_6 olefins useful as octane-enhancers in motor fuel.

The codimerization of ethylene and butadiene and the oligomerization reactions of dienes are discussed in Sections 4.4 and 4.5.

Aluminum-Catalyzed Oligomerization of Ethylene

Ethylene readily inserts into Al—H and Al—C bonds to form C_2–C_{40} alkylaluminum compounds. These compounds are intermediates in the commercial production of linear α-olefins and α-alcohols [29,30]:

$$\mathrm{H{-}Al} \xrightarrow{nC_2H_4} \mathrm{H{\left(CH_2CH_2\right)}_n Al} \begin{cases} \xrightarrow{-\mathrm{AlH}} \mathrm{H(CH_2CH_2)_{n-1}CH{=}CH_2} \\ \xrightarrow[\mathrm{H_2O}]{[\mathrm{O}]} \mathrm{H(CH_2CH_2)_nOH} \end{cases}$$

The synthesis of olefins is catalytic because β-hydrogen abstraction from the growing alkyl chain regenerates an Al—H bond that can start growth of a new alkyl chain. The linear olefins produced in this way are intermediates in the synthesis of fatty acid esters, aldehydes, and alcohols by the carbonylation reactions described in Chapter 5.

In contrast to the catalytic α-olefin synthesis, the alcohol synthesis uses the alkylaluminum compound stoichiometrically. An R_3Al compound prepared by ethylene oligomerization is oxidized with air to $(RO)_3Al$, which is hydrolyzed to produce the linear alcohol. Such "fatty alcohols" are biodegradable and are used in the manufacture of detergents. The process competes economically with the hydroformylation of olefins, even though it uses the aluminum compound stoichiometrically rather than as a catalyst.

The chemistry of ethylene addition to Al—C and Al—H bonds [31] dictates the manner in which ethylene oligomerization is carried out commercially. Trialkylaluminum compounds catalyze the reaction of ethylene, aluminum, and hydrogen to form triethylaluminum (which is extensively dimerized in the liquid state):

$$2Al + 3H_2 + 6C_2H_4 \xrightarrow{R_3Al} 2AlEt_3 \rightleftharpoons Al_2Et_6$$

The insertion of ethylene does not stop when $Al—C_2H_5$ groups are formed. Continued insertion into the Al—C bonds produces alkylaluminum compounds in which the alkyl groups are polyethylene chains. The length of the chains is governed by reaction conditions but generally represents a statistical distribution of sizes based on the amount of ethylene available.

Practical syntheses of terminal olefins employ reaction temperatures at which β-hydrogen elimination is frequent. Thus chain growth and termination occur at comparable rates:

$$>Al—Et \xrightarrow{nC_2H_4} >Al—(CH_2CH_2)_nEt \xrightarrow{\Delta} >Al—H + CH_2{=}CH(C_2H_4)_nH$$

$$>Al—H \xrightarrow{C_2H_4} >Al—Et$$

The reaction is carried out typically at 200–250° and 130–250 atmospheres pressure [29]. The high pressure of ethylene prevents the α-olefin products from reinserting into the growing alkyl chains by a simple mass action effect. The predominance of ethylene insertion gives linear olefins with 4–8 carbon atoms as the major products.

The alcohol synthesis process carried out by Conoco [30,32] employs a temperature of only 115–130° in the chain growth step to avoid olefin formation by β-hydrogen abstraction. Ethylene and triethylaluminum are reacted at about 120° and 135 atmospheres to form a statistical mixture of $RR'R''Al$ compounds with most alkyl chain lengths in the C_6–C_{14} range (all even numbered). This mixture is oxidized with dry air at about 35° to form the corresponding $(RO)(R'O)(R''O)Al$ mixture. Hydrolysis with sulfuric acid yields a mixture of fatty alcohols. In laboratory preparations of alcohols from R_3Al compounds, trimethylamine oxide is a convenient reagent for oxidation of the R—Al bond [33].

Shell Higher Olefins Process

A nickel-catalyzed oligomerization of ethylene is used to prepare linear α-olefins on a large scale [34]. As in the aluminum-catalyzed oligomerization, insertion of ethylene into M—H and M—C bonds forms metal alkyls with a statistical distribution of chain lengths. β-Hydrogen abstraction produces olefin and regenerates an Ni—H species to repeat the chain growth sequence:

$$\text{Ni—H} \xrightarrow{nC_2H_4} \text{Ni}(C_2H_4)_n\text{H} \xrightarrow{-\text{NiH}} H_2C{=}CH(C_2H_4)_{n-1}H$$

The mechanism of this process is like that discussed for olefin dimerization in the next subsection. A nickel hydride catalyst is generated by reduction of a nickel salt in the presence of a chelating ligand such as diphenylphosphinoacetic acid [35]:

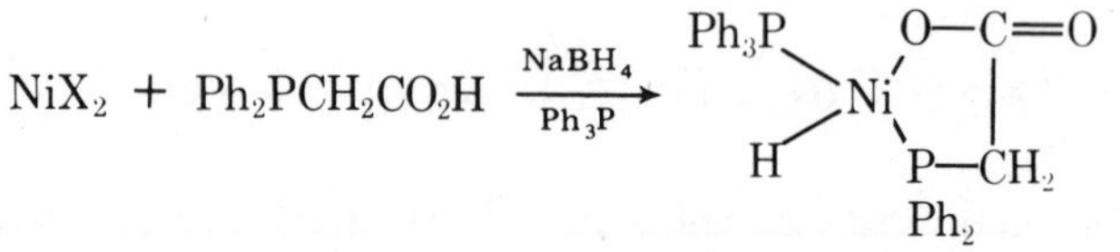

$$NiX_2 + Ph_2PCH_2CO_2H \xrightarrow[Ph_3P]{NaBH_4} (Ph_3P)(H)Ni(O{-}C{=}O)(P{-}CH_2)Ph_2$$

In practice a catalyst of this sort is allowed to react with ethylene in a solvent such as ethylene glycol at about 100° and 40 atmospheres pressure. A rapid reaction occurs to form a mixture of linear α-olefins. The chain lengths are typically distributed as follows [36]:

C_4-C_8	41%
C_{10}-C_{18}	40.5%
C_{20+}	18.5%

The C_{10}-C_{18} olefins, for which ready markets exist, are separated by distillation and used as such. The higher and lower boiling products are then used in a complex sequence of catalytic reactions (Figure 4.3).

The low-boiling and high-boiling olefins are isomerized separately over heterogeneous catalysts to produce internal olefins. This step is necessary because the next step is olefin metathesis, which often does not work well with terminal olefins (Section 9.2). In the process variant shown in the figure, the low- and high-boiling internal olefins react over a heterogeneous catalyst such as MoO_3 on Al_2O_3 to produce a broad range of internal olefins. Owing to the prevalence of reactions such as:

$$CH_3CH{=}CHCH_3 + C_{10}H_{21}CH{=}CHC_{10}H_{21} \rightleftharpoons 2\ CH_3CH{=}CHC_{10}H_{21}$$

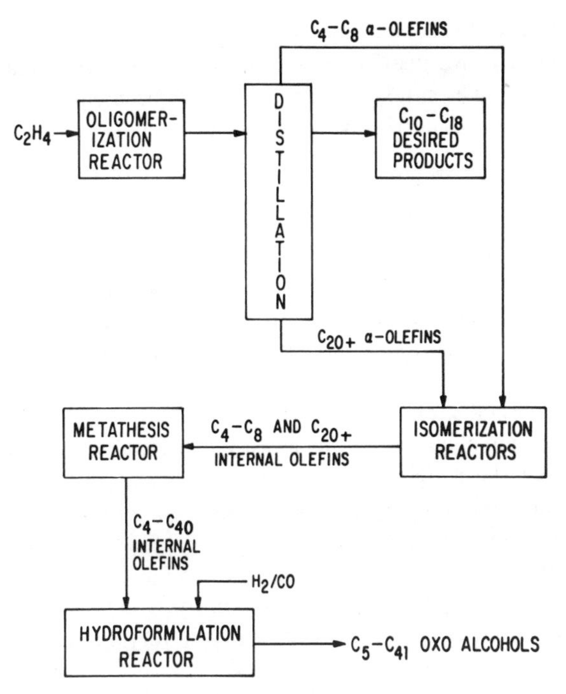

Figure 4.3 A block diagram of the Shell Higher Olefins Process to illustrate complete utilization of ethylene [36].

much of the product is in the useful C_{10}–C_{18} range. Hydroformylation is performed with a cobalt catalyst that converts internal olefins to terminal alcohols (Section 5.5).

$$CH_3CH{=}CHC_{10}H_{21} \rightleftharpoons H_2C{=}CHC_{11}H_{23} \xrightarrow{H_2/CO} H(O{=})CCH_2CH_2C_{11}H_{23} \xrightarrow{H_2} HO(CH_2)_{14}H$$

In this way a high proportion of the ethylene is converted to useful fatty alcohols.

Olefin Dimerization

The dimerization of ethylene to butenes is the simplest and one of the best studied oligomerization reactions [37,38]. Industrially, it is not very significant, because the butenes are generally cheaper than ethylene. However, *in situ* formation of 1-butene during ethylene polymerization may be a convenient way to produce branched polyethylene with properties approaching those of low-density polyethylene. Another potentially useful reaction is the dimerization of propylene, which converts this inexpensive olefin to useful C_6 compounds such as dimethylbutenes and methylpentenes [39]. The dimerization of acrylonitrile to 1,4-dicyanobutene has been extensively investigated as a synthesis of hexamethylenediamine for nylon production [40,41]. The catalytic approaches have been only modestly successful, but electrolytic hydrodimerization is used commercially [42]:

$$2CH_2{=}CHCN + 2e^- + 2H^+ \rightarrow NC(CH_2)_4CN$$

The first simple olefin dimerization was discovered in the course of Ziegler's study of ethylene oligomerization to long-chain α-olefins by alkylaluminum compounds [31]. It was observed that traces of nickel from reactor corrosion diverted the ethylene oligomerization to the production of 1-butene. This chance discovery led to the invention of the Ziegler catalysts for polymerization, hydrogenation, and isomerization of olefins. As discussed earlier, this chemistry provides the basis for the Shell α-olefins process.

The dimerization of simple olefins has been studied extensively as a model for 1,4-hexadiene synthesis, (Section 4.4). Greatest attention has been given to the nickel [38] and rhodium [43] complexes that are commercially attractive for preparation of *trans*-1,4-hexadiene. However, very effective catalysts for olefin dimerization also arise from reaction of alkylaluminum compounds with cobalt salts and with titanium(IV) complexes. Recently, a tantalum catalyst has been discovered to dimerize ethylene and propylene by a very different mechanism from that of the Ziegler-type catalysts [44].

The dimerization of propylene with nickel-based Ziegler catalysts has been explored extensively, and a process based on this chemistry has been tested in a pilot plant [45]. In large-scale operations the catalyst is prepared as it is used by mixing $NiCl_2$, triethylaluminum, and butadiene in chlorobenzene to give a C_{12} π-allyl complex of nickel (Section 4.5). A phosphine is added to the solution, and the mixture is fed to a continuous reactor along with liquid propylene at 15 atmospheres pressure. Rapid dimerization occurs at 30–40°C to give a mixture of *n*-hexenes, 2-methylpentenes, and 2,3-dimethylbutenes in 85–90% yield. The distribution of the isomeric C_6 products depends on the nature of the phosphine used in catalyst preparation. At low temperatures the proportion of methylpentenes can be as high as 80% if PMe_3 or $Ph_2PCH_2PPh_2$ is used. However, with sterically bulky phosphines such as $C_2H_5P(t\text{-}Bu)_2$, 70–80% of the mixture is dimethylbutenes.

Effective catalysts can be prepared from combinations of π-allylnickel halides with phosphines and Lewis acids such as $AlCl_3$ or $EtAlCl_2$ [45]. Similarly, zero-valent nickel complexes such as bis(1,5-cyclooctadiene)nickel and $Ni[P(OPh)_3]_4$ react with Lewis acids to give catalysts for dimerization of propylene. The same catalyst systems also bring about the dimerization of ethylene to butenes and co-dimerization of ethylene and propylene to pentenes. The products are generally the thermodynamically favored internal isomers because the dimerization catalysts also catalyze double-bond migration (Section 3.3).

The active catalyst for olefin dimerization in all these systems is probably a nickel hydride complex [38]. The Ni—H function can be formed by β-hydrogen abstraction from alkyl intermediates formed by interaction of an alkyl or allylnickel complex with propylene:

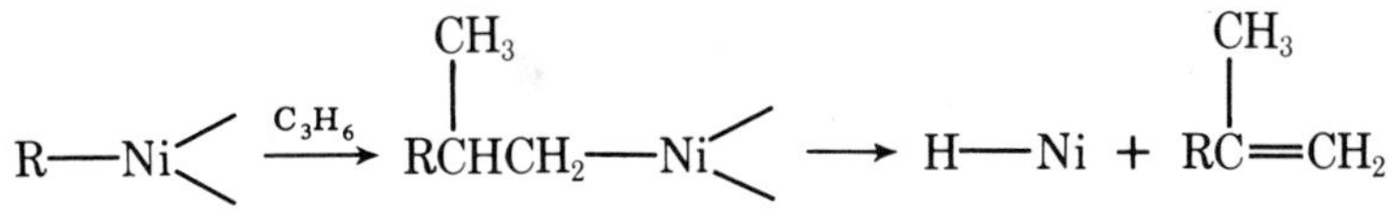

$$\text{R—Ni}\langle \xrightarrow{C_3H_6} \text{R}\overset{CH_3}{\overset{|}{\text{C}}}\text{HCH}_2\text{—Ni}\langle \longrightarrow \text{H—Ni} + \text{R}\overset{CH_3}{\overset{|}{\text{C}}}\text{=CH}_2$$

The other coordination sites about the nickel are occupied by phosphine, olefin, and halide ligands as in **8**. Similarly, the active catalyst in the well-studied rhodium system [43] may resemble **9**. The formation of **9** is discussed in the section on hexadiene synthesis (Section 4.4).

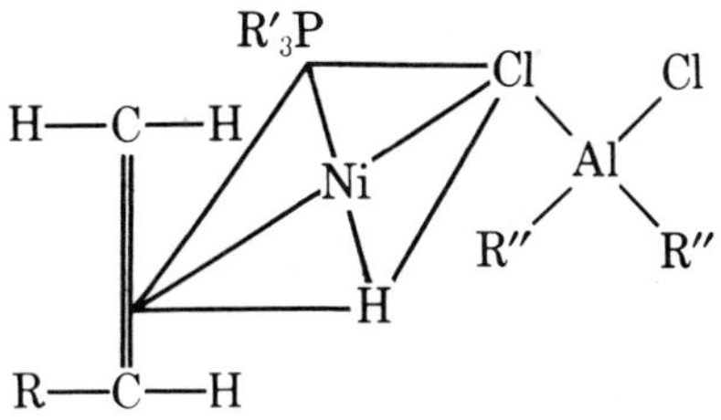

8

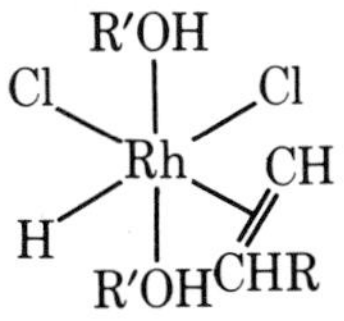

9

These complexes correspond to the olefin hydride complexes shown in the 2-o'clock position of the catalytic cycle of Figure 4.4.

This mechanism employs the familiar steps of olefin coordination to a metal hydride, insertion into the M—H bond, and subsequent insertion into an M—C bond, just as in olefin polymerization catalysts. However, the olefin dimerization catalysts effect β-hydrogen abstraction from the growing alkyl chain after almost every M—C insertion. As a result, olefin dimers predominate in the product. Only small quantities of trimers and tetramers form under normal operating conditions.

The ethylene dimerization of Figure 4.4 can produce only a single product because ethylene insertion into an M—H or M—C bond has no regioselective aspect.

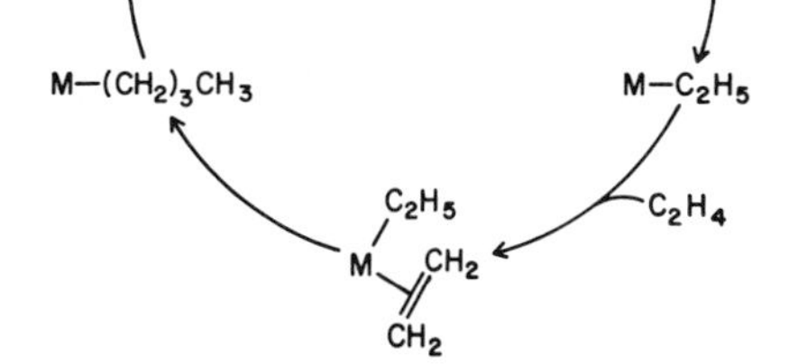

Figure 4.4 Catalytic cycle for dimerization of ethylene to 1-butene by a metal hydride catalyst (M = Ni, Rh).

With propylene, however, both insertion steps can produce isomers:

$$M\text{—}H \xrightarrow{C_3H_6} MCHMe_2 \xrightarrow{C_3H_6} M\text{—}CH(Me)CH_2CHMe_2$$

$$MCHMe_2 \xrightarrow{C_3H_6} M\text{—}CH_2CH(Me)CHMe_2$$

$$M\text{—}H \xrightarrow{C_3H_6} MCH_2Et \xrightarrow{C_3H_6} M\text{—}CH(Me)CH_2CH_2Et$$

$$MCH_2Et \xrightarrow{C_3H_6} M\text{—}CH_2CH(Me)CH_2Et$$

The relative frequencies of the various insertion modes determine the product distribution. A very crowded catalyst site produced by bulky ligands favors formation of alkyl groups with RCH_2 joined to the metal. This effect enhances production of the branched-chain olefins.

An entirely different mechanism has been proposed for another catalyst that is extremely selective for the dimerization of propylene to 2,3-dimethyl-1-butene [44]. A stable alkylidenetantalum complex (Chapter 9) reacts with propylene in two stages, as shown in Figure 4.5. At 0° the alkylidene group is removed from **10**, and an unstable metallocycle (**11**) forms. At 35°C, the metallocycle serves as a catalyst for the dimerization of propylene with more than 90% selectivity to the dimethylbutene. In the catalyst cycle, a 1,3-hydrogen shift across the face of the ring converts the metallocycle to the olefin complex (**12**). Dissociation of the dimeric olefin and coordination of propylene give the bis(propylene) complex (**13**). (Dissociation and association of olefinic ligands may occur simultaneously.) Reorganization of bonds in **13** regenerates the metallocycle (**11**).

The tantalum catalyst converts ethylene to 1-butene efficiently but is less effective for dimerization of 1-butene and higher olefins. Because it does not involve hydride intermediates that might catalyze double-bond migration, the terminal olefin products are stable in the reaction mixture.

This metallocyclic mechanism is probably fairly common for olefin dimerization, but its significance has not been appreciated until recently. The dimerization of ethylene to 1-butene or cyclobutane has been observed with metallocyclic nickel and titanium complexes [46]. The formation of a metallacyclopentane from two moles of ethylene closely resembles the coupling of two moles of butadiene to a metallocycle, which is the first step in oligomerization of the diene by nickel or titanium catalysts (Section 4.5).

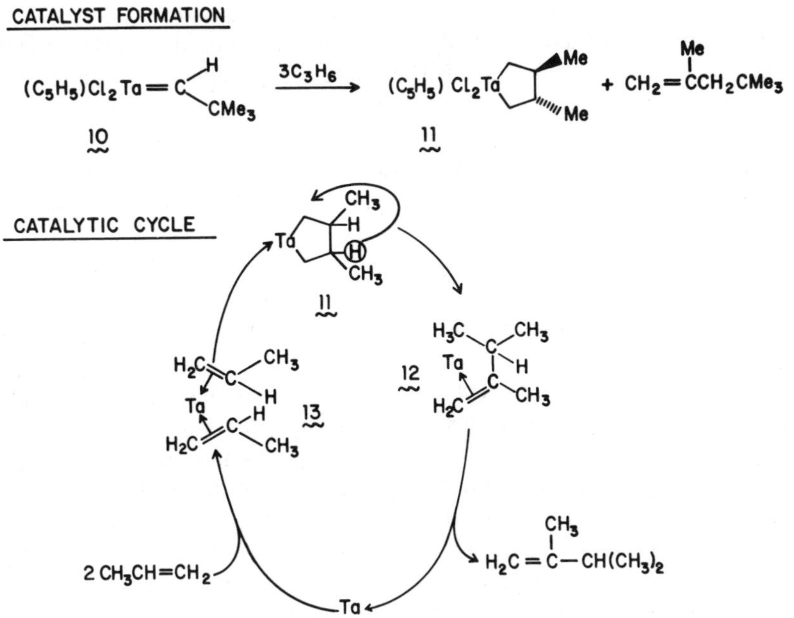

CATALYST FORMATION

CATALYTIC CYCLE

Figure 4.5 A metallocyclic mechanism for propylene dimerization. Nonparticipating ligands are omitted from intermediates in the cycle.

4.4 1,4-HEXADIENE

Codimerization of ethylene and butadiene with a rhodium catalyst is used to produce *trans*-1,4-hexadiene, which is a comonomer in ethylene-propylene-diene elastomers. Ziegler catalysts based on nickel, cobalt and iron salts are also very effective for this codimerization [47]. The nickel catalysts produce largely the industrially interesting *trans* isomer, whereas the cobalt and iron catalysts give the *cis* isomer cleanly. Since many of these catalysts also bring about olefin isomerization, care is required to remove the product from the reaction mixture before it isomerizes to the more stable conjugated 2,4-hexadiene. Similar nickel-based Ziegler catalysts can be used to prepare ethylidenenorbornene [48], which is also widely used in ethylene-propylene-diene elastomers. Ethylene reacts with norbornadiene in two metal-catalyzed steps:

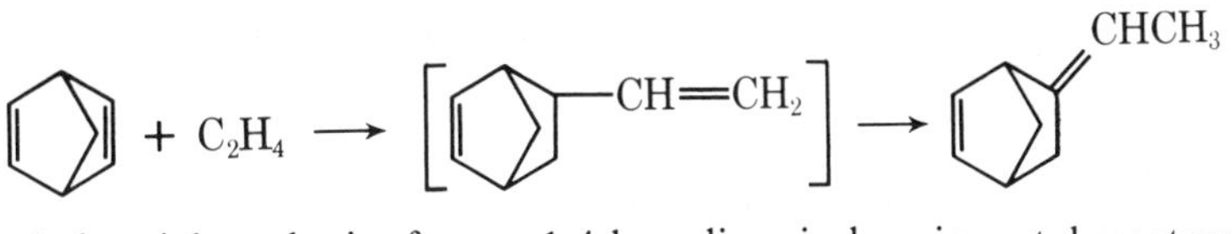

$$\text{norbornadiene} + C_2H_4 \longrightarrow \left[\text{norbornene}{-}CH{=}CH_2\right] \longrightarrow \text{norbornene}{=}CHCH_3$$

The industrial synthesis of *trans*-1,4-hexadiene is done in metal reactors under pressure, but the codimerization of ethylene and butadiene can be studied conveniently in glass pressure bottles in the laboratory [49]. Addition of an equimolar mixture of ethylene and butadiene to a methanol solution of commercial $RhCl_3 \cdot 3H_2O$ leads to a slow reaction that accelerates with time. The initial induction period during which Rh(III) is reduced to Rh(I) can be eliminated by adding a preformed rhodium(I) catalyst such as $[RhCl(C_2H_4)_2]_2$. The product from the 1:1 mole ratio of ethylene and butadiene is a mixture of hexadienes that forms by isomerization of the initially formed *trans*-1,4-isomer. Isomerization can, however, be suppressed by maintaining high concentrations of butadiene [47], that is, operating at low conversion.

Most hexadiene syntheses with the nickel- and cobalt-based Ziegler catalysts are performed in metal pressure reactors. Typically [50], $NiCl_2(PBu_3)_2$ and $Al_2Cl_2(i\text{-}Bu)_4$ are mixed in tetrachloroethylene that has been presaturated with butadiene and ethylene. The two olefins react rapidly at 65° to give mixed hexadienes (mostly *trans*-1,4) and some 3-methyl-1,4-pentadiene, which are isolated by distillation. Instead of using a nickel(II) complex with a reducing agent, nickel(0) complexes can be used to codimerize ethylene and butadiene if acidic cocatalysts are supplied [51]. A mixture of bis(1,5-cyclooctadiene)nickel, $C_6F_5PPh_2$, $EtAlCl_2$, and Et_2AlOEt gives extremely high yields of *trans*-1,4-hexadiene [52]. The aluminum cocatalyst can be replaced by a protonic acid. A solution of $Ni[P(OEt)_3]_4$ and H_2SO_4 in methanol effects hexadiene synthesis under conditions in which several intermediates can be observed [53]. As noted in Section 2.4, such combinations of acid and Ni(0) generate a nickel hydride that is believed to be the true catalyst.

All the nickel and rhodium catalysts are based on metal hydride complexes. In the rhodium system several steps are involved in catalyst formation during the induction period [43]:

$$RhCl_3 + C_2H_5OH \rightarrow \text{“RhCl”} + HCl + CH_3CHO$$

$$\text{“RhCl”} + HCl \rightarrow \text{“HRhCl}_2\text{”} \xrightarrow{\text{ligands}} HRhCl_2\ (\text{ligand})_3$$

The rhodium hydride enters the catalytic cycle of Figure 4.6 at the top. Reaction with butadiene gives a *syn*-π-crotyl complex (**14**) that coordinates a molecule of ethylene to form **15.** Insertion of ethylene at the less hindered end of the crotyl ligand forms a *trans*-4-hexenylrhodium complex (**16**). β-Hydrogen elimination yields a Rh—H complex of *trans*-1,4-hexadiene. Dissociation of the product completes the catalytic cycle [49].

With the rhodium and the $[HNiL_4]^+$ catalysts it seems likely that the stereochemistry about the double bond in the product is determined by an isomerization of the π-crotyl intermediate [47,53]. The initial insertion of butadiene into the metal-hydrogen bond is believed to give an *anti*-crotyl intermediate:

H—Ni ⟶ Ni ⇌ Ni

This isomer would yield *cis*-1,4-hexadiene as the final product, but isomerization to the *syn*-crotyl complex is faster than ethylene insertion [53].

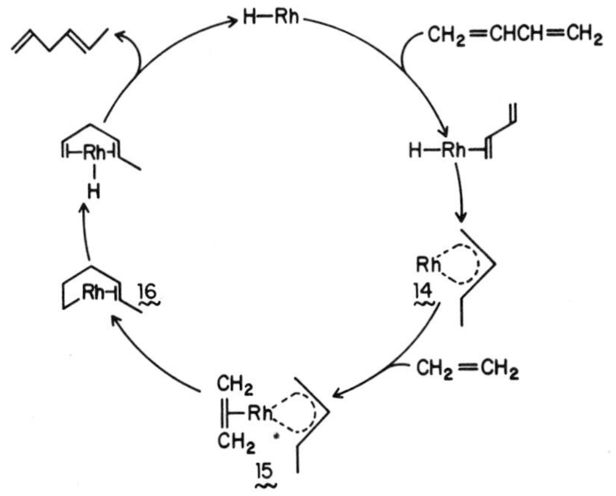

H—Rh
$CH_2=CHCH=CH_2$
H—Rh
Rh
14
$CH_2=CH_2$
CH_2
Rh
CH_2
15
Rh
16
Rh
H

Figure 4.6 Mechanism of rhodium-catalyzed codimerization of ethylene and butadiene. Presumably several intermediates are formally six-coordinate with coordination sites filled by alcohol, chloride, and olefin ligands, which are omitted for clarity.

In the Ziegler systems, product stereochemistry is believed to arise from monodentate or bidentate coordination of the incoming butadiene ligand, just as in butadiene polymerization. With nickel catalysts in the presence of excess phosphine ligand, monodentate coordination gives the *syn*-crotyl intermediate, which yields *trans*-1,4-hexadiene:

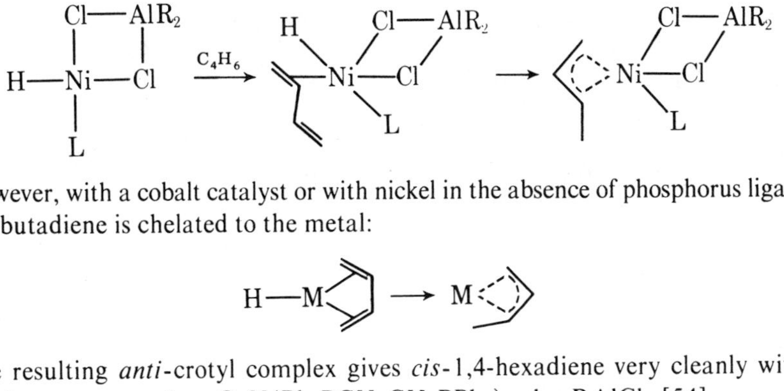

Cl—AlR_2
H—Ni—Cl
L
$\xrightarrow{C_4H_6}$
H
Cl—AlR_2
Ni—Cl
L
$\longrightarrow$
Cl—AlR_2
Ni—Cl
L

However, with a cobalt catalyst or with nickel in the absence of phosphorus ligands, the butadiene is chelated to the metal:

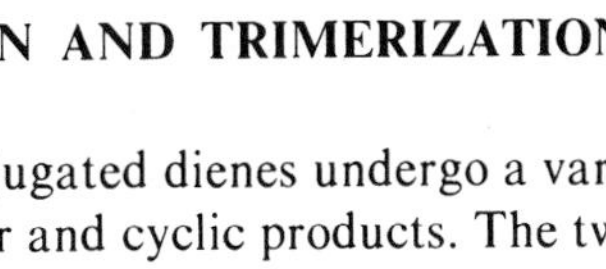

H—M $\longrightarrow$ M

The resulting *anti*-crotyl complex gives *cis*-1,4-hexadiene very cleanly with a catalyst system such as $CoH(Ph_2PCH_2CH_2PPh_2)_2$ plus $RAlCl_2$ [54].

4.5 DIMERIZATION AND TRIMERIZATION OF DIENES

Butadiene and other conjugated dienes undergo a variety of oligomerization reactions to give both linear and cyclic products. The two following are used commercially:

- Cyclodimerization of butadiene with a nickel catalyst to produce 1,5-cyclooctadiene, which is used in preparation of flame retardants such as tetrabromocyclooctane.

- Cyclotrimerization of butadiene to give 1,5,9-cyclododecatriene, an intermediate in the manufacture of dodecanedioic acid (Chapter 10) and lauryllactam [55].

Other potential industrial applications include dimerization of butadiene to linear octatrienes and cyclodimerization of isoprene to dimethylcyclooctadienes. Some of these hydrocarbons would be difficult to synthesize by conventional methods.

Diene Dimerization

Ziegler catalysts like those used for synthesis of polybutadiene and 1,4-hexadiene can be modified to produce both linear and cyclic dimers and trimers of butadiene. Nickel catalysts are the most versatile and can be used to make almost any of the products by suitable choice of ligands and reaction conditions [38]. Emphasis here is centered on 1,5-cyclooctadiene (COD) and 1,3,7-octatriene (OT). Other accessible dimers

include 4-vinylcyclohexene and 1,2-divinylcyclobutane. Many substituted butadienes also dimerize to give analogous cyclooctadienes and octatrienes [56]. The isoprene dimers and trimers are especially interesting for synthesis of terpenoid and sesquiterpenoid compounds of biological interest. The dimethylcyclooctadienes can be used as intermediates in the synthesis of fragrances [57]. Low-valent palladium catalysts are often advantageous for preparation of the linear dimers such as 2,7-dimethyl-2,4,6-octatriene [58].

The cyclodimerization of butadiene to give 1,5-cyclooctadiene is catalyzed by a zerovalent nickel complex that contains one mole of a triaryl phosphite ligand [59]. In large-scale syntheses nickel(II) acetylacetonate is reduced with an organoaluminum compound in the presence of the phosphite. This catalyst differs from that used in hexadiene synthesis or butadiene polymerization in that the aluminum compound need not be a Lewis acid. An equally useful catalyst is obtained by reacting the triaryl phosphite with $Ni(COD)_2$ in liquid butadiene. The most effective catalysts are based on bulky ligands such as tris(*o*-phenylphenyl) phosphite. The catalyst containing this ligand rapidly dimerizes butadiene at 80° and 1 atmosphere pressure in a hydrocarbon solvent such as cyclooctadiene. The product is 96% 1,5-cyclooctadiene, along with small amounts of trimers and 4-vinylcyclohexene. The latter product is the normal thermal dimer of butadiene, although it is not commonly formed below 150°. When butadiene is dimerized at low temperatures and low conversions, 1,2-divinylcyclobutane may be isolated in yields up to 40%. It readily isomerizes to COD and vinylcyclohexene in the presence of the catalyst.

The same catalysts used to cyclodimerize butadiene produce linear dimers when a slightly acidic coreactant is added to the reaction mixture [38]. The coreactant may be ROH, R_2NH, HCN, or active methylene compounds. It may be incorporated in the product or it may promote formation of an octatriene. For example, phenols can give either octatrienes [60] or phenoxyoctadienes [61] as major products:

$$2\ \text{butadiene} \xrightarrow[m\text{-cresol}]{Ni^0} \text{octatriene} + \text{COD}$$

$$2\ \text{butadiene} + \text{PhOH} \xrightarrow{Ni[P(OPh)_3]_4} \text{PhO-octadiene}$$

In general, however, zerovalent palladium catalysts are more effective for production of these linear dimers. With $Pd(PPh_3)_2$ (maleic anhydride) as a catalyst, isoprene gives 2,7-dimethyl-1,3,7-octatriene exclusively [58].

The activity of these catalysts is modified by the presence of carbon dioxide. The CO_2-promoted reaction of $Pt(PPh_3)_3$ gives octatrienes [62]. With $Pd(diphos)_2$ it gives both octatrienes and octadiene lactones [63]. The presence of CO_2 can also induce formation of octadienols from butadiene and water [64]. It has been speculated that CO_2 and water combine under pressure to give carbonic acid, which adds to an octatriene precursor to give a water-sensitive octadienyl carbonate. In the nickel system a CO_2 insertion product of an isoprene dimer complex has been characterized [65].

The mechanism of butadiene dimerization has been studied thoroughly [38,66,67]. It is likely that the same intermediates are involved in formation of the linear and cyclic dimers. Some of these intermediates also occur in the pathway to 1,5,9-cyclododecatriene, a cyclic trimer of butadiene (below). It appears that two molecules of butadiene coordinate to a zerovalent nickel atom that bears one phosphorus ligand, as shown in Figure 4.7. The critical step in dimer formation is coupling the two monodentate butadiene ligands to an octadienyl ligand in complex **17.** On the basis of its proton nmr spectrum the dimer-ligand is believed to bond to the metal through a π-allyl function, as well as a σ-allyl. However, the spectroscopically detectable σ,π form (**17**) is tautomeric with three isomeric σ-allylic structures (**18–20**). Each isomer can undergo reductive elimination of two Ni—C bonds to generate a different cyclic dimer and reform the "L—Ni" catalyst. In practice all three processes occur, and each can predominate with proper choice of reaction conditions and the ligand L. The formation of 1,3,7-octatriene can be rationalized by β-hydrogen elimination from the tautomer (**18**) that ordinarily gives rise to divinylcyclobutane:

LNi(H) (**18**) ⟶ LNi(H) ⟶ LNi +

This rationalization oversimplifies the situation because it ignores the role of the

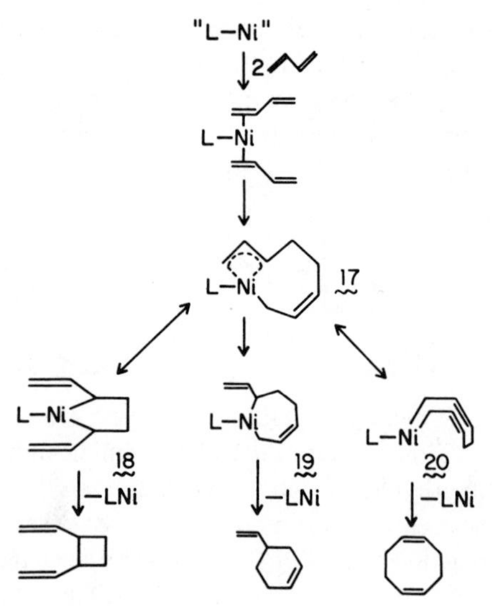

Figure 4.7 Pathways for conversion of butadiene to cyclic dimers (L = R_3P or $(RO)_3P$) [38].

weak acid cocatalyst. It has been suggested [38] that the cocatalyst protonates one allyl function to achieve the Ni—C bond cleavage shown as the second step earlier.

Cyclododecatriene Synthesis

Butadiene and substituted dienes trimerize readily under the influence of soluble transition metal catalysts [68]. The catalyst used commercially, a mixture of $TiCl_4$ and an alkylaluminum chloride, gives the *cis, trans, trans* isomer (ctt). Zerovalent nickel catalysts produce the all-*trans* isomer (ttt). The course of the nickel-catalyzed reaction

ctt

ttt

is very sensitive to the nature of the ligands on the metal, since, as described earlier, it can produce polymer or either of two butadiene dimers as major products. In addition it can catalyze cotrimerization of butadiene and ethylene to form either *cis, trans*-1,5-cyclodecadiene or 1,4,9-decatriene [69]:

$2C_4H_6 + C_2H_4$ —"Ni°"→ (cis, trans-1,5-cyclodecadiene)

$2C_4H_6 + C_2H_4$ —"Ni°" + L→ (1,4,9-decatriene)

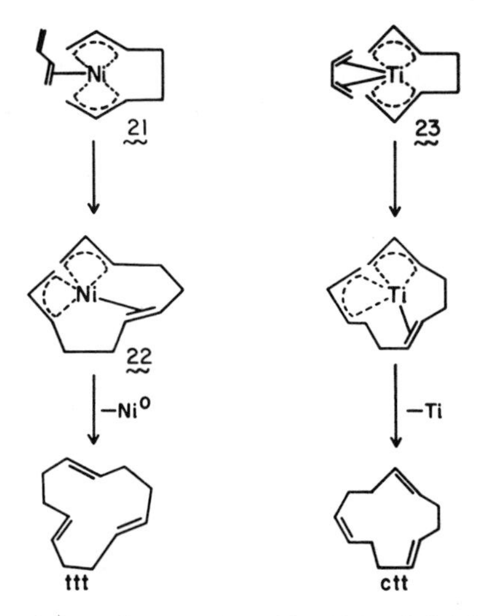

Figure 4.8 Possible schemes for butadiene reaction with a bis(π-allylic) ligand to give all-*trans* or *cis, trans, trans*-1,5,9-cyclododecatriene.

The commercial synthesis of cyclododecatriene is based on the observation [70] that butadiene trimerizes rapidly in the presence of a catalyst almost identical to that used for polymerization of ethylene. In the industrial process [71], the catalyst is prepared by mixing $TiCl_4$ with excess $Al_2Cl_3Et_3$ in benzene. Butadiene and freshly mixed catalyst solution are fed to a reactor in which trimerization occurs in a few minutes at about 70° and 1 atmosphere pressure. After catalyst deactivation, distillation gives a 75–90% yield of 1,5,9-cyclododecatriene, which is almost entirely the *cis, trans, trans* isomer. Minor amounts of the all-*trans* isomer, polybutadiene, 1,5-cyclooctadiene, and 4-vinylcyclohexene also form.

The cyclotrimerization of butadiene with nickel catalysts has been studied more thoroughly [38,72], even though it is not used industrially. When nickel(II) acetylacetonate is reduced with $Al(C_2H_5)_2(OC_2H_5)$ in the presence of butadiene, a highly reactive complex is formed. This compound, sometimes dubbed "naked nickel," catalyzes trimerization of butadiene to *trans, trans, trans* 1,5,9-cyclododecatriene. Zerovalent nickel complexes, for example, $Ni(COD)_2$, which lack phosphorus ligands, are also very effective for this purpose.

It seems likely that the initial steps in trimerization are identical to those in diene dimerization. Two butadiene ligands couple to form complex **17** of Figure 4.7 or its isomer, in which the double bond has a *trans*-configuration. In the absence of phosphorus ligands, L is another molecule of butadiene. This third butadiene inserts into an Ni—C bond of **17** or its tautomer (**21**), shown in Figure 4.8, to give a nickel complex (**22**), which bears a bis(allylic) C_{12} ligand. Complex **22** can be isolated at low temperature [72]. On warming in the presence of butadiene, the ends of the allyl functions couple to form the cyclic triene, initially as a Ni(0) complex. The *trans* configurations of the C=C bonds in the triene derive from the *syn* conformations of the bis(π-allyl) intermediates. It is tempting to speculate that the monodentate bonding of the butadiene ligand in **21** dictates formation of the *syn* allyl intermediate. In contrast the additional vacant orbitals of a titanium catalyst permit bidentate coordination of the diene, as shown in **23.** This structure should yield an *anti* allyl ligand, which in turn generates a *cis* double bond in the final product (ctt).

The active species in the titanium catalyst system is not well characterized. Chemical evidence [73] and analogy to olefin polymerization catalysts suggest that the Ti species in Figure 4.8 are Ti(III) compounds. Other ligands about the metal might include an ion such as $[Cl_2AlEt_2]^-$. However, bis(benzene)titanium(0) catalyzes the trimerization when used in combination with $Et_2Al_2Cl_4$ [74]. This result suggests a Ti(0) $\rightleftharpoons$ Ti(II) cycle exactly analogous to that observed with nickel. The $Ti(C_6H_6)_2$ catalyst ordinarily gives *cis,trans,trans*-1,5,9-cyclododecatriene, but addition of triphenylphosphine yields the all-*trans* isomer. This result supports the proposal that the stereochemistry of the product is dictated by the number of orbitals available for coordination of the third butadiene molecule.

4.6 ADDITIONS TO OLEFINS AND DIENES

Addition of H—X functions (X = CN, SiR_3, OR, NR_2) to olefins and dienes is closely related to hydrogenation in concept and mechanism. The addition of HCN to butadiene is a major industrial synthesis of adiponitrile, a key intermediate in production of 6,6-nylon (Section 11.3). Hydrosilylation has extensive application in synthesis of silicone polymers. Soluble metal complexes also catalyze addition of alcohols, phenols, and amines to dienes [38] but generally do not catalyze additions of ammonia and water to simple olefins.

Hydrocyanation

The adiponitrile synthesis commercialized by Du Pont requires addition of two moles of hydrogen cyanide to butadiene with good regioselectivity.

$$\text{CH}_2{=}\text{CHCH}{=}\text{CH}_2 \xrightarrow{\text{HCN}} \text{CH}_3\text{CH}{=}\text{CHCH}_2\text{CN} \rightleftharpoons \text{CH}_2{=}\text{CHCH}_2\text{CH}_2\text{CN} \xrightarrow{\text{HCN}} \text{NC(CH}_2)_4\text{CN}$$

Several catalyst systems, including cuprous halides [75] and nickel(0) complexes [76], can effect the first addition. Reportedly, in practice, butadiene and HCN react in the presence of a zerovalent nickel compound stabilized by triaryl phosphite ligands to give a 2:1 mixture of pentenenitriles and 2-methyl-3-butenenitrile [77]. The same complex catalyzes isomerization of the branched isomer to the desired linear product:

$$\text{CH}_2{=}\text{CH—}\underset{\text{CH}_3}{\underset{|}{\text{CH}}}\text{—CN} \rightleftharpoons \text{CH}_3\text{CH}{=}\text{CHCH}_2\text{CN}$$

The $Ni[P(OAr)_3]_4$ complex also effects addition of the second mole of HCN to mixed pentenenitriles to produce adiponitrile with good selectivity. According to the patent literature cited by Luedeke [77], reaction temperatures in all steps are modest, and pressures are just high enough to condense unreacted butadiene. Lewis acids are reported to promote the nickel-catalyzed reactions.

Although little has been reported on the mechanism of adiponitrile synthesis, substantial information on related systems has accumulated [78]. The hydrocyanation of an olefin by a low-valent metal complex is suggested to occur by a sequence of reactions analogous to the steps involved in hydrogenation (ligands omitted for simplicity) [79]:

$$\mathrm{M + HCN \rightleftharpoons H{-}M{-}CN}$$

$$\mathrm{H{-}M{-}CN + C{=}C \longrightarrow H{-}M(C{=}C){-}CN}$$

$$\mathrm{H{-}M(C{=}C){-}CN \longrightarrow H{-}\overset{|}{\underset{|}{C}}{-}\overset{|}{\underset{|}{C}}{-}M{-}CN}$$

$$\mathrm{H{-}\overset{|}{\underset{|}{C}}{-}\overset{|}{\underset{|}{C}}{-}M{-}CN \longrightarrow H{-}\overset{|}{\underset{|}{C}}{-}\overset{|}{\underset{|}{C}}{-}CN + M}$$

The addition of HCN to zerovalent complexes such as NiL_4 gives $HNiL_3CN$ complexes, which often dissociate a ligand to give coordinatively unsaturated $HNiL_2CN$ [80]. This species presumably coordinates a mole of olefin such as a pentenenitrile. A study of the hydrocyanation of 1-hexene [81] suggests that olefin insertion into the Ni—H bond is the step that determines whether linear or branched product is formed. This study indicates that Lewis acid cocatalysts play a role in the product-determining step. The final reductive elimination of alkyl and cyanide entities to form the nitrile product may be promoted by the phosphite ligand. Triethyl phosphite assists the elimination of phenyl and cyano ligands from a somewhat analogous complex [82]:

$$\mathrm{PhNiCN(PR_3)_2 \xrightarrow{(EtO)_3P} PhCN + Ni(0)\ complexes}$$

Hydrosilylation

The addition of an Si—H bond to a C=C function has been explored intensively as a route to alkylsilanes [83–85]. In addition to its use in laboratory syntheses the hydrosilylation reaction is used in many ways in the manufacture of silicone polymers. Probably the broadest application is the "curing" of silicone rubbers, a step that converts a syrupy polymer to a gum rubber or a plastic polymer to a hard material such as dental cement. This toughening process is accomplished by forming crosslinks between polymer chains. Commonly, an SiH function of one chain is added to a vinyl group of another chain [86]:

$$(\text{—O})_2\text{Si(R)H} + \text{H}_2\text{C=C(H)—Si(R)(O—)}_2 \longrightarrow (\text{—O})_2\text{Si(R)—CH}_2\text{CH}_2\text{—Si(R)(O—)}_2$$

A variety of catalysts are used for different applications, but soluble platinum complexes and platinum-on-alumina heterogeneous catalysts are common.

The vinyl silanes that provide the crosslinking sites are often made by adding an Si—H function to acetylene. Long-chain alkyl substituents are introduced by adding an Si—H group to a C=C bond of a terminal olefin. For addition of silane Si—H bonds to unactivated olefins, the usual catalyst of choice is chloroplatinic acid, $H_2PtCl_6{\cdot}6H_2O$, usually designated as Speier's catalyst [83]. Although dozens of transition metal complexes catalyze the addition, the stable, easily available platinum compound is preferred. It catalyzes Si—H addition in the presence of many kinds of functional groups. A strong steric influence is noted, since the silicon almost always attaches to the less crowded end of a C=C bond. Terminal olefins are hydrosilylated in preference to internal olefins. In fact internal olefins often isomerize to form terminal products [87] as in:

$$\text{3-heptene} \rightleftharpoons \text{1-heptene} \xrightarrow{\text{HSiR}_3} \text{n-C}_7\text{H}_{15}\text{SiR}_3$$

If the silane is optically active, it retains its configuration. Recently, asymmetric induction has been observed in hydrosilylation of olefins and ketones with catalysts that bear chiral ligands [88].

In a typical hydrosilylation [87] the olefin and the silane are mixed with a solution of Speier's catalyst (10^{-5} mole Pt/mole Si) in a polar solvent such as 2-propanol. After a brief induction period, a vigorous exothermic reaction occurs. The mixture is commonly heated to complete the reaction, and the product is isolated by distillation. The induction period can be reduced or eliminated by use of a lower valent platinum catalyst such as $[Pt(C_2H_4)Cl_2]_2$ or $Pt(C_2H_4)(PPh_3)_2$. It seems likely that catalyst formation from H_2PtCl_6 involves reduction of the platinum to an oxidation state of 0 or 2+. The silane is a likely reducing agent.

Accumulated evidence suggests that hydrosilylation occurs by a mechanism like that shown in Figure 4.9 [85]. A zerovalent platinum compound "Pt^0" coordinates an olefin and oxidatively adds H—SiR_3 to form the platinum hydride complex (**24**). As in hydrogenation mechanisms the sequence of the two steps could be reversed. Insertion of the olefin into the Pt—H bond yields the alkyl silyl platinum(II) complex (**25**). Reductive elimination of Pt—C and Pt—Si bonds regenerates the Pt^0 catalyst and gives the observed alkylsilane. Similar mechanisms

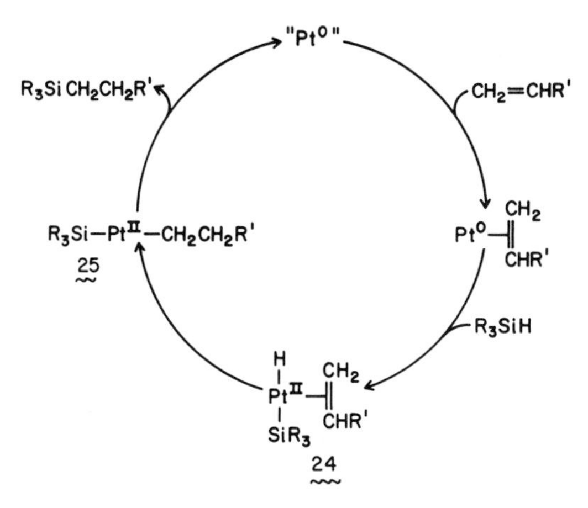

Figure 4.9 Hydrosilylation of a terminal olefin by a zerovalent platinum complex such as $Pt(PPh_3)_2(C_2H_4)$. The phosphine ligands are omitted for clarity.

can be written for catalysis of Si—H addition by other low-valent metal catalysts such as $Ir(CO)(Cl)(PPh_3)_2$.

A minor application of hydrosilylation is synthesis of solvent-resistant silicone rubbers for use as liners in self-sealing fuel tanks for military aircraft. Resistance to gasoline or jet fuel is attained by incorporation of polar cyanoethyl groups as substituents on the siloxane polymer. For example, cyanoethylation of $HSiMeCl_2$ gives a monomer that can be copolymerized with more normal silicone monomers:

$$HSiMeCl_2 + CH_2{=}CHCN \longrightarrow$$

$$Cl_2MeSiCH_2CH_2CN \xrightarrow{R_2Si(OH)_2} \left[\begin{array}{c} R \\ | \\ -Si-O- \\ | \\ R \end{array} \right]_m \left[\begin{array}{c} CH_3 \\ | \\ -Si-O- \\ | \\ CH_2CH_2CN \end{array} \right]_n$$

The reaction of acrylonitrile with the silane is catalyzed by a mixture of cuprous chloride, tetramethylethylenediamine, and a trialkylamine [89]. The chelating diamine probably solubilizes the copper(I) salt. When all three components are present, 75–80% yields of the β-cyanoethylsilane result.

GENERAL REFERENCES

Boor, J., *Ziegler-Natta Catalysts and Polymerizations*, Academic Press, 1978.

Albright, L. F., *Processes for Major Addition Type Plastics and Their Monomers*, McGraw-Hill, 1974.

Keii, T., *Kinetics of Ziegler-Natta Polymerization*, Chapman and Hall, 1972.

Coordination Polymerization, Ed., J. C. W. Chien, Academic Press, 1975.

The Stereo Rubbers, Ed., W. M. Saltman, Wiley, 1977.

Caunt, A. D., "Ziegler Polymerization," *Catalysis,* **1,** 234 (1977).

Sittig, M., *Polyolefin Production Processes,* Noyes Data Corp., Park Ridge, N.J., 1976.

Ballard, D. G. H., "Transition Metal Alkyl Compounds as Polymerization Catalysts," *J. Polym. Sci., Polym. Chem. Ed.,* **13,** 2191 (1975).

SPECIFIC REFERENCES

1. *Chem. Eng. News* (Dec. 5, 1977), p. 21; I. J. Levine and F. J. Karol, U.S. Patent 4,011,382 (1977).
2. A. Clark, *Catal. Rev.,* **3,** 145 (1969).
3. F. J. Karol, G. L. Karapinka, C. Wu, A. W. Dow, R. N. Johnson, and W. L. Carrick, *J. Poly. Sci.,* **A1, 10,** 2621 (1972); G. L. Karapinka, U.S. Patent 3,709,853 (1973).
4. W. L. Carrick, *Macromol. Syn.,* **2,** 33 (1966).
5. R. Spitz, A. Revillon, and A. Guyot, *J. Catal.,* **35,** 335, 345 (1974); A. Zecchina, E. Garrone, et al, *J. Phys. Chem.,* **79,** 966, 972, 978, 984 (1975).
6. E. A. Youngman and J. Boor, *Macromol. Rev.,* **2,** 33 (1967).
7. E. J. Vandenberg, *Macromol. Syn.,* **5,** 95 (1974).
8. R. J. Kern, H. Schnecko, W. Lintz, and L. Kollar, *Macromol. Syn., Coll.,* **1,** 425 (1977).
9. R. J. Kern, *Macromol. Syn., Coll.,* **1,** 1 (1977).
10. P. J. T. Tait, in R. N. Howard, Ed., *Developments in Polymerization,* Vol. 2, Applied Science Publishers, 1979.
11. P. Cossee, *J. Catal.,* **3,** 80 (1964); E. J. Arlman, *ibid,* **5,** 178 (1966).
12. D. R. Armstrong, P. G. Perkins, and J. J. P. Stewart, *J. Chem. Soc. Dalton,* 1972 (1972).
13. O. Novaro, E. Blaisten-Barojas, E. Clementi, G. Giunchi, and M. E. Ruiz-Vizcaya, *J. Chem. Phys.,* **68,** 2337 (1978).
14. G. Crespi, A. Valvasorri, and U. Flisi, "Olefin Copolymers," in W. M. Saltman, Ed., *The Stereo Rubbers,* Wiley, 1977, pp. 365–427.
15. S. Cesca, *Macromol. Rev.,* **10,** 1 (1975).
16. R. German, G. Vaughan, and R. Hank, *Rubber Chem. Tech.,* **40,** 569 (1967).
17. G. Henrici-Olivé and S. Olivé, *Angew. Chem. Int. Ed.,* **10,** 776 (1971).
18. A. G. Evans, J. C. Evans, and E. H. Moon, *J. Chem. Soc. Dalton,* 2390 (1974).
19. W. Cooper, "Polydienes by Coordination Catalysts," in W. M. Saltman, Ed., *The Stereo Rubbers,* Wiley, 1977, pp. 21–75.
20. E. W. Duck and J. M. Locke, "Polydienes by Anionic Catalysts," in W. M. Saltman, Ed., *The Stereo Rubbers,* Wiley, 1977, p. 139.
21. W. M. Saltman and E. Schoenberg, *Macromol. Syn.,* **2,** 50 (1966).
22. J. S. Lasky, *Macromol. Syn., Coll.,* **1,** 141 (1977).
23. C. F. Gibbs, V. L. Folt, E. J. Carlson, S. E. Horne, and H. Tucker, British Patent 916,383 (1963).
24. M. Gippin, *Macromol. Syn.,* **2,** 42 (1966).
25. G. Henrici-Olivé, S. Olivé, and E. Schmidt, *J. Organomet. Chem.,* **39,** 201 (1972).
26. Ph. Teyssié and F. Dawans, in W. M. Saltman, Ed., *The Stereo Rubbers,* Wiley, 1977, p. 79.
27. J. P. Durand, F. Dawans, and Ph. Teyssié, *J. Polym. Sci.,* **A-1, 8,** 979 (1970).
28. Ph. Teyssié, Proceedings of the 1st International Symposium on Homogeneous Catalysis, Corpus Christi, TX, Nov. 29–Dec. 1, 1978.

29. K. L. Lindsay, "Alpha-Olefins," in J. J. McKetta and W. A. Cunningham, Eds., *Encyclopedia of Chemical Processing and Design,* Vol. 2, Marcel Dekker, 1977, p. 482.
30. A. Lundeen and R. Poe, "Alpha-Alcohols," in J. J. McKetta and W. A. Cunningham, Eds., *Encyclopedia of Chemical Processing and Design,* Vol. 2, Marcel Dekker, 1977, p. 465.
31. K. Ziegler, in *Organometallic Chemistry,* H. Zeiss, Ed., Reinhold, 1960, pp. 194–195, 229–231.
32. A. J. Lundeen and J. E. Yates, U.S. Patent 3,450,735 (1969).
33. G. W. Kabalka and R. J. Newton, *J. Organomet. Chem.,* **156,** 65 (1978).
34. *Chem. Week,* 70 (Oct. 23, 1974); *Chem. Market. Reporter* (April 18, 1977).
35. R. S. Bauer, H. Chung, P. W. Glockner, and W. Keim, U.S. Patent 3,644,563 (1972); R. F. Mason, U.S. Patent 3,737,475 (1973).
36. F. H. Kowaldt, Thesis, Aachen (1977).
37. G. Lefebvre and Y. Chauvin, in R. Ugo, Ed., *Aspects of Homogeneous Catalysis,* Vol. 1, Carlo Manfredi, 1970, p. 108.
38. P. W. Jolly and G. Wilke, *The Organic Chemistry of Nickel,* Vol. 2, Academic Press, 1975.
39. G. Henrici-Olivé and S. Olivé, *Transition Met. Chem.,* **1,** 109 (1976).
40. M. Hidai and A. Misono, "Dimerization of Acrylic Compounds," in R. Ugo, Ed., *Aspects of Homogeneous Catalysis,* Vol. 2, D. Reidel, 1974, p. 159.
41. W. Strohmeier and A. Kaiser, *J. Organomet. Chem.,* **114,** 273 (1976).
42. F. Beck, *Angew Chem. Int. Ed.,* **11,** 760 (1972).
43. R. Cramer, *J. Am. Chem. Soc.,* **87,** 4717 (1965).
44. S. J. McLain and R. R. Schrock, *J. Am. Chem. Soc.,* **100,** 1315 (1978).
45. B. Bogdanovic, H. Biserka, H. G. Karmann, H. G. Nüssel, D. Walter, and G. Wilke, *Ind. Eng. Chem.,* **62,** 34 (Dec. 1970).
46. R. H. Grubbs, A. Miyashita, et al., *J. Am. Chem. Soc.,* **100,** 1300, 2418, 7416 (1978).
47. A. C. L. Su, "Catalytic Codimerization of Ethylene and Butadiene," in *Adv. Organomet. Chem.,* **17,** 269 (1978).
48. H. M. J. C. Creemers, U.S. Patent 3,767,717 (1973).
49. R. Cramer, *J. Am. Chem. Soc.,* **89,** 1633 (1967).
50. R. G. Miller, T. J. Kealy, and A. L. Barney, *J. Am. Chem. Soc.,* **89,** 3756 (1967).
51. R. Cramer, U.S. Patents 4,025,570 and 4,028,429 (1977).
52. A. C. L. Su and J. W. Collette, *J. Organomet. Chem.,* **36,** 177 (1972).
53. C. A. Tolman, *J. Am. Chem. Soc.,* **92,** 6777 (1970).
54. M. Iwamoto and S. Yuguchi, *J. Org. Chem.,* **31,** 4290 (1966); A. Miyake, G. Hata, M. Iwamoto, and S. Yuguchi, Proceedings of the 7th World Petroleum Congress, Mexico City, 1967, p. 37.
55. W. Griehl and D. Ruestem, *Ind. Eng. Chem.,* **62,** 16 (Mar. 1970).
56. P. Heimbach, "Nickel Catalyzed Syntheses of Methyl-Substituted Cyclic Olefins," in R. Ugo, Ed., *Aspects of Homogeneous Catalysis,* Vol. 2, D. Reidel, 1974, p. 79.
57. A. J. de Jong, Ger. Offenleg. 2,704,547 (1977).
58. A. D. Josey, *J. Org. Chem.,* **39,** 139 (1974); U.S. Patent 3,925,497 (1975).
59. W. Brenner, P. Heimbach, H. Hey, E. W. Müller, and G. Wilke, *Liebigs Ann. Chem.,* **727,** 161 (1969).
60. J. Feldman et al, American Chemical Society, *Petrol. Div. Preprints* **9** (4) A55 (1964); U.S. Patents 3,284,529 (1966); 3,480,685 (1969).
61. F. J. Weigert and W. C. Drinkard, *J. Org. Chem.,* **38,** 335 (1973).
62. J. F. Kohnle, L. H. Slaugh, and K. L. Nakamaye, *J. Am. Chem. Soc.,* **91,** 5904 (1969).
63. Y. Sasaki, Y. Inoue, and H. Hashimoto, *Chem. Comm.,* 605 (1976).

64. K. E. Atkins, R. M. Manyik, and G. L. O'Connor, U.S. Patent 3,992,456 (1976).

65. P. W. Jolly, S. Stobbe, G. Wilke, R. Goddard, C. Krüger, J. C. Sekutowski, and Y. H. Tsay, *Angew. Chem. Int. Ed.*, **17,** 124 (1978).

66. P. W. Jolly, I. Tkatchenko, and G. Wilke, *Angew. Chem. Int. Ed.*, **10,** 329 (1971).

67. C. R. Graham and L. M. Stephenson, *J. Am. Chem. Soc.*, **99,** 7098 (1977).

68. H. Breil, P. Heimbach, M. Kröner, H. Müller, and G. Wilke, *Makromol. Chem.*, **69,** 18 (1963).

69. P. Heimbach and G. Wilke, *Liebigs Ann. Chem.*, **727,** 183 (1969).

70. G. Wilke, *Angew. Chem. Int. Ed.*, **2,** 105 (1963).

71. W. Ring and J. Gaube, *Chem. Ing. Tech.*, **38,** 1041 (1966).

72. B. Bogdanovic, P. Heimbach, M. Kröner, and G. Wilke, *Liebigs Ann. Chem.*, **727,** 143 (1969).

73. L. I. Zakharkin and V. M. Akhmedov, *Zhur. Org. Khim.*, **2,** 998 (1966).

74. V. M. Akhmedov, M. T. Anthony, M. L. H. Green, and D. Young, *J. Chem. Soc. Dalton,* 1419 (1975).

75. D. Y. Waddan, U.S. Patent 3,869,501 (1975).

76. P. Arthur and B. C. Pratt, U.S. Patent 2,571,099 (1951); W. C. Drinkard and R. V. Lindsey, U.S. Patent 3,496,215 (1970).

77. V. D. Luedeke, "Adiponitrile," in J. J. McKetta and W. A. Cunningham, Eds., *Encyclopedia of Chemical Processing and Design,* Vol. 2, Marcel Dekker, 1977, p. 146.

78. E. S. Brown, "Addition of Hydrogen Cyanide to Olefins," in I. Wender and P. Pino, Eds., *Organic Syntheses via Metal Carbonyls,* Vol. 2, Wiley-Interscience, 1977, p. 655.

79. E. S. Brown, *Aspects of Homogeneous Catalysis,* **2,** 57 (1974).

80. J. D. Druliner, A. D. English, J. P. Jesson, P. Meakin, and C. A. Tolman, *J. Am. Chem. Soc.*, **98,** 2156 (1976).

81. B. W. Taylor and H. E. Swift, *J. Catal.*, **26,** 254 (1972).

82. G. Favero, M. Gaddi, A. Morvillo, and A. Turco, *J. Organomet. Chem.*, **149,** 395 (1978).

83. J. L. Speier, "Homogeneous Catalysis of Hydrosilation by Transition Metals," *Adv. Organomet. Chem.*, **17,** 407 (1979).

84. E. Lukevics, Z. V. Belyakova, M. G. Pomerantseva, and M. G. Voronkov, *Organomet. Chem. Rev.*, **5,** 1 (1977).

85. J. F. Harrod and A. J. Chalk, "Hydrosilation Catalyzed by Group VIII Complexes," in I. Wender and P. Pino, Eds., *Organic Syntheses via Metal Carbonyls,* Vol. 2, Wiley-Interscience, 1977, p. 673.

86. R. N. Meals, *Pure Appl. Chem.*, **13,** 141 (1966).

87. J. C. Saam and J. L. Speier, *J. Am. Chem. Soc.*, **80,** 4104 (1958).

88. I. Ojima, K. Yamamoto, and M. Kumada, "Asymmetric Hydrosilylation by Means of Homogeneous Catalysts with Chiral Ligands," in *Aspects of Homogeneous Catalysis,* **3,** 186 (1977).

89. B. A. Bluestein, *J. Am. Chem. Soc.*, **83,** 1000, (1961); U.S. Patent 2,971,970 (1961).

5 REACTIONS OF CARBON MONOXIDE

The organic chemistry of carbon monoxide embraces some of the most important applications of homogeneous catalysis. Even though reactions such as hydroformylation of olefins have been used industrially since the early 1940s [1], these processes continue to grow in importance. This growth is expected to continue, even when conventional organic feedstocks are scarce, because carbon monoxide is available from many sources. Reforming of natural gas and naphtha is now the major source of "synthesis gas," a mixture of carbon monoxide and hydrogen. However, available technology also permits use of coal or heavy petroleum fractions as feedstocks for production of synthesis gas. It is likely that coal-based synthesis gas will be a major source of chemicals 15–20 years hence.

Projected changes in relative costs and availability of feedstocks suggest that many chemicals currently produced from ethylene will be made from synthesis gas in the future. As described in Chapter 12, the substitution of synthesis gas for ethylene as a feedstock for acetic acid production is well under way [2], and a similar trend may develop for vinyl acetate. An analogous substitution may also occur for ethylene glycol production. Union Carbide is developing a synthesis gas-based process (Section 5.9) that could possibly displace conventional ethylene oxidation technology (Section 6.4).

5.1 COORDINATION OF CARBON MONOXIDE

Carbon monoxide is an extremely effective ligand for stabilization of low-valent metal complexes. In fact zerovalent carbonyl complexes of all the first-row transition metals from vanadium to nickel are well characterized [3]. The compositions of these compounds are generally predicted by the "18-electron rule" (Section 2.1). It forecasts monomeric compounds from metals in even-numbered groups of the periodic table, for example, $Cr(CO)_6$, $Fe(CO)_5$, and $Ni(CO)_4$. Odd-numbered

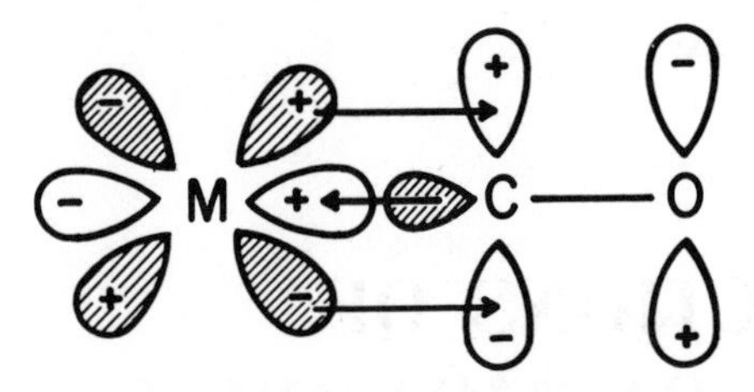

Figure 5.1 Interaction of metal and carbon monoxide orbitals in a metal carbonyl complex.

groups form paramagnetic $[V(CO)_6]$ or dimeric carbonyl compounds such as $Mn_2(CO)_{10}$ and $Co_2(CO)_8$. The transition metals of the second and third rows are prone to form metal-metal bonded cluster compounds. Cluster compounds stabilized by CO as a ligand are known for most transition metals but are especially abundant in Group VIII. The metal carbonyl clusters seem to offer exciting new possibilities in catalysis, since, at least in a formal sense, they combine characteristics of mononuclear carbonyls with those of a metal surface [4].

The study of the metal carbonyls has been stimulated by their importance as catalysts for reactions of carbon monoxide. Even though CO is widely used as a feedstock in manufacture of organic compounds, it is not very reactive by itself. Powerful electrophiles and nucleophiles attack the CO molecule, but a catalyst is generally required to bring about reactions with olefins, alcohols, and hydrogen. In practice the catalysts are generally low-valent complexes of Group VIII metals. Cobalt and rhodium compounds are especially important from an industrial viewpoint.

The catalytic activation of carbon monoxide seems to involve changes in the molecular orbitals induced by coordination to a metal atom. Carbon monoxide is an extremely weak *sigma* donor ligand, but, even more than the isoelectronic dinitrogen molecule, it can bond quite tenaciously to a metal. The general bonding pattern illustrated in Figure 5.1 closely resembles that described for the M—N≡N bond in Section 2.2. The weak ligand-to-metal *sigma* donor interaction is reinforced by back donation from filled metal *d* orbitals to vacant ligand antibonding orbitals. The two bonding modes are synergistic in transferring electron density between the ligand and the metal [5]. The most stable M—CO bonds seem to result when a strong electron donor ligand such as a phosphine is *trans* to the carbonyl. In this position the R_3P and CO ligands share metal orbitals, and electron density is transferred from P to CO.

The complex interaction between the CO ligand and the metal ion modifies the reactivity of the carbon monoxide molecule. A CO ligand in a cationic carbonyl complex such as $[Mn(CO)_6]^+$ becomes susceptible to attack by external nucleophiles such as amines, alkoxides, and carbanions. The most important modification from a practical viewpoint is, however, susceptibility to attack by other ligands coordinated to the same metal ion. This insertion or alkyl migration process described here is a key step in all the practical CO-based syntheses discussed in this chapter.

5.2 CO INSERTION PROCESSES

Nearly all the reactions discussed in this chapter involve the "insertion" of a carbon monoxide into a C—M *sigma*-bond:

$$M(CR_3)(C{\equiv}O) \rightleftharpoons M(\square)\text{—}C(=O)\text{—}CR_3$$

As indicated in the equation, the reaction occurs within the coordination sphere of the metal. The alkyl group migrates to the carbon of the CO ligand and leaves behind a vacant coordination site. Because of this mechanistic aspect, the reaction should be termed "alkyl migration," but insertion is more commonly used. The reaction is readily reversible if some additional ligand is not present to occupy the coordination site vacated by the alkyl group. The reverse reaction, decarbonylation, has some synthetic utility, as discussed in Section 5.6. The stereochemistry of a chiral alkyl group is generally preserved during its migration in a carbonylation or decarbonylation process.

The insertion process has been difficult to study in detail, especially with the catalytically active cobalt and rhodium complexes, which involve rather labile intermediates. The most detailed studies of insertion have been done with the pseudooctahedral $RMn(CO)_5$ compounds, which are more tractable than the catalytic $RCo(CO)_4$ and $RRh(CO)(PPh_3)_3$ species. Other studies have used relatively stable $RCo(CO)_3(PR'_3)$ complexes and the kinetically sluggish iridium analogs of the rhodium catalysts. Collectively, the results of these investigations support the general statements made above [6].

The CO insertion and decarbonylation reactions with methylmanganese compounds are instructive. The reaction of methylmanganese pentacarbonyl with a ligand such as triphenylphosphine or ^{13}CO occurs in two steps [7]:

$$CH_3Mn(CO)_5 \rightleftharpoons CH_3COMn(CO)_4 \xrightarrow{L} cis\text{-}CH_3COMn(CO)_4L$$

Even in the presence of excess labeled CO, the acetyl group contains no ^{13}C, because the methyl migrates to a CO already present on the metal. The first observable product is the *cis*-substituted acetyl derivative. Evidently the intermediate five-coordinate acetyl complex has enough configurational stability to permit the incoming ligand to seek out the coordination site vacated by the methyl group. The initially observed *cis*-acetyl complex later isomerizes to the *trans* isomer by CO dissociation and reassociation without involvement of the acetyl carbonyl group. The motion of the CH_3 and CO ligands during the insertion or migration process has been calculated in some detail [8].

5.3 ACETIC ACID SYNTHESIS

Acetic acid and acetic anhydride are major industrial chemicals used in the manufacture of vinyl acetate, cellulose acetate, pharmaceuticals, dyes, and pesticides [2]. Acetic acid is a major solvent in the oxidation of *p*-xylene to terephthalic acid, and acetate esters are used extensively as solvents. Acetic acid and its derivatives are made in three major ways, all of which employ homogeneous catalysts:

Oxidation of ethylene via acetaldehyde (Section 6.1).

Oxidation of butane or naphtha (Section 10.5).

Carbonylation of methanol.

Currently, these three processes are used on comparable scales in the United States.

Acetic acid has been made by carbonylation of methanol for almost 20 years [2], but this reaction has become more attractive with the development of a catalyst that operates under low pressure. The original process developed by BASF [1,9] uses a cobalt carbonyl catalyst with an iodine compound as a cocatalyst. Cobaltous iodide is commonly used for *in situ* generation of $Co_2(CO)_8$ and HI. As is common with cobalt-catalyzed carbonylation, severe conditions (about 210°/700 atm) are required to give commercially acceptable reaction rates. The newer Monsanto process [10], catalyzed by a rhodium salt with an iodide cocatalyst, will produce acetic acid even at atmospheric pressure, although higher pressures are used in practice. Both catalysts are effective for the conversion of higher alcohols to the homologous carboxylic acids. The cobalt system has been used commercially in Europe for synthesis of propionic acid from ethanol, although carboxylation of ethylene (Section 5.4) may be more economical. Alkyl halides can be used in place of alcohols. Reportedly, phenylacetic acid is now made by carbonylation of benzyl chloride in the presence of base [11]:

$$PhCH_2Cl + CO + 2OH^- \rightarrow PhCH_2COO^- + Cl^- + H_2O$$

The commercial synthesis of acetic acid with the rhodium catalyst is carried out by reacting methanol with CO at about 180° and 30–40 atmospheres pressure with about 10^{-3} molar catalyst concentration [10]. The reaction produces acetic acid or its methyl ester with greater than 99% selectivity. In comparison with the cobalt system, the presence of hydrogen in the gas stream is less apt to cause hydrogenation of CH_3—M and CH_3COM intermediates to give methane or acetaldehyde.

The rhodium catalyst may be almost any soluble rhodium compound, but commercial derivatives such as $RhCl_3 \cdot 3H_2O$ are rapidly converted to $[Rh(CO)_2I_2]^-$ under reaction conditions. Many iodine compounds can be used as cocatalysts, but it seems likely that much of the cocatalyst is present in the plant streams as methyl iodide. Both rhodium and iodine are sufficiently expensive that nearly quantitative recycle is necessary for economic reasons. The rhodium complexes are nonvolatile

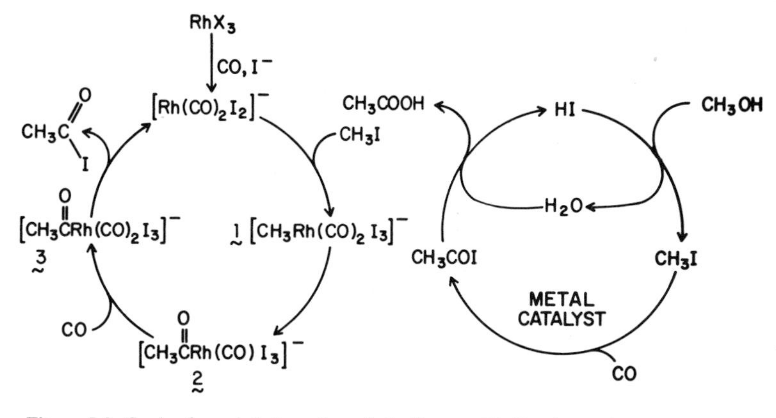

Figure 5.2 Cycles for catalytic action of rhodium and iodine in acetic acid synthesis.

and reportedly stay in the reaction loop for years. The volatile iodine compounds are more troublesome, especially since iodine contamination of the product is undesirable. As with the BASF catalyst, the use of iodide in an acidic reaction medium requires a highly corrosion-resistant metal reactor.

The role of iodine in the two catalyst systems is the conversion of methanol to the more electrophilic methyl iodide. The iodine is used in a catalytic sense, as shown in the right hand loop in Figure 5.2. This iodine cycle interacts with a cycle involving the metal carbonyl catalyst. The rhodium cycle [12] shown in the figure begins with oxidative addition of methyl iodide to $[Rh(CO)_2I_2]^-$, a formal transition from Rh(I) to Rh(III). Although oxidative additions to rhodium(I) complexes are well known, this one is extraordinarily fast [13]. Evidently the extra electron density in the anionic complex makes it a powerful nucleophile.

The methylrhodium complex(**1**) formed by the oxidative addition is kinetically unstable and rapidly isomerizes to an isolable acetyl monocarbonyl complex (**2**) [12]. This isomerization of **1** to **2** is a nice illustration of the stepwise character of the methyl migration ("insertion") reaction. The five-coordinate acetyl derivative (**2**) reacts with CO to produce a labile six-coordinate acetyl complex (**3**). In the absence of water or alcohol, it eliminates acetyl iodide spontaneously and regenerates the $[Rh(CO)_2I_2]^-$ catalyst. The acetyl iodide can reenter the iodine cycle to form acetic acid and HI by hydrolysis. Under other reaction conditions acetic anhydride or methyl acetate form by partial hydrolysis or by methanolysis. In the commercial reaction system a second mode of cleavage of the acetylrhodium complex is available:

$$[CH_3CORh(CO)_2I_3]^- \xrightarrow{H_2O} [HRh(CO)_2I_3]^- \xrightarrow{-HI} [Rh(CO)_2I_2]^-$$
$$+$$
$$CH_3COOH$$

The relative rates of the two acetyl cleavage mechanisms have not been reported.

5.4 CARBOXYLATION OF OLEFINS

Carboxylic acid derivatives form in many catalytic reactions of olefins, CO, and hydrogen donors. These syntheses are catalyzed by standard carbonylation catalysts such as $Co_2(CO)_8$ [1]. Such processes complement the acetic acid synthesis from methanol because they provide routes to higher aliphatic acids from readily available terminal olefins. As in the acetic acid synthesis the products are formed from an acyl metal complex, as shown by the synthesis of propionic acid derivatives from ethylene:

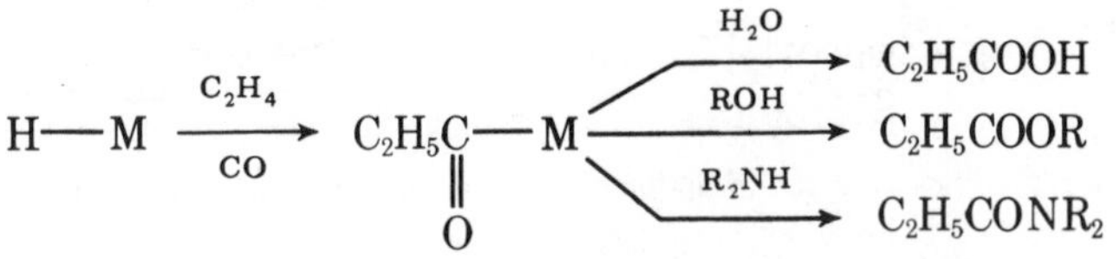

$$H{-}M \xrightarrow[CO]{C_2H_4} C_2H_5\overset{}{\underset{\displaystyle \overset{\|}{O}}{C}}{-}M \begin{cases} \xrightarrow{H_2O} C_2H_5COOH \\ \xrightarrow{ROH} C_2H_5COOR \\ \xrightarrow{R_2NH} C_2H_5CONR_2 \end{cases}$$

Industrial applications of carboxylation have been limited because good alternative processes exist for production of most large-volume acids. For example, the widely used 2-ethylhexanoic acid ("isooctanoic") is commonly made by oxidation of the corresponding aldehyde. However, some propionic acid is made from ethylene, as shown above, and fatty esters are made from olefins such as 1-hexene and 1-octene. The interest in the latter processes has been enhanced by the development of long-lived synthetic lubricants based on esters such as pentaerythrityl heptanoate. In a closely related reaction, acrylates have been made commercially by reaction of acetylene, carbon monoxide, and alcohols (Section 8.4).

Propionic acid can be made directly by reaction of ethylene, CO, and water ("hydrocarboxylation"), but the higher acids are produced as methyl or ethyl esters. The hydrocarboxylation reaction requires a solvent or a phase transfer agent to bring the olefin and water in contact. A drawback arising from the presence of water is the formation of hydrogen via the shift reaction (Section 5.8): $CO + H_2O \rightleftharpoons CO_2 + H_2$. As in acetic acid synthesis, hydrogen leads to byproduct formation by hydrogenation or hydroformylation of the olefin. These side reactions can be tolerated to some extent for cheap olefins such as ethylene but not for the more expensive higher olefins.

All the commercial carboxylation processes use cobalt carbonyl catalysts [1], but palladium complexes such as $PdCl_2(PPh_3)_2$ may have commercial potential because they are effective at lower pressures [14,15]. In a typical carboalkoxylation with a cobalt catalyst, a methanol solution of $Co_2(CO)_8$ and 1-pentene is heated to 140–170° under 100–200 atmospheres Co pressure for several hours (80–90% conversion of olefin). The resulting product is a solution of 70% methyl hexanoate and 30% branched esters along with minor byproducts [16,17]. Direct synthesis of carboxylic acids is performed similarly, except that the solvent is aqueous dioxane or acetone. With $PdCl_2(PPh_3)_2$ as the catalyst the carbonylation is done at 80–110° and pressures as low as about 35 atmospheres [18]. The yield of methyl octanoate from 1-heptene under such conditions is less than 50% because the selectivity for

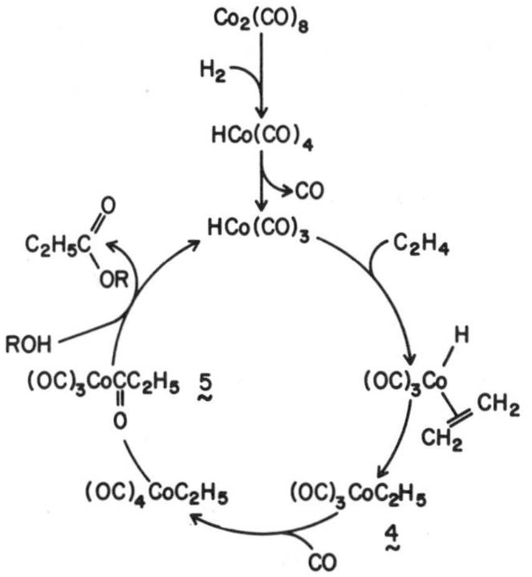

Figure 5.3 Major steps in the cobalt carbonyl-catalyzed synthesis of propionic acid or its esters (R = H or alkyl).

formation of the linear product is only about 60%. If, however, the reaction is run in the presence of excess $SnCl_2$, the selectivity rises to 87%, and the yield of linear ester, to 76% [18]. The branched ester is the major product when the $PdCl_2(PPh_3)_2$-catalyzed carbonylation is done in the presence of HCl. For example, propylene, CO, and methanol yield predominantly methyl isobutyrate [14], a compound of interest as a possible precursor of methyl methacrylate.

The mechanism of the cobalt-catalyzed carboxylation is sketched in Figure 5.3. Its main features are common to most carbonylation reactions catalyzed by $Co_2(CO)_8$ [19]. The reaction of $Co_2(CO)_8$ with adventitious hydrogen or with an alcohol forms $HCo(CO)_4$. This "hydride" is actually a strong acid in polar solvents. It is quite unstable and readily decomposes to cobalt metal in a CO-free atmosphere. The high pressures generally required in cobalt carbonyl-catalyzed reactions are needed to stabilize $HCo(CO)_4$ and the intermediates in the catalyst cycle.

It is commonly assumed that the catalyst cycle is entered by dissociation of a CO ligand to give $HCo(CO)_3$, as shown in the figure. There is, however, some evidence that the replacement of CO by olefin may occur via an associative substitution. The cycle in the figure indicates coordination of ethylene and insertion of the olefin into the Co—H bond to give the coordinatively unsaturated ethyl complex (**4**). The subsequent steps are coordination of CO and insertion of CO into the Co—C_2H_5 bond to give a labile acyl compound (**5**). All the steps in the cycle to this point are common to carboxylation and to the hydroformylation reaction discussed in Section 5.5. The two reactions differ mainly in the mode of cleavage of the acyl group from the metal.

Grossly, the acyl cleavage with an alcohol or with water proceeds as shown in Figure 5.3 to give an ester or a carboxylic acid and to regenerate the catalytic

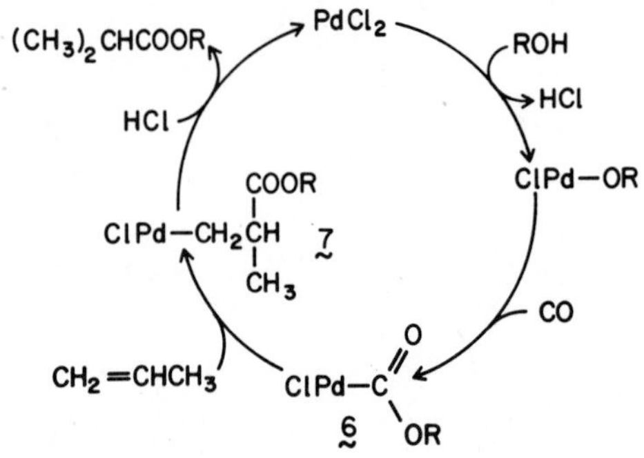

Figure 5.4 A proposed mechanism for isobutyrate synthesis from propylene. Neutral ligands have been omitted.

$HCo(CO)_3$. The detailed mechanism of cleavage is not established, but studies on the more stable acyl manganese carbonyl derivatives are suggestive. Both with water and with alcohols, the initial reaction seems to involve the acid- or base-catalyzed addition of the O—H bond to the acyl carbonyl function [20]:

$$CH_3C(=O)\text{—}Mn(CO)_5 \xrightarrow{ROH} CH_3C(OR)(OH)\text{—}Mn(CO)_5 \longrightarrow CH_3C(=O)OR + HMn(CO)_5$$

Presumably, a similar carbonyl addition occurs with acyl cobalt carbonyl complexes. The transfer of hydrogen from OH to metal is analogous to the facile β-hydrogen transfer from CH to metal that occurs in alkyl derivatives of the transition metals.

The mechanism of carbonylation with the palladium catalysts is less well defined. Two major mechanisms have been proposed. One involves formation of a palladium hydride that acts as a catalyst in very much the same way as $HCo(CO)_4$ does. The hydride mechanism seems to be generally accepted for the $PdCl_2(PPh_3)_2/SnCl_2$ system [18]. The tin chloride assists in formation of a hydride by creation of labile $SnCl_3^-$ ligands:

$$PdCl_2L_2 \overset{SnCl_2}{\rightleftharpoons} Pd(SnCl_3)_2L_2 \overset{H_2}{\rightleftharpoons} HPd(SnCl_3)L_2 + H^+ + SnCl_3^-$$

This assistance to hydride formation is much like that proposed for the $H_2PtCl_6/SnCl_2$ hydrogenation catalyst (Section 3.4). However, the $SnCl_3^-$ ligand probably also has other roles. It affects the direction of addition of the Pd—H bond to the olefin, as judged by its effect on the ratio of linear to branched esters.

Another mechanism for the palladium system has been suggested to account for the predominance of branched products in the absence of $SnCl_2$ [15,16]. The version of this mechanism sketched in Figure 5.4 accounts for the role of an acid as a cocatalyst in the reaction. In this scheme a complex such as $PdCl_2(PPh_3)_2$ reacts with the alcohol to form a labile alkoxy compound. Coordination of carbon monoxide and insertion of CO into the Pd—O bond forms the carboalkoxy complex (**6**). (Palladium compounds that contain the $—COOCH_3$ ligand are well characterized [21].) Insertion of propylene into the Pd—C bond should yield the $Pd—CH_2$ derivative (**7**) when the coordination sphere is crowded with bulky ligands such as triphenylphosphine.

The major role of the acid, HCl, in this scheme is to cleave the Pd—C bond in **7** before β-hydrogen elimination can occur. The acid cleavage yields the observed alkyl isobutyrate and regenerates $PdCl_2(PPh_3)_2$. If β-hydrogen abstraction occurs, the organic product is a methacrylate ester, but the catalyst is reduced to metallic palladium or a Pd(O) complex.

5.5 HYDROFORMYLATION

The oldest and largest homogeneous catalytic reaction of olefins is hydroformylation, the addition of CO and hydrogen to produce an aldehyde. The most prominent example is manufacture of butyraldehyde from propylene and synthesis gas [1]:

$$CH_3CH{=}CH_2 + H_2 + CO \rightarrow CH_3CH_2CH_2CHO$$

The aldehyde is a versatile chemical intermediate produced on a scale of about 3 million tons per year worldwide. Some butyraldehyde is hydrogenated to 1-butanol, but the greater part is self-condensed to form 2-ethylhexyl derivatives [22]:

$$CH_3CH_2CH_2CHO \xrightarrow[-H_2O]{\text{base}} CH_3(CH_2)_2CH{=}C(CHO)(C_2H_5)$$

$$CH_3CH_2CH_2CHO \xrightarrow{H_2} \text{1-butanol}$$

$$CH_3(CH_2)_2CH{=}C(CHO)(C_2H_5) \xrightarrow{H_2} CH_3(CH_2)_3CH(CHO)(C_2H_5)$$

$$CH_3(CH_2)_3CH(CHO)(C_2H_5) \xrightarrow{H_2} \text{2-ethylhexanol}$$

$$CH_3(CH_2)_3CH(CHO)(C_2H_5) \xrightarrow{O_2} \text{2-ethylhexanoic acid ("isooctanoic acid")}$$

Much of the 2-ethylhexanol is converted to phthalate esters for use as plasticizers. Butanol and other short-chain alcohols made by hydroformylation technology are used extensively as solvents.

A second major application of hydroformylation is the synthesis of fatty alcohols from terminal olefins such as 1-octene. The process sometimes combines hydroformylation and reduction steps, since cobalt carbonyl catalysts are effective for both reactions:

$$RCH{=}CH_2 \xrightarrow[CO]{H_2} RCH_2CH_2CHO \xrightarrow{H_2} R(CH_2)_3OH$$

The linear alcohols prepared this way, like those made by ethylene telomerization (Section 4.3), are used in synthesis of biodegradable detergents and for production of adipate esters. The latter serve as high-temperature lubricants and as plasticizers.

In addition to the industrial applications, hydroformylation is a useful laboratory synthesis of aldehydes from both terminal and internal olefins. Such syntheses can be done at atmospheric pressure by using $HCo(CO)_4$ as a stoichiometric hydroformylation reagent or by using rhodium-based catalysts such as $HRh(CO)(PPh_3)_3$ [23]. Hydroformylation of internal olefins such as 2-butene generates a chiral center. When the aldehyde synthesis is done with a rhodium complex that bears chiral ligands, asymmetric induction occurs and yields optically active aldehydes. The asymmetric hydrogenation catalysts formed from $[Rh(olefin)_2]^+$ salts and the chiral ligand DIOP are quite effective for this purpose.

DIOP = Ph_2PCH_2–*CH(–O–)–*CH(–O–)–CH_2PPh_2, the two O atoms joined to $C(CH_3)_2$

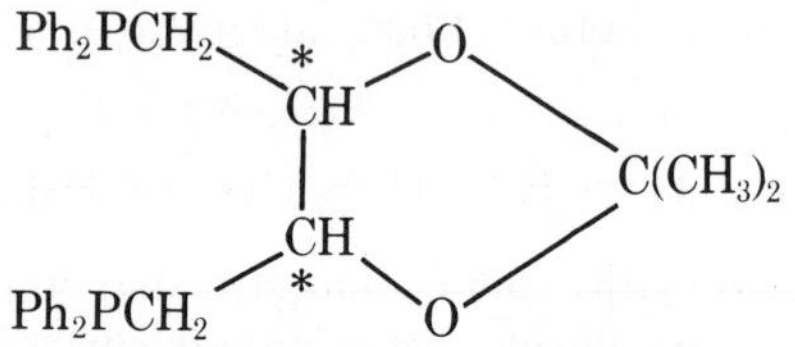

Industrial hydroformylations take place with two types of catalysts. Simple cobalt carbonyl catalysts are used for roughly 90% of all production, both of aldehydes and of alcohols. A phosphine-modified cobalt catalyst has been developed by Shell specifically for linear alcohol synthesis. A new rhodium-based system has been commercialized recently by Union Carbide. The cobalt and rhodium systems are discussed separately here.

Cobalt Catalysts

In 1938 Roelen, at Ruhrchemie AG, discovered that cobalt-containing heterogeneous catalysts effect the addition of carbon monoxide and hydrogen to ethylene to yield aldehydes and ketones [1]. It was soon recognized that the actual catalyst is the soluble cobalt carbonyl formed by reductive carbonylation of cobalt oxide. This chemistry was soon developed into an industrial synthesis of aldehydes and alcohols.

The usual commercial process for manufacture of butyraldehyde does not differ greatly from that developed in the 1940s. Typically, the catalyst is prepared in the hydroformylation reactor by treating finely divided cobalt metal or a cobalt(II) salt with synthesis gas (1 : 1 H_2:CO). This treatment produces the following reactions:

$$Co^{2+} \xrightarrow[CO]{H_2} Co^0 \xrightarrow{CO} Co_2(CO)_8 \xrightarrow{H_2} HCo(CO)_4$$

Addition of propylene to this catalyst/solvent/synthesis gas mixture at 120–140° and 200 atmospheres pressure [23] forms a mixture of products. The main component is *n*-butyraldehyde, which is formed in 60–70% yield. Isobutyraldehyde is the major impurity, but C_4 alcohols and dipropyl ketones are also present.

The high pressure in the reactor is necessary to stabilize the cobalt carbonyl intermediates. However, the requirement for a high temperature reflects the conditions needed to prepare the catalyst. If $HCo(CO)_4$ is prepared in advance in a separate reactor, the hydroformylation proceeds smoothly at 90–120°. Under the milder conditions, catalyst selectivity is better and butyraldehyde forms in 72–80% yield [24].

The simple cobalt carbonyl catalysts are reasonably satisfactory for industrial hydroformylation processes, but several problems exist. The carbonyls $HCo(CO)_4$ and $Co_2(CO)_8$ are unstable and volatile. These properties make them difficult to separate from aldehyde products during the product purification and catalyst recycle steps. More significantly, the limited selectivity for the desired linear aldehydes and alcohols increases the consumption of starting materials and requires large distillation facilities for product purification. Equally seriously, the severe reaction conditions impose economic penalties for high-pressure reactor investment and for construction and operation of the gas compressor.

Two new catalyst systems described here provide better product selectivity and lower pressure reaction conditions [25]. Rhodium catalysts give high yields of linear aldehydes under mild conditions. Phosphine-modified cobalt carbonyl catalysts are advantageous for direct synthesis of linear alcohols.

Typically, the synthesis of long-chain alcohols from terminal olefins takes place in two stages [1]. For example, 1-octene is hydroformylated to nonanal at 120–170° and 200–300 atmospheres pressure. The nonanal is then hydrogenated with a heterogeneous catalyst in a separate step. In principle the two reactions can be combined by increasing the H_2:CO ratio. However, simple cobalt carbonyls are only weak catalysts for aldehyde hydrogenation and lead to several side reactions. The hydroformylation and aldehyde hydrogenation steps can, however, be combined satisfactorily by addition of a trialkylphosphine ligand to the standard cobalt system. In the process developed by Shell [26] the active catalyst appears to be $HCo(CO)_3(PBu_3)$. This tributylphosphine-modified catalyst is less active for hydroformylation than $HCo(CO)_4$ and yields an inferior rate even at 180°. It is, however, much more active for hydrogenation and gives a linear: branched alcohol ratio of about 7:1 compared to 4:1 typical of unmodified cobalt hydroformylation catalysts [1]. The $HCo(CO)_3(PBu_3)$ catalyst is also more stable and is used at about 100 atmospheres' pressure versus 200–300 atmospheres. This stability also simplifies catalyst recycle because the alcohols can be distilled from the catalyst. A limitation of this process is that some olefin is lost through hydrogenation to alkane.

Cobalt-catalyzed hydroformylation was one of the first homogeneous catalytic reactions to be characterized from a mechanistic viewpoint [27]. Recent infrared studies of the reaction have tended to confirm that the intermediates under typical high-pressure conditions resemble those observed earlier at atmospheric pressure [28,29]. Except for the final product-forming step the chemistry is identical to that in the carbonylation reaction shown in Figure 5.3. The $Co_2(CO)_8$ catalyst is almost completely hydrogenated to $HCo(CO)_4$ at 150° and 290 atmospheres' H_2/CO pressure [28]. The hydride reacts with an olefin to form an olefin complex, an alkyl derivative (**4**), and an acyl derivative (**5**), just as shown in the figure. With unsymmetrical olefins such as propylene the direction of olefin insertion into the Co—H bond determines whether a linear or a branched aldehyde is formed:

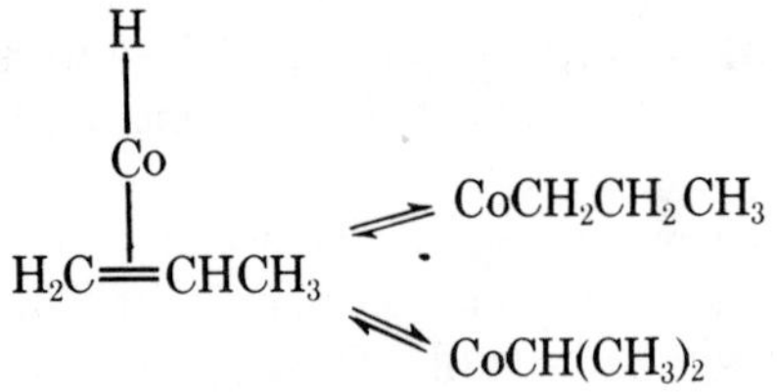

$$\underset{H_2C{=}CHCH_3}{\overset{\overset{H}{|}}{\underset{|}{Co}}} \rightleftharpoons CoCH_2CH_2CH_3 \qquad \underset{H_2C{=}CHCH_3}{\overset{\overset{H}{|}}{\underset{|}{Co}}} \rightleftharpoons CoCH(CH_3)_2$$

A key intermediate in the overall process is the coordinatively unsaturated acylcobalt compound (**8**). It can undergo several reactions with CO and hydrogen:

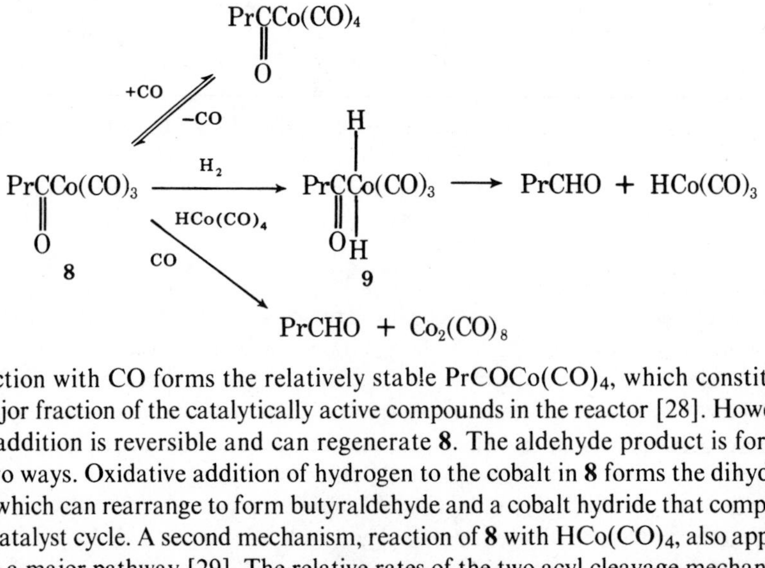

$$\underset{\mathbf{8}}{PrC(=O)Co(CO)_3} \underset{-CO}{\overset{+CO}{\rightleftharpoons}} PrC(=O)Co(CO)_4$$

$$\underset{\mathbf{8}}{PrC(=O)Co(CO)_3} \xrightarrow{H_2} \underset{\mathbf{9}}{PrC(=O)Co(H)_2(CO)_3} \longrightarrow PrCHO + HCo(CO)_3$$

$$\underset{\mathbf{8}}{PrC(=O)Co(CO)_3} \xrightarrow[CO]{HCo(CO)_4} PrCHO + Co_2(CO)_8$$

Reaction with CO forms the relatively stable $PrCOCo(CO)_4$, which constitutes a major fraction of the catalytically active compounds in the reactor [28]. However, CO addition is reversible and can regenerate **8**. The aldehyde product is formed in two ways. Oxidative addition of hydrogen to the cobalt in **8** forms the dihydride (**9**), which can rearrange to form butyraldehyde and a cobalt hydride that completes the catalyst cycle. A second mechanism, reaction of **8** with $HCo(CO)_4$, also appears to be a major pathway [29]. The relative rates of the two acyl cleavage mechanisms are not known, but this step is often rate determining in aldehyde synthesis.

A very similar catalytic cycle is involved in the hydroformylation catalyzed by mixtures of $Co_2(CO)_8$ and trialkylphosphines. While complex mixtures of cobalt species form in this system at low pressures of CO and H_2, the major species at 190° and 15–40 atmospheres is $HCo(CO)_3(PBu_3)$ [28,30]. Even in the presence of excess olefin the known alkyl and acyl derivatives are not observed. This result suggests that olefin insertion into the Co—H bond may be rate determining, as well as product determining. The predominance of the *n*-propyl derivative may result from crowding in the coordination sphere of the cobalt caused by the bulky phosphine ligand. Both steric and electronic effects are probably involved in the stabilization of these complexes by the trialkylphosphine ligand. This stabilization permits operation at high temperatures and low CO pressures, conditions that favor hydrogenation of the aldehyde products to alcohols [30].

Rhodium Catalysts

Research in industrial laboratories in the 1950s showed that many rhodium complexes catalyze hydroformylation of olefins under very mild conditions. Despite their high activity, however, the simple rhodium compounds were not attractive, because they gave mostly branched aldehydes, for example, isobutyraldehyde from propylene. It was later found that addition of phosphorus ligands such as triphenylphosphine or triphenyl phosphite gave active catalysts with excellent selectivity for formation of the desired linear aldehydes [1,23,25]. This modified rhodium technology was put in commercial production by Union Carbide in 1976 [31,32]. Simultaneously with the commercial development Wilkinson and his coworkers studied the mechanism of hydroformylation with $HRh(CO)(PPh_3)_3$ as a catalyst [33]. These studies have produced a good understanding of the role of the catalyst.

The commercial catalyst may be prepared by reaction of high-surface-area metallic rhodium (such as rhodium on charcoal) with synthesis gas in the presence of triphenylphosphine or a phosphite. In the presence of this catalyst propylene reacts with synthesis gas at 10–20 atmospheres pressure and about 100° to give predominantly *n*-butyraldehyde. Little or no aldehyde is hydrogenated to C_4 alcohols, although some olefin is lost through hydrogenation to propane. The selectivity to the desired linear aldehyde is very high, greater than 90% when excess phosphorus ligand is present. The ligand is important to stabilize the catalyst during product recovery, as well as to direct the reaction to formation of the desired product. Since both selectivity and stability are favored by a large excess of ligand, it has been suggested that molten triphenylphosphine (m.p. 79°) is the ideal solvent for the reaction. Indeed the catalytic rhodium complex is very selective and is stable almost indefinitely in this medium [34].

The mechanism of hydroformylation with the modified rhodium catalysts (Figure 5.5) is quite similar to that of the cobalt catalysts. Many of the intermediates isolated in the course of mechanism studies at atmospheric pressure [33] have also been observed by infrared studies at pressures typical of commercial operations [35]. The immediate catalyst precursor is $HRh(CO)(PPh_3)_3$. This compound is much more stable than $HCo(CO)_4$ but requires the presence of some excess ligand to stabilize it during product distillation and catalyst recycle. As in the cobalt system, ligand dissociation is necessary to create a coordinatively unsaturated complex (**10**) that can bind the olefin. In this case one of the bulky phosphine ligands probably dissociates, as shown in the figure. Propylene enters the vacant coordination site in **10** and inserts into the Rh—H bond to form a propyl or isopropyl group. The effect of the phosphine ligand concentration on the product ratio probably operates at the olefin insertion step. If, as shown in the figure, the

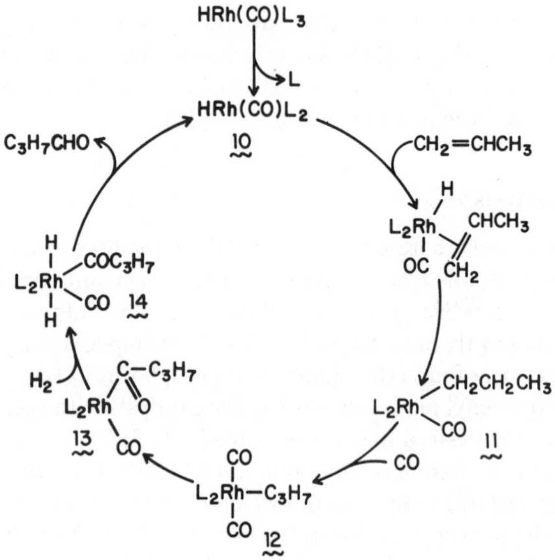

Figure 5.5 Mechanism of hydroformylation of propylene by a phosphine-modified rhodium catalyst (L = triphenylphosphine).

olefin π-complex bears two large phosphine ligands, crowding in the coordination sphere of the metal will favor formation of the *n*-propyl derivative (**11**). If, however, one of the phosphine ligands has been replaced by a much smaller carbon monoxide ligand, there will be space to form the bulky isopropyl group.

The propyl complex (**11**) is coordinatively unsaturated and readily binds another CO ligand to give the dicarbonyl complex (**12**). Carbon monoxide insertion (or propyl migration) generates a σ-acyl compound (**13**) that is again coordinatively unsaturated. Oxidative addition of H_2 and reductive elimination of acyl and hydride ligands from **14** yield the aldehyde and regenerate the catalytically active rhodium hydride. The major uncertainties in the catalytic cycle concern the number of phosphine and CO ligands in the intermediates.

Despite the advantages of the modified rhodium catalysts (mainly high selectivity and low-pressure operations), the high cost of rhodium inventory and recycle may limit their use [36]. The rhodium catalysts will probably be used in the future when high selectivity to linear aldehydes is required. For applications in which some branched products are acceptable or desirable, simple cobalt catalysts will be preferred. The phosphine-modified cobalt catalysts have some advantages for production of linear alcohols.

5.6 DECARBONYLATION

The CO insertion step in hydroformylation is reversible and provides an interesting synthesis of alkanes from aldehydes:

$$RCHO \xrightarrow{\text{catalyst}} RH + CO$$

Similarly, many acyl chlorides decarbonylate to give chloroalkanes or chloroarenes when treated with soluble catalysts. Although these reactions are not universal and do not have commercial utility, they have several applications in laboratory syntheses of complex organic molecules. Decarbonylation is a useful way to remove formyl groups in steroid molecules. Synthesis applications have been reviewed recently [37]. In addition to the aldehyde and acyl chloride reactions there have been reports of decarbonylation of carboxylic acid anhydrides, ketones, and ketenes.

Decarbonylation is usually effected by heating the aldehyde or acyl chloride with a metal complex in a high boiling solvent such as benzonitrile. Wilkinson's catalyst, $RhCl(PPh_3)_3$, is one of the most active CO-abstracting reagents. This compound can be used stoichiometrically to remove CO from the organic compound under very mild conditions [38]:

$$C_6H_{13}CHO + RhCl(PPh_3)_3 \xrightarrow[25°]{CH_2Cl_2} C_6H_{14} + RhCl(CO)(PPh_3)_2 + PPh_3$$

At such a low temperature the rhodium carbonyl product is stable. At high temperatures, however, CO can be expelled from the complex and catalytic decarbonylation occurs [38]:

$$PhCOCl \xrightarrow[180°]{RhCl(PPh_3)_3} PhCl + CO$$

For catalytic decarbonylation the moderately air-stable carbonyl complex $RhCl(CO)(PPh_3)_2$ seems just as effective as Wilkinson's catalyst and is more convenient to handle. When high reaction temperatures cannot be tolerated, a rhodium complex with chelating phosphine ligands such as the commercially available $Ph_2PCH_2CH_2PPh_2$ can be used catalytically [39].

Zerovalent palladium complexes are also effective catalysts for decarbonylation, but equally good results are often obtained with palladium(II) chloride or supported metallic palladium catalysts [37]. For example, an 88% yield of toluene is obtained when *p*-tolualdehyde is boiled with 5% palladium-on-carbon for 30 minutes [40].

Although decarbonylation of aromatic aldehydes and acyl chlorides is straightforward and generally proceeds in good yield, several complications are observed with aliphatic derivatives. Alkanoyl chlorides often give olefins rather than alkyl chlorides as products. Isomerization of the alkyl groups sometimes occurs. For example, palmitoyl chloride gives mixtures of pentadecenes and isomeric C_{15} chlorides [41]:

$$CH_3(CH_2)_{14}COCl \xrightarrow[-CO]{Pd(acac)_2}$$

$$\underset{+\text{ isomers}}{CH_3(CH_2)_{12}CH{=}CH_2} + \underset{\text{isomers}}{C_{15}H_{31}Cl} + \underset{\text{isomers}}{C_{15}H_{31}COCl}$$

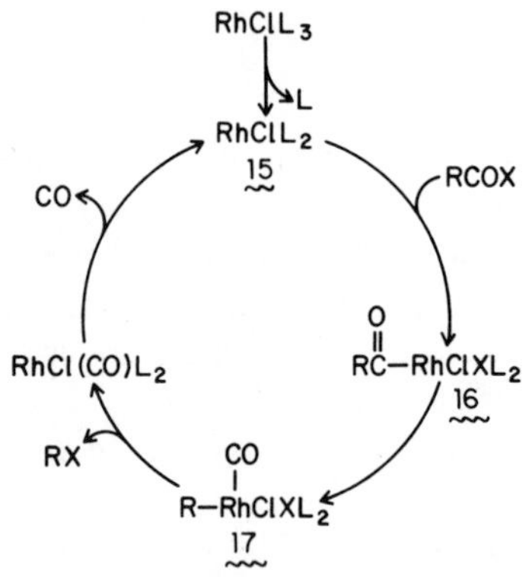

Figure 5.6 Simplified scheme for catalytic decarbonylation; L = CO or PPh_3, X = H or Cl.

The acyl chlorides recovered from such reactions contain grossly altered alkyl groups. Nonanoyl chloride gives 2-methyloctanoyl, 2-ethylheptanoyl, and 2-propylhexanoyl chlorides.

The mechanism generally accepted for decarbonylation [37,42] is illustrated in Figure 5.6 for the reaction catalyzed by Wilkinson's catalyst or $RhCl(CO)(PPh_3)_2$. The initial step with either aldehydes or acyl chlorides is oxidative addition to a coordinatively unsaturated rhodium(I) complex (**15**). The oxidative addition products (**16**) of acyl chlorides are five-coordinate acylrhodium(III) compounds, which are often isolable [42,43]. The aldehyde addition products are usually not stable and undergo further reaction immediately. However, the product from 8-quinolinecarboxaldehyde and $RhCl(PPh_3)_3$ is stabilized by chelation and has been isolated [44]. It has the following structure:

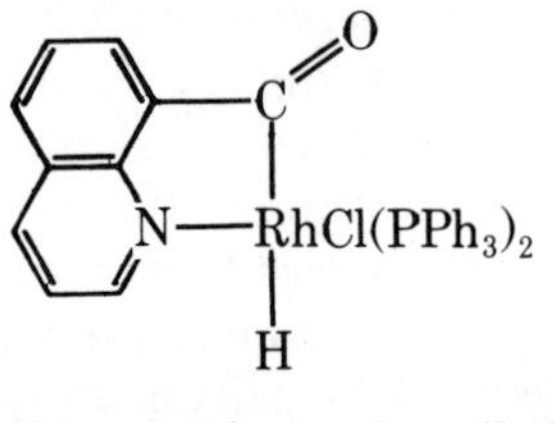

Both types of oxidative addition products give alkyl or aryl rhodium carbonyl complexes by CO elimination from the acyl—Rh bond. This reaction is the reverse of CO insertion and is most correctly considered as a migration of the alkyl group from CO to Rh. The final organic products arise from coupling of the alkyl and H or Cl ligands of the alkyl complex (**17**). As noted previously, the rhodium(I) product of this reductive elimination step is stable at low temperature but eliminates CO at high temperature to complete the catalytic cycle. The side reactions observed with aliphatic aldehydes and acyl chlorides probably result from rearrangement of the alkylrhodium intermediate (**17**).

5.7 ISOCYANATES FROM NITRO COMPOUNDS

Aromatic nitro compounds react with carbon monoxide in the presence of palladium catalysts to give the corresponding isocyanates:

$$ArNO_2 + 3\,CO \rightarrow ArN{=}C{=}O + 2CO_2$$

This unusual reaction is potentially important because diisocyanates such as TDI and MDI are used to prepare polyurethane foams and elastomers (Chapter 11). The reductive carbonylation reaction shown here has many advantages over the usual two-step isocyanate synthesis based on phosgene:

$$ArNO_2 \xrightarrow{H_2} ArNH_2 \xrightarrow{COCl_2} ArNCO + 2HCl$$

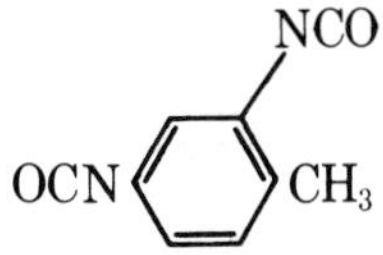

TDI

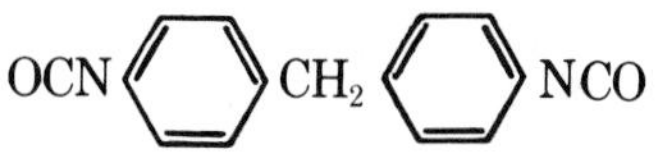

MDI

Arco recently announced plans to construct a plant to produce MDI by reductive carbonylation [45]. It seems likely that they chose to make MDI because it has good market potential and because it has only one NCO group per ring. Attempts to produce TDI and other isocyanates with two NCO groups per ring usually give low yields in the reductive carbonylation process.

The reductive carbonylation of a nitro compound is effected by heating it under carbon monoxide pressure in the presence of a palladium salt or metallic Pd or Rh [46]. For example, when an acetonitrile solution of nitrobenzene is treated with carbon monoxide at 700 atmospheres pressure and 200° in the presence of $PdCl_2$, phenyl isocyanate is formed in 55% yield [47]. When the solvent is an alcohol, the products are carbamate esters [48], which are useful polyurethane precursors like isocyanates. For example, nitrobenzene in methanol gives a methyl ester:

$$PhNO_2 + 3CO + MeOH \xrightarrow{PdCl_2} PhNHC(=O)OMe + 2CO_2$$

It seems likely that the palladium catalyst alternates between Pd(O) and Pd(II) with nitrobenzene as the oxidant and CO as the reductant. Metal salts such as $CoCl_2$ assist the reaction by promoting the redox cycle [49,50].

The mechanism of this reductive carbonylation of nitro compounds is not well established. The palladium salt may be involved primarily in the reduction of the nitro compound. Carbon monoxide reduces $PdCl_2$ to Pd(O) readily in protonic media [51]:

$$Pd^{II} + CO \rightleftharpoons Pd^{II}\text{—}CO \xrightarrow{H_2O} Pd^{II}\text{—}C(=O)\leftarrow OH_2 \longrightarrow Pd^0 + CO_2 + 2H^+$$

A palladium(0) complex should function nicely to abstract oxygen from an aromatic nitro compound to form a nitroso compound. Further O abstraction would yield an arylnitrene, ArN⁚, perhaps stabilized as a metal complex. It has been shown that nitrenes generated by decomposition of aryl azides react rapidly with CO to form isocyanates [52]. Another possible mechanism would involve cooxidation of $ArNH_2$ and CO by a palladium salt [53]:

$$Pd^{II}\text{—}CO \leftarrow NH_2Ar \longrightarrow Pd^0 + O{=}C{=}NAr + 2H^+$$

5.8 CO OXIDATION AND THE SHIFT REACTION

In most commercial production of hydrogen from hydrocarbons, a mixture of carbon monoxide and hydrogen is produced by steam reforming or partial oxidation. The so-called shift reaction:

$$CO + H_2O \rightleftharpoons CO_2 + H_2$$

is then performed to maximize hydrogen yield. Ordinarily it is done in two stages, first at 350–450° over a chromium-based catalyst that gives a rapid conversion of most of the CO. A second catalyst based on copper and zinc is then used at 200–300°, at which temperature the equilibrium constant is more favorable for complete conversion [54].

Recently there have been several reports of catalysis of the shift reaction by soluble complexes such as $Ru_3(CO)_{12}$, $Pt(PR_3)_3$, $[Rh(CO)_2I_2]^-$, and $[HFe(CO)_4]^-$ [55–58]. Although the reactions are too slow and too cumbersome from an engineering viewpoint for practical use, these studies can provide valuable insights into the mode of operation of the commercial heterogeneous catalysts. The rhodium system [56] is one of the best understood. A solution of $[Rh(CO)_2I_2]^-$ in aqueous acetic acid containing HCl converts CO to equimolar amounts of hydrogen and carbon dioxide at 80–90° and ½ atmosphere pressure. The rate is low (5–9 cycles per day), but the catalyst seems relatively stable. The reaction becomes rapid at higher temperatures. In fact it becomes a serious problem in carbonylation reactions in the presence of water because the hydrogen produced consumes olefin by hydrogenation (Section 5.4).

It appears that this catalyst functions through a two-electron redox cycle. The CO oxidation appears to occur by hydrolysis of a high-valent metal carbonyl complex such as rhodium(III):

$$Rh^{III}\text{—}C{\equiv}O \leftarrow OH_2 \longrightarrow Rh^{I} + CO_2 + 2H^+$$

In this respect the shift reaction closely resembles the stoichiometric oxidation of CO by $PdCl_2$ [51]. In both cases the high-valent metal ion appears to activate the CO by withdrawal of electron density. Partial positive charge on the carbonyl carbon atom renders it susceptible to attack by nucleophiles such as water.

The soluble rhodium catalysts, in contrast to $PdCl_2$, are able to complete a catalytic cycle by reoxidation of the low-valent hydrolysis product [56]. A rhodium(I) carbonyl complex is oxidized by protons:

$$Rh^{I}(CO) + 2H^{+} \rightarrow Rh^{III}(CO) + H_2$$

In this way the protons generated during CO oxidation are converted to the desired product, H_2. It seems quite plausible that similar redox cycles occur on the surface of the solid metal oxides used as commercial shift catalysts.

5.9 CO HYDROGENATION

As noted at the beginning of this chapter, rising prices and limited supplies of conventional petrochemical feedstocks are a major incentive to use synthesis gas for the production of organic chemicals. This incentive has stimulated much research on selective hydrogenation of carbon monoxide to useful products. Three processes for which soluble catalysts might be used commercially are the syntheses of ethylene glycol, ethanol, and vinyl acetate. In addition to these potential commercial applications, there is much research on finding homogeneous catalytic equivalents of the Fischer-Tropsch reaction in which CO is hydrogenated to hydrocarbons over a heterogeneous catalyst [59,60]. The three potential industrial processes are discussed individually.

Ethylene Glycol

Union Carbide has announced semiworks development of a process in which synthesis gas is converted to ethylene glycol by a soluble catalyst [31]. The process has many difficulties associated with it, especially high reaction pressures and difficult product separations, but it has major economic potential if these problems can be solved. A variety of rhodium complexes catalyze the transformation of synthesis gas (1:1 H_2:CO) to a mixture of methanol, ethylene glycol, propylene glycol, and glycerol. The conditions are severe (210–250° and 500–3400 atmospheres pressure), but yields of ethylene glycol range up to 70% [61]. Similar results are obtained with mononuclear complexes such as $Rh(CO)_2(acac)$ or with the carbonyl clusters $Rh_4(CO)_{12}$ or $Rh_6(CO)_{16}$. The infrared spectrum of the reaction mixture at 180° under high pressure shows a pattern of C≡O stretching frequencies assigned to a cluster anion, $[Rh_{12}(CO)_{30}]^{2-}$. It has not, however, been demonstrated that the cluster compound is a catalytic intermediate. Cobalt compounds such as cobalt(II) acetate also catalyze the hydrogenation of CO to ethylene glycol [62], but they are usually less active than the rhodium catalysts and are less selective for glycol formation. Presumably, the cobalt salts are converted to cobalt carbonyl complexes that also catalyze methanol synthesis and homologation, as discussed here.

Little is known about the mechanism of this reaction. One possibly related observation is that several early transition metal complexes stoichiometrically reduce CO to a two-carbon ligand. With $Zr(C_5Me_5)_2H_2$ [63] the following sequence of reactions is believed to occur (C_5Me_5 ligands omitted):

$$H{-}Zr{-}H \xrightarrow{CO} H{-}Zr(CO){-}H\ (\mathbf{18}) \rightleftharpoons \left[H{-}Zr{-}C(H){=}O\right]\ (\mathbf{19}) \rightarrow H{-}Zr{-}O{-}CH{=}CH{-}O{-}Zr{-}H\ (\mathbf{20})$$

Coordination of CO to the dihydride gives the CO complex (**18**), which may be in equilibrium with the formyl complex (**19**). Dimerization of **19** would yield the enedioxy compound (**20**), a potential glycol precursor. The formyl complex also gives rise to a methoxy compound $(C_5Me_5)_2Zr(H)(OCH_3)$ under other conditions. While there is no direct evidence to link the zirconium and rhodium chemistry, the pathway via a formyl complex seems quite plausible for both methanol and glycol syntheses. A number of metal carbonyl compounds have been reduced to formyl derivatives [64]. For example, iron carbonyl gives an isolable formyl anion:

$$Fe(CO)_5 + [BH(OR)_3]^- \longrightarrow [HC({=}O)Fe(CO)_4]^- + B(OR)_3$$

Alcohol Synthesis

As mentioned earlier, $Co_2(CO)_8$ catalyzes the hydrogenation of CO to methanol and higher alcohols:

$$CO + 2H_2 \rightarrow CH_3OH \xrightarrow{CO/H_2} C_2H_5OH + H_2O$$

A 1:1 H_2:CO mixture in dioxane at 182° and 300 atmospheres yields methanol as the primary product, along with some methyl formate [65]. As the reaction proceeds, a second faster process produces ethanol, 1-propanol, and their formate esters. The methanol synthesis is not especially interesting from an industrial viewpoint, because copper-based heterogeneous catalysts produce methanol rapidly and cleanly at pressures below 100 atmospheres. However, the homologation of methanol to mixtures of higher alcohols may be useful for the production of synthetic fuels.

It has been proposed [65] that methanol synthesis occurs by a radical process mediated by mononuclear cobalt complexes. The rate-determining step in this scheme is production of a formyl radical from cobalt carbonyl hydride:

$$HCo(CO)_4 + CO \rightleftharpoons H\dot{C}O + \dot{C}o(CO)_4$$

The formyl radical or formaldehyde derived from it would then be hydrogenated by $HCo(CO)_4$ in much the same way that aldehydes are reduced to alcohols in direct alcohol synthesis by hydroformylation.

The methanol homologation reaction has been known for many years [66] and has been reinvestigated recently [67]. When methanol is treated with synthesis gas at 160–180° and 300 atmospheres pressure in the presence of $Co_2(CO)_8$ or a cobalt(II) salt, ethanol and higher linear alcohols and formate esters are produced. Typically, about half of the methanol is converted to higher alcohols with 35–40% selectivity to ethanol. The reaction is promoted by iodide ion, rather like acetic acid synthesis from methanol. It appears that facilitation of C—O bond cleavage is essential to success, just as in acetic acid synthesis. The proposed mechanism [67] involves acid-promoted nucleophilic attack on methanol. ($HCo(CO)_4$ is a strong acid in polar media.)

$$\overset{H^+\ \downarrow}{CH_3\text{—}OH} \xrightarrow{\ \ } H_2O + CH_3X \qquad (X^- \text{ attacks } CH_3)$$

$X = I$ or $Co(CO)_4$

The $CH_3Co(CO)_4$, which is formed directly or via CH_3I, reacts with CO and hydrogen to form acetaldehyde and ethanol, sequentially.

Somewhat similar chemistry seems to be involved in the homologation of dimethyl ether or methyl acetate to ethyl acetate, which is catalyzed by ruthenium salts and iodide ion [68]. The reaction may be sequential:

$$CH_3OCH_3 \xrightarrow{CO} CH_3O\underset{\underset{O}{\|}}{C}CH_3 \xrightarrow[H_2]{CO} CH_3CH_2O\underset{\underset{O}{\|}}{C}CH_3$$

In contrast to the cobalt-catalyzed homologation, acetaldehyde is not observed as a byproduct.

Vinyl Acetate Synthesis

Recent patents issued to Halcon Corporation describe CO chemistry that could lead to a synthesis gas-based process for manufacture of vinyl acetate. In the basic reaction, methyl acetate, which can be derived from CO and H_2 as described in Section 5.3, reacts with synthesis gas to form ethylidene diacetate:

$$CH_3OAc \xrightarrow{CO/H_2} CH_3CH(OAc)_2 \xrightarrow{-HOAc} H_2C{=}CHOAc$$

The diacetate can be converted to vinyl acetate and acetic acid in high yield by passage over a metal oxide catalyst at high temperature.

The reaction of methyl acetate with CO is catalyzed by a rhodium or palladium salt in combination with an iodide [69]. For example, $RhCl_3 \cdot 3H_2O$, together with 3-picoline and methyl iodide, brings about the reaction of a 1:2 $CO:H_2$ mixture with methyl acetate at 150° and about 140 atmospheres pressure. The product contains 44% ethylidene diacetate, 6.5% acetic anhydride, and 0.6% acetaldehyde.

The catalyst is very much like that used in the Monsanto acetic acid process (Section 5.3). The chemistry seems to combine elements of that process with the reactions involved in methanol homologation:

$$CH_3OAc \xrightarrow{HI} CH_3I \xrightarrow{[Rh]} CH_3RhI \xrightarrow{CO} CH_3C(=O)RhI$$

$$CH_3C(=O)RhI \xrightarrow{AcOH} (CH_3C(=O))_2O + [Rh] + HI$$

$$CH_3C(=O)RhI \xrightarrow{H_2} CH_3CHO + [Rh] + HI$$

$$CH_3CHO + (CH_3CO)_2O \rightleftharpoons CH_3CH(OCOCH_3)_2$$

As in the processes described previously, it appears necessary to convert the $CH_3—O$ function of methyl acetate to $CH_3—I$ to react with a low-valent metal complex, designated as [Rh]. Oxidative addition produces a $CH_3—Rh$ bond that readily undergoes CO insertion of form an acetyl compound, which is a key intermediate. Reaction with acetic acid forms the anhydride. Reaction with hydrogen produces acetaldehyde, just as in the rhodium-catalyzed hydroformylation process. Combination of the anhydride and the aldehyde would yield ethylidene diacetate. Although no information about the mechanism of the diacetate synthesis has been published, this scheme accounts for the products very nicely.

GENERAL REFERENCES

Falbe, J., *Carbon Monoxide in Organic Synthesis,* Springer-Verlag, 1970.

Wender, I., and P. Pino, Eds., *Organic Syntheses via Metal Carbonyls,* Vol. 1, 1968; Wiley-Interscience, Vol. 2, 1977.

Stone, F. G. A., and R. West, Eds., *Advances in Organometallic Chemistry,* Vol. 17, *Catalysis and Organic Syntheses,* Academic Press, 1979.

SPECIFIC REFERENCES

1. J. Falbe, *Carbon Monoxide in Organic Synthesis,* Springer-Verlag, 1970.
2. R. P. Lowry and A. Aguilo, *Hydrocarbon Proc.,* 103 (Nov. 1974).
3. F. Calderazzo, R. Ercoli, and G. Natta, "Metal Carbonyls: Preparation, Structure and Properties,"

in I. Wender and P. Pino, Eds, *Organic Syntheses via Metal Carbonyls,* Vol. 1, Wiley-Interscience, 1968.

4. E. L. Muetterties, *Science,* **196,** 839 (1977).
5. K. G. Caulton, R. L. DeKock, and R. F. Fenske, *J. Am. Chem. Soc.,* **92,** 515 (1970).
6. F. Calderazzo, *Angew Chem. Int. Ed.,* **16,** 299 (1977).
7. K. Noack, M. Ruch, and F. Calderazzo, *Inorg. Chem.,* **7,** 345 (1968); K. Noack and F. Calderazzo, *J. Organomet. Chem.,* **10,** 101 (1967).
8. H. Berke and R. Hoffmann, *J. Am. Chem. Soc.,* **100,** 7224 (1978).
9. P. Ellwood, *Chem. Eng.,* 148 (May 19, 1969).
10. J. F. Roth, J. H. Craddock, A. Hershman, and F. E. Paulik, *Chem. Tech.,* **1,** 600 (1971); H. D. Grove, *Hydrocarbon Proc.,* 76 (Nov. 1972); F. E. Paulik, U.S. Patent 3,769,329 (1973).
11. L. Cassar and M. Foa, *J. Organomet. Chem.,* **134,** C15 (1977); U.S. Patent 4,128,572 (1978).
12. D. Forster, *Ann. N.Y. Acad. Sci.,* **295,** 79 (1977); *Adv. Organomet. Chem.,* **17,** 255 (1979).
13. D. Forster, *J. Am. Chem. Soc.,* **97,** 951 (1975).
14. K. Bittler, N. von Kutepow, D. Neubauer, and H. Reis, *Angew. Chem. Int. Ed.,* **7,** 329 (1968).
15. D. M. Fenton, *J. Org. Chem.,* **38,** 3192 (1973).
16. P. Pino, F. Piacenti, and M. Bianchi, "Hydrocarboxylation of Olefins," in I. Wender and P. Pino, Eds., *Metal Syntheses via Metal Carbonyls,* Vol. 2, Wiley-Interscience, 1977, p. 233.
17. F. Piacenti, M. Bianchi, and R. Lazzaroni, *Chim. Ind.* (*Milan*), **50,** 318 (1968).
18. J. F. Knifton, *J. Org. Chem.,* **41,** 2885 (1976).
19. R. F. Heck and D. S. Breslow, *J. Am. Chem. Soc.,* **85,** 2779 (1963).
20. R. W. Johnson and R. G. Pearson, *Inorg. Chem.,* **10,** 2091 (1971).
21. F. Rivetti and U. Romano, *J. Organomet. Chem.,* **154,** 323 (1978).
22. C. E. Loeffler, L. Strautzenberger, and J. D. Unruh, "Butyraldehydes and Butyl Alcohols," in J. J. McKetta and W. A. Cunningham, Eds., *Encyclopedia of Chemical Processing and Design,* Vol. 5, Marcel Dekker, 1977, p. 358.
23. P. Pino, F. Piacenti, and M. Bianchi, "Reactions of Carbon Monoxide and Hydrogen with Olefinic Substrates: The Hydroformylation Reaction," in I. Wender and P. Pino, Eds., *Organic Syntheses via Metal Carbonyls,* Vol. 2, Wiley-Interscience, 1977, p. 43.
24. R. Kummer, H. J. Nienburg, H. Hohenschutz,and M. Strohmeyer in D. Forster and J. F. Roth Eds., *Homogeneous Catalysis—II, Advances in Chemistry Series* **132,** American Chemical Society, (1974), p. 19.
25. F. E. Paulik, *Catal. Rev.,* **6,** 49 (1972).
26. L. H. Slaugh and R. D. Mullineaux, U.S. Patents 3,239,569 and 3,239,570 (1966).
27. D. S. Breslow and R. F. Heck, *J. Am. Chem. Soc.,* **83,** 4023 (1961).
28. R. Whyman, *J. Organomet. Chem.,* **81,** 97 (1974).
29. N. H. Alemdaroglu, J. L. M. Penninger, and E. Oltay, *Monatsh. Chem.,* **107,** 1153 (1976).
30. G. F. Pregaglia, A. Andreeta, G. F. Ferrari, and R. Ugo, *J. Organomet. Chem.,* **30,** 387 (1971).
31. R. L. Pruett, *Ann. N.Y. Acad. Sci.,* **295,** 239 (1977).
32. R. Fowler, H. Connor, and R. A. Baehl, *Chem. Tech.,* 772 (1976); *Hydrocarbon Proc.,* **55,** 247 (Sept. 1976).
33. D. Evans, G. Yagupsky, and G. Wilkinson, *J. Chem. Soc. A,* 2660 (1968); G. Yagupsky, C. K. Brown, and G. Wilkinson, *ibid,* 1392, 2753 (1970).
34. R. Fowler, British Patent 1,387,657 (1975); G. Wilkinson, U.S. Patent 4,108,905 (1978).
35. D. E. Morris and H. B. Tinker, *Chem. Tech.,* **2,** 554 (1972).

36. B. Cornils, R. Payer, and K. C. Traenckner, *Hydrocarbon Proc.*, 83 (June 1975).
37. J. Tsuji, "Decarbonylation Reactions Using Transition Metal Compounds," in I. Wender and P. Pino, Eds., *Organic Syntheses via Metal Carbonyls*, Vol. 2, Wiley-Interscience, 1977, p. 595; J. Tsuji and K. Ohno, *Synthesis*, 157 (1969).
38. K. Ohno and J. Tsuji, *J. Am. Chem. Soc.*, **90**, 99 (1968).
39. D. H. Doughty and L. H. Pignolet, *J. Am. Chem. Soc.*, **100**, 7083 (1978).
40. J. O. Hawthorn and M. H. Wilt, *J. Org. Chem.*, **25**, 2215 (1960).
41. T. A. Foglia, P. A. Barr, and M. J. Idacavage, *J. Org. Chem.*, **41**, 3452 (1976).
42. J. K. Stille, et al, *J. Am. Chem. Soc.*, **96**, 1508, 1514, 1518 (1974).
43. K. S. Y. Lau, Y. Becker, F. Huang, N. Baenziger, and J. K. Stille, *J. Am. Chem. Soc.*, **99**, 5664 (1977).
44. J. W. Suggs, *J. Am. Chem. Soc.*, **100**, 640 (1978).
45. *Chem. Eng. News*, 12 (Oct. 10, 1977); *Chem. Week*, 28 (July 26, 1978).
46. W. B. Hardy and R. P. Bennett, *Tetrahedron Lett.*, 961 (1967); British Patent 1,025,436 (1966).
47. W. W. Prichard, U.S. Patent 3,576,836 (1971).
48. B. A. Mountfield, British Patent 993,704 (1965); J. G. Zajacek and J. J. McCoy, U.S. Patent 3,993, 685 (1976).
49. G. F. Ottmann, E. H. Kober, and D. F. Gavin, U.S. Patents 3,481,968 (1969) and 3,523,966 (1970).
50. H. Tietz, K. Unverferth, and K. Schwetlick, *Z. Chem.*, **17**, 368 (1977); **18**, 98 (1978).
51. V. A. Golodov, Yu. L. Sheludyakov, R. I. Di, and V. K. Fokanov, *Kinet. Katal.*, **18**, 234 (1977).
52. R. P. Bennett and W. B. Hardy, *J. Am. Chem. Soc.*, **90**, 3295 (1968).
53. E. W. Stern and M. L. Spector, *J. Org. Chem.*, **31**, 596 (1966).
54. J. A. Kent, Ed., *Riegel's Handbook of Industrial Chemistry*, 7th ed., Van Nostrand Reinhold, 1974, p. 87.
55. R. M. Laine, R. G. Rinker, and P. C. Ford, *J. Am. Chem. Soc.*, **99**, 252 (1977).
56. C. H. Cheng, D. E. Hendriksen, and R. Eisenberg, *J. Am. Chem. Soc.*, **99**, 2791 (1977).
57. R. B. King, C. C. Frazier, R. M. Hanes, and A. D. King, *J. Am. Chem. Soc.*, **100**, 2925 (1978).
58. T. Yoshida, Y. Ueda, and S. Otsuka, *J. Am. Chem.Soc.*, **100**, 3941 (1978).
59. G. Henrici-Olivé and S. Olivé, *Angew. Chem. Int. Ed.*, **15**, 136 (1976).
60. C. Masters, *Adv. Organomet. Chem.*, **17**, 61 (1979).
61. R. E. Pruett and W. E. Walker, U.S. Patent 3,957,857 (1976).
62. W. F. Gresham, U.S. Patent 2,636,046 (1953).
63. J. M. Manriquez, D. R. McAlister, R. D. Sanner, and J. E. Bercaw, *J. Am. Chem. Soc.*, **100**, 2716 (1978).
64. C. P. Casey and S. M. Neumann, *J. Am. Chem. Soc.*, **98**, 5395 (1976).
65. J. W. Rathke and H. M. Feder, *J. Am. Chem. Soc.*, **100**, 3623 (1978).
66. I. Wender, R. Levine, and M. Orchin, *J. Am. Chem. Soc.*, **71**, 4160 (1949).
67. M. Novotny and L. R. Anderson, Seminar, Catalysis Club of Philadelphia, Chester, PA (May 25, 1978).
68. G. Braca, G. Sbrana, G. Valenti, G. Andrich, and G. Gregorio, *J. Am. Chem. Soc.*, **100**, 6238 (1978).
69. N. Rizkalla and C. N. Winnick, Ger. Offenleg. 2,610,035 (1976).

6 OXIDATION OF OLEFINS AND DIENES

The oxidation of olefins to alcohols, aldehydes, and epoxides is a large and growing part of the industrial application of homogeneous catalysis. Twenty years ago development of the Wacker process for oxidation of ethylene to acetaldehyde was a milestone in replacement of acetylene as a major feedstock for production of organic chemicals. Now, as the Wacker process approaches obsolescence, new processes such as the Halcon glycol acetate synthesis and the Oxirane propylene oxide synthesis seem to have great potential for growth.

Two families of olefin oxidation catalysts are discernible. One is based on the almost unique ability of palladium(II) salts to oxidize olefins to aldehydes, ketones, enol esters, and other products. The other family, which includes the Halcon and Oxirane processes, involves a metal or metalloid ion in one step in a complex series of organic reactions.

The palladium-catalyzed oxidations are potentially useful for many transformations of olefins. However, only the Wacker processes for acetaldehyde from ethylene and acetone from propylene are used on a significant scale in the homogeneous catalytic mode. Similar chemistry can be used to produce vinyl acetate, allyl acetate, glycol acetates, and 1,4-diacetoxy-2-butene, but a heterogeneous palladium catalyst is usually used for these reactions. Despite the engineering advantages of the heterogeneous catalysts, studies of the homogeneous palladium-catalyzed acetoxylation reactions have contributed much to our understanding of the commercial processes.

The palladium-catalyzed oxidations are among the best studied homogeneous catalytic processes. Intimate details of the mechanisms are controversial, but the scope and character of these reactions have been well explored. Excellent reviews of organopalladium chemistry and its catalytic uses are available [1–4].

In contrast to the widespread study of palladium-catalyzed oxidations the development of the molybdenum and tellurium chemistry associated with the Oxirane and Halcon processes seems to have been empirical. Much of the definition of the scope and practice of these reactions is in the patent literature. Few reviews are available and studies of mechanism are quite limited. The lack of academic attention may reflect the newness and the complexity of these reactions. The organometallic chemist may also find these processes unappealing because the Mo and Te atoms play such a limited part. There seems to be no convincing evidence for organometallic intermediates. In this respect these processes resemble the cobalt-catalyzed hydrocarbon oxidations discussed in Chapter 10.

6.1 WACKER ACETALDEHYDE SYNTHESIS

The synthesis of acetaldehyde by oxidation of ethylene, generally known as the Wacker process, was a major step in displacement of acetylene as a starting material in manufacture of organic chemicals. Acetylene-based acetaldehyde (Section 8.3) was a major intermediate for production of acetic acid and of butyraldehyde, itself a major intermediate (Section 5.5). The cost was, however, high because a large energy input is required to produce acetylene. When an efficient process for synthesis of acetaldehyde from ethylene was developed, its adoption was rapid. Ironically this development originated in the laboratories of the Consortium für Elektrochemische Industrie, a branch of Wacker-Chemie [5], which was established to promote use of acetylene. Although the Wacker process itself is now being displaced by CO-based technology (Chapter 5), it is still a major application of homogeneous catalysis.

The invention of the Wacker process was a triumph of common sense. It had been known since 1894 [6] that ethylene is oxidized to acetaldehyde by palladium chloride as a stoichiometric reagent (eq. 1). It was not until 1956, however, that this reaction was combined with the known reactions (2) and (3) to yield a catalytic acetaldehyde synthesis (eq. 4).

$$C_2H_4 + PdCl_2 + H_2O \rightarrow CH_3CHO + Pd^0 + 2\,HCl \tag{1}$$

$$Pd^0 + 2CuCl_2 \rightarrow PdCl_2 + Cu_2Cl_2 \tag{2}$$

$$Cu_2Cl_2 + 2\,HCl + \tfrac{1}{2}\,O_2 \rightarrow 2\,CuCl_2 + H_2O \tag{3}$$

$$C_2H_4 + \tfrac{1}{2}\,O_2 \rightarrow CH_3CHO \tag{4}$$

This sequence of reactions permits the precious metal salt to be used in a catalytic sense.

In commercial practice two modes of operation have developed [5,7]. The single-stage Hoechst process involves feeding an oxygen-ethylene mixture to an aqueous solution of $PdCl_2$ and $CuCl_2$ at 120–130° and 4 atmospheres' pressure. The acetaldehyde is fractionated from the gases as they exit from the reactor. In a two-stage process ethylene is stoichiometrically oxidized to acetaldehyde by a $PdCl_2$-$CuCl_2$ solution in one reactor. The organic-free aqueous solution leaving this reactor contains $PdCl_2$ and Cu_2Cl_2. The copper salt is reoxidized with air in

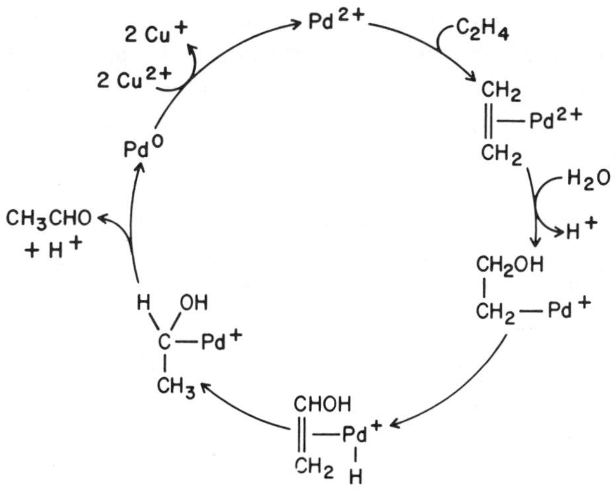

Figure 6.1 Major features of the oxidation of ethylene by palladium salts in water.

a separate reactor before being recycled to the synthesis unit. Both processes give 95% yields of acetaldehyde, along with small amounts of 2-chloroethanol, ethyl chloride, acetic acid, chloroacetaldehydes, and acetaldehyde condensation products. The two-stage process requires a higher capital investment than the single-stage process but permits the use of air rather than oxygen as the oxidant. It also avoids the explosion hazards involved in mixing oxygen and ethylene.

Recently halide-free catalyst systems for ethylene oxidation have been developed [8]. Such systems typically comprise a palladium salt along with a heteropoly acid salt such as a phosphomolybdovanadate. An obvious advantage is reduction in halide-promoted corrosion of process equipment, but it is not clear that these systems will be used commercially.

The Wacker chemistry can also be used to oxidize higher olefins. Terminal olefins yield methyl ketones. For example, one of the commercial syntheses of acetone is the palladium-catalyzed oxidation of propylene [9,10]. Although this reaction is rapid and clean, rates and yields decrease with increasing alkyl chain length. In a standard set of experiments [11] propylene was oxidized in 5 minutes at 20° to give a 90% yield of acetone. In contrast 1-decene required 1 hour at 70° and gave only a 34% yield of 2-decanone. Cyclic olefins react easily to give fairly good yields of cyclic ketones, but other internal olefins give mixtures of products.

The nature of the products in these palladium-catalyzed oxidations is strongly dependent on reaction conditions. The oxidation of ethylene in acetic acid can be directed to give vinyl acetate (Section 6.2), ethylene glycol acetates, or 2-chloroethyl acetate as major products [4]. Similarly, in methanol, either methyl vinyl ether or $CH_3CH(OCH_3)_2$ can be made to predominate. Even in water [12], 2-chloroethanol becomes a major product at high chloride and copper concentrations.

The major features of the mechanism of ethylene oxidation (Figure 6.1) seem to be well established, but a lively controversy about the detailed mechanism has developed recently. There is general agreement that the initial step is replacement of a chloride ion in $[PdCl_4]^{2-}$ (shown as Pd^{2+} in the figure) by ethylene. Additionally, it is agreed that other chloride ions are replaced by water and hydroxide ligands. As a consequence the rate of the reaction is sharply decreased by high chloride concentrations.

The major point of disagreement is the mechanism of the addition of OH^- to the coordinated ethylene to give a hydroxyethylpalladium intermediate. This intermediate has never been observed but is commonly assumed to be present in the reaction mixture. Until recently, the OH addition was considered to be the result of migration of a coordinated OH^- ligand from palladium to carbon, as shown in equation 5.

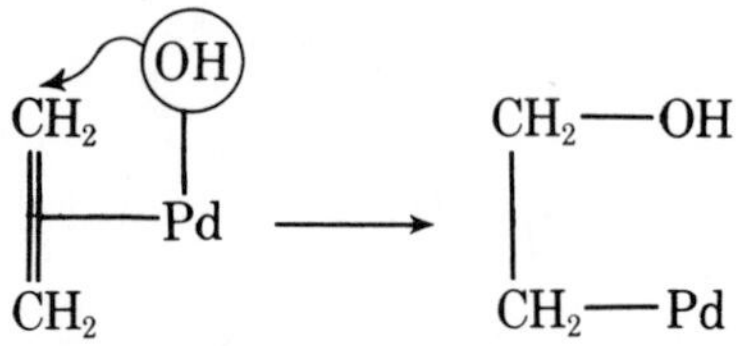

$$CH_2{=}CH_2\text{—}Pd(OH) \longrightarrow HOCH_2\text{—}CH_2\text{—}Pd \qquad (5)$$

$$H_2O + CH_2{=}CH_2\text{—}Pd \xrightarrow{-H^+} HO\text{—}CH_2\text{—}CH_2\text{—}Pd \qquad (6)$$

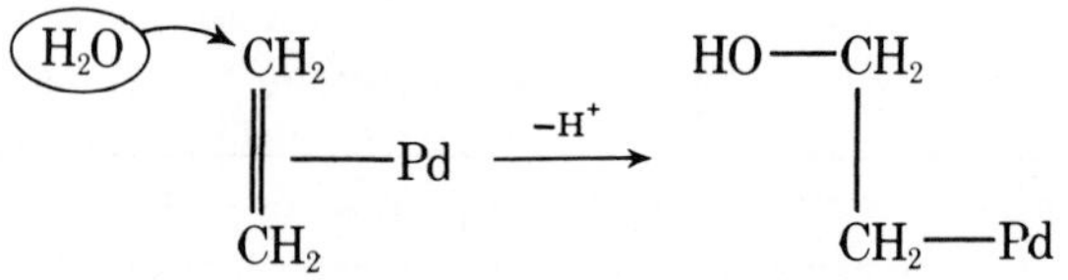

This migration (or ethylene insertion into a Pd—OH bond) is consistent with the observed kinetics. However, recent stereochemical studies [13,14] of addition to coordinated *cis*- or *trans*-CHD=CHD indicate attack by an external nucleophile (eq. 6). It is quite possible that both mechanisms (*cis* and *trans* addition) operate and that reaction conditions determine which predominates.

One remarkable aspect of the ethylene oxidation is that no H/D exchange with solvent occurs. When the reaction takes place in D_2O, no deuterium appears in the product. Evidently the hydrogen migrations involved in transformation of the $HOCH_2CH_2Pd$ intermediate to CH_3CHO occur entirely within the coordination sphere of the palladium. β-Hydrogen elimination to give a vinyl alcohol complex as shown in Figure 6.1 seems a likely pathway. Analogous platinum complexes of vinyl alcohol are well characterized [15]. Readdition of the hydride ligand to the coordinated vinyl alcohol gives the α-hydroxyethylpalladium complex shown at the 8-o'clock position in the figure. A β-hydrogen elimination from OH completes acetaldehyde formation and results in formation of zerovalent palladium or a palladium hydride. Either species is rapidly reoxidized by copper(II) in a chloride-containing medium.

6.2 ACETOXYLATION OF OLEFINS AND DIENES

The reaction of an olefin with a palladium(II) salt in acetic acid is a versatile synthesis of vinylic and allylic acetates. These reactions are generally used industrially as heterogeneous processes, but extensive development effort has been applied to the analogous soluble processes. Some of the most promising reactions are the following:

$$CH_2{=}CH_2 \rightarrow CH_2{=}CHOAc$$
$$CH_2{=}CH_2 \rightarrow HOCH_2CH_2OAc$$

$$CH_3CH{=}CH_2 \rightarrow CH_2{=}CHCH_2OAc$$

$$CH_2{=}CHCH{=}CH_2 \rightarrow AcOCH_2CH{=}CHCH_2OAc$$

The chemistry of these processes, which are discussed in the following sections, closely parallels that of the Wacker process.

Vinyl Acetate and Glycol Acetates from Ethylene

The reaction of ethylene with palladium salts in acetic acid can give several different products under different reaction conditions (Figure 6.2). The original version of the reaction was a stoichiometric oxidation of ethylene with $PdCl_2$ and sodium acetate in acetic acid [16]. The reaction proceeded at room temperature to give vinyl acetate as a major product, but yields were low. Roughly 10 grams of $PdCl_2$ were needed to make 1 gram of vinyl acetate. Such an inefficient use of a precious metal compound obviously had no direct industrial application, but it pointed the way to catalytic processes in which air or oxygen was the ultimate oxidant.

$$CH_2{=}CH_2 + HOAc + \tfrac{1}{2}\,O_2 \rightarrow CH_2{=}CHOAc + H_2O$$

The two liquid-phase catalytic processes that evolved parallel the one- and two-stage acetaldehyde processes. In the one-stage process [17], a suspension of $PdCl_2$, $Cu(OAc)_2$, KCl, and KOAc in acetic acid is saturated with ethylene at 40–45 atmospheres and 120°. Small amounts of oxygen are then added as required

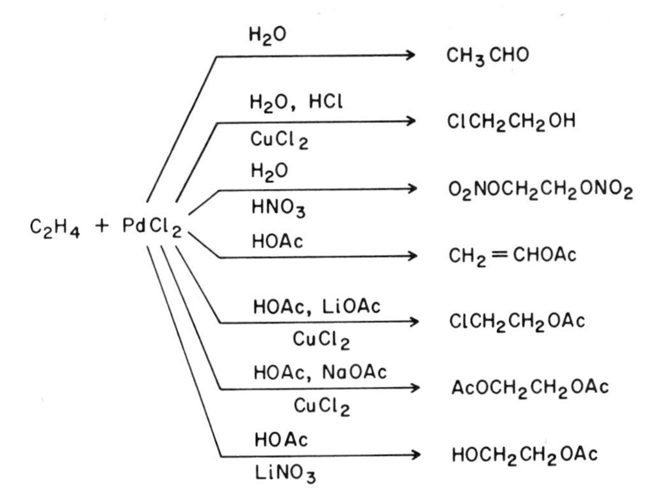

Figure 6.2 Major products from ethylene oxidation in water or acetic acid.

to produce reaction. This process, in a continuous version, gives high yields of vinyl acetate, along with some acetaldehyde. The aldehyde probably forms by reaction of vinyl acetate with water, which is a coproduct in the reaction. In the two-stage process, ethylene is stoichiometrically oxidized to vinyl acetate by $PdCl_2$ and $CuCl_2$ (or benzoquinone) in one reactor. The volatile vinyl acetate is separated by distillation. The reduced working solution is sent to a second reactor where it is reoxidized with air before it is recycled to the vinyl acetate synthesis unit.

Neither liquid-phase process is industrially attractive despite attempts to commercialize both versions. The one-stage process is hazardous because oxygen and ethylene are mixed under pressure. Careful control is needed to avoid explosive mixtures. The two-stage process encounters serious mechanical problems in pumping copper(I) chloride/acetate slurries between reactors. In addition the product distribution is changed by the presence of $CuCl_2$, as we shall see. Both processes use extremely corrosive working solutions and require exotic materials of construction.

The solution to these problems was development of a heterogeneous catalyst that converted a gaseous mixture of ethylene, oxygen, and acetic acid vapor to vinyl acetate. The composition of the catalysts that were developed for this purpose suggests that the chemistry is similar to that in the liquid phase. Typically, a porous silica is impregnated with Na_2PdCl_4 and $HAuCl_4$, and the catalyst is reduced to give a well-dispersed palladium-gold alloy. This catalyst is impregnated with potassium acetate before use. In use at 140–170° and about 5–10 atmospheres pressure, such a catalyst gives a 96% yield of vinyl acetate (based on ethylene), along with small amounts of carbon dioxide [18]. It is tempting to speculate that the pores of the silica catalyst support are saturated with acetic acid under the reaction conditions and that the vinyl acetate synthesis occurs in solution in an acetic acid film.

Both the liquid- and vapor-phase processes convert ethylene and volatile carboxylic acids to vinyl carboxylates in good yield [4]. For example, isobutyric acid gives vinyl isobutyrate in 96% yield with the heterogeneous catalyst [18]. Nonvolatile acids can be converted to vinyl esters easily by ester interchange with vinyl acetate:

$$C_2H_3OAc + RCOOH \rightleftharpoons C_2H_3OOCR + HOAc$$

Both palladium(II) and mercury(II) salts are good catalysts for this reaction [19–21].

In contrast to the ease with which acids can be varied in vinyl carboxylate synthesis, the range of operable olefins is very limited. Most olefins other than ethylene give allylic rather than vinylic acetates when they are oxidized in acetic acid under standard vinyl acetate synthesis conditions. Stoichiometric oxidations may, however, give enol acetates. The oxidation of propylene by $PdCl_2$ in acetic acid without added acetate gives 96% 2-propenyl acetate, whereas oxidation in the presence of 0.9 M sodium acetate yields 94% allyl acetate [22].

In the attempts to develop a liquid-phase vinyl acetate process it was found that changes in the oxidant can completely change the course of ethylene oxidation. In the presence of large quantities of $CuCl_2$ (Cu : Pd = 200) and lithium acetate the major product is 2-chloroethyl acetate, along with some ethylene glycol diacetate [23]. Little vinyl acetate is formed. The glycol diacetate becomes the major product with $CuCl_2$ in the presence of sodium acetate. Potassium acetate makes vinyl acetate the predominant product.

Substitution of nitrate for acetate in these anion-promoted reactions has two major effects. Nitrate can replace copper(II) as an oxidant and it changes the major product to 2-hydroxyethyl acetate (ethylene glycol monoacetate). Reaction of ethylene with an acetic acid solution of $LiNO_3$ and $PdCl_2$ at 50° and 1 atmosphere pressure gives predominantly monoacetate, along with a little diacetate and acetaldehyde [24]. These products are easily hydrolyzed to ethylene glycol, and so this reaction has received much attention as a potential glycol synthesis process. Recently, however, development of a nonprecious metal catalyst system for glycol acetate synthesis (Section 6.4) has reduced interest in the palladium-catalyzed reaction [25].

The great range of ethylene oxidation products summarized in Figure 6.2 probably arises by closely related mechanisms [2]. It is generally agreed that ethylene coordination to palladium sensitizes the olefin to attack by a nucleophile, acetate ion in the case of acetoxylation. As in acetaldehyde synthesis the reaction with the nucleophile could occur within the coordination sphere of the palladium. Recent work suggests, however, attack on coordinated ethylene by an external nucleophile. Stereochemical studies in acetic acid have not resolved this point [26], but, in methanol, coordinated *cis*-CHD=CHD yields a modestly stable 2-methoxyethyl palladium complex [27]. The stereochemistry of this adduct is consistent with *trans* (or *exo*) attack by uncoordinated methanol or methoxide. By analogy the initial reaction in acetic acid is proposed to occur as shown in Figure 6.3. The situation in halide-free acetic acid oxidations is further complicated by the observation that $Na_2Pd(OAc)_4$, $Na_2Pd_2(OAc)_6$, and $Na_2Pd_3(OAc)_8$ oxidize ethylene at different rates [22]. The binuclear complex is most reactive.

The 2-acetoxyethylpalladium intermediate probably yields all the observed ethylene oxidation products, as shown in Figure 6.3. In the absence of an added oxidizing agent, β-hydrogen elimination forms a vinyl acetate complex. When the enol acetate dissociates from the metal, the unstable palladium hydride product decomposes to metallic palladium unless an oxidant intercepts it. The β-hydrogen transfer appears to be reversible. Both 1,1- and 1,2-dideuteroethylenes yield the same mixture of six monodeuterated and dideuterated vinyl acetates [26]. This result is consistent with intramolecular H/D scrambling via rapid β-hydrogen elimination and readdition.

The β-acetoxyethyl complex also appears to be the precursor of the 2-substituted ethyl acetates that form when an oxidant such as Cu^{2+} or NO_3^- is added. Evidently the Pd—C bond is cleaved by the oxidant and the alkyl fragment reacts with the predominant nucleophile (OAc, Cl, or NO_3 ion) in the system. The simplest explanation would be a one-electron oxidation to form a $AcOCH_2CH_2\cdot$ radical, but the stereochemical specificity of the reaction suggests a more complex process [28]. Oxidation of *cis*-1-decene-d_1 with $CuCl_2$ gives mostly *erythro*-$C_8H_{17}CH(OAc)$-CHDCl, consistent with "backside" attack of Cl^- or $CuCl_2$ on the Pd—C bond

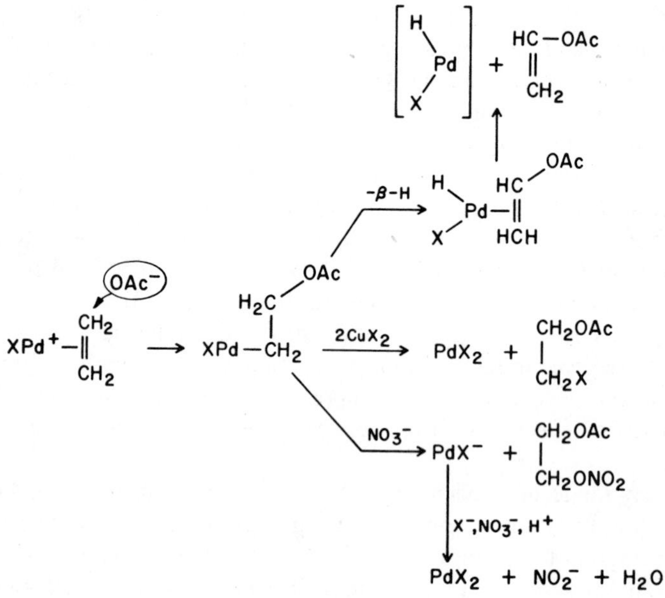

$$XPd^+\text{—}\overset{CH_2}{\underset{CH_2}{\|}} \xrightarrow{OAc^-} XPd\text{—}CH_2CH_2OAc$$

$$XPd\text{—}CH_2CH_2OAc \xrightarrow{-\beta\text{-}H} H(X)Pd\text{—}\|(HC\text{—}OAc)(HCH) \rightarrow [HPdX] + HC(OAc)\text{=}CH_2$$

$$XPd\text{—}CH_2CH_2OAc \xrightarrow{2CuX_2} PdX_2 + CH_2OAc\text{—}CH_2X$$

$$XPd\text{—}CH_2CH_2OAc \xrightarrow{NO_3^-} PdX^- + CH_2OAc\text{—}CH_2ONO_2$$

$$PdX^- \xrightarrow{X^-, NO_3^-, H^+} PdX_2 + NO_2^- + H_2O$$

Figure 6.3 Formation of ethylene oxidation products in acetic acid (X = Cl, OAc).

[29]. Similarly, the oxidation by nitrate can be viewed as a nucleophilic displacement of XPd^- by NO_3^-, as shown in the figure. This explanation is undoubtedly oversimplified. It is likely that electron transfer from Pd to NO_3^- precedes rather than follows Pd—C bond cleavage. In the nitrate-promoted oxidation the relative yields of monoacetates and diacetates may reflect the relative rates of nitrate and acetate attack. This ratio varies extensively with reaction conditions [30]. With $LiNO_3$ and $PdCl_2$ in acetic acid, 98% $HOCH_2CH_2OAc$ is obtained.

Allylic Oxidation

π-Allyl complexes of palladium are remarkably stable and form under many different conditions. Isobutylene gives a π-methallylpalladium complex, along with a simple olefin complex [31,32]:

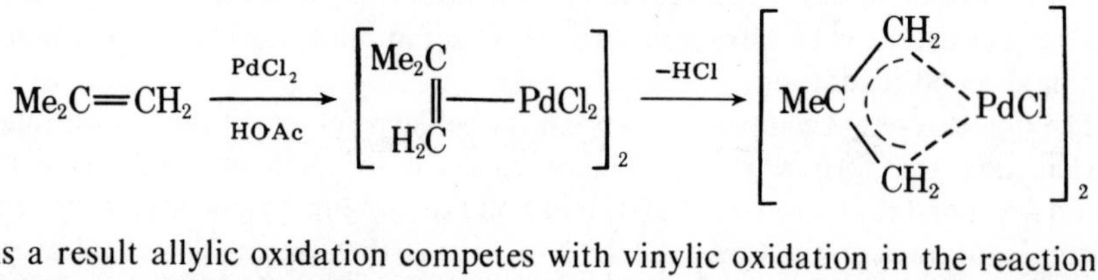

$$Me_2C\text{=}CH_2 \xrightarrow[HOAc]{PdCl_2} \left[\overset{Me_2C}{\underset{H_2C}{\|}}\text{—}PdCl_2\right]_2 \xrightarrow{-HCl} \left[MeC\overset{CH_2}{\underset{CH_2}{\langle\ \rangle}}PdCl\right]_2$$

As a result allylic oxidation competes with vinylic oxidation in the reaction of propylene and higher olefins. Stoichiometric oxidation of propylene in unbuffered acetic acid gives 96% 2-propenyl acetate, but allyl acetate forms in 94% yield in the presence of sodium acetate [22].

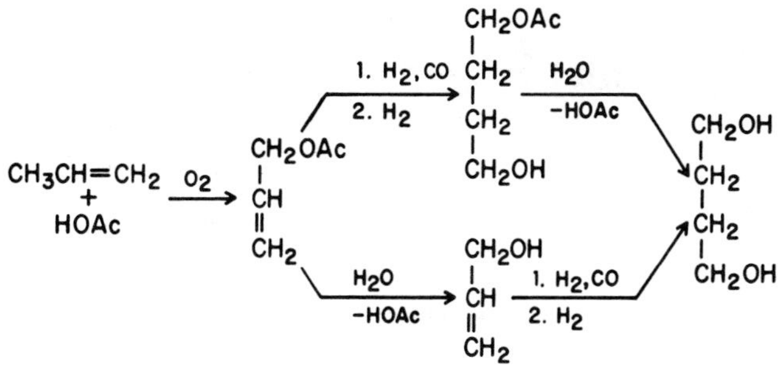

$CH_3CH{=}CH_2 + HOAc \xrightarrow{O_2} CH_2OAc{-}CH{=}CH_2$

$CH_2OAc{-}CH{=}CH_2 \xrightarrow{1.\ H_2, CO;\ 2.\ H_2} CH_2OAc{-}CH_2{-}CH_2{-}CH_2OH \xrightarrow[-HOAc]{H_2O} CH_2OH{-}CH_2{-}CH_2{-}CH_2OH$

$CH_2OAc{-}CH{=}CH_2 \xrightarrow[-HOAc]{H_2O} CH_2OH{-}CH{=}CH_2 \xrightarrow{1.\ H_2, CO;\ 2.\ H_2} CH_2OH{-}CH_2{-}CH_2{-}CH_2OH$

Figure 6.4 Propylene-based processes for synthesis of 1,4-butanediol.

Cooxidation of propylene and acetic acid vapor over a standard heterogeneous catalyst for vinyl acetate synthesis gives allyl acetate in high yield [33–35]. This finding has been used in two potentially attractive processes for synthesis of 1,4-butanediol and tetrahydrofuran (Figure 6.4). One approach has been direct hydroformylation of allyl acetate [34] with a conventional cobalt-based catalyst (Section 5.5). A second route involves hydrolysis of allyl acetate to allyl alcohol [35] and hydroformylation of the alcohol [36] with $HRh(CO)(PPh_3)_3$ as the catalyst. The 1,4-butanediol produced is a valuable polymer intermediate (for polybutylene terephthalate) and can be cyclized to tetrahydrofuran by treatment with acid. These processes, as well as one based on butadiene (below), have attracted interest because starting material costs in the conventional acetylene-based synthesis (Section 8.2) have escalated dramatically [37].

The simplest mechanism for allylic oxidation begins with coordination of the olefin to a palladium(II) ion to form a simple *pi* complex. The actual oxidation step is elimination of a proton from an allyl position to give a π-allyl complex, as shown for isobutylene earlier. Allylic oxidation products form by reactions of the π-allyl ligand. Solvolysis of π-allylpalladium chloride in NaOAc-buffered acetic acid forms allyl chloride and allyl acetate [4]. The latter product predominates in halide-free systems like the surface of the catalyst used in the vapor-phase synthesis. Hydrolysis of π-methallylpalladium chloride gives a mixture of products that include isobutyraldehyde and methacrolein [31]. The unsaturated aldehyde predominates when an oxidizing agent such as excess MnO_2 is present [38].

In all the palladium-catalyzed oxidations the kinetics of various reaction paths are very sensitive to reaction conditions. Olefin *pi* complexes can either react with a nucleophile directly to form vinylic oxidation products, or they may lose allylic protons to form allylic products. Acidity and ionic strength of the reaction medium seem to be major factors in deciding the course of the reaction.

Diene Oxidation

The oxidation of butadiene by palladium salts gives a variety of products (Figure 6.5) under different reaction conditions [39], just as in ethylene oxidation. The stoichiometric oxidation of butadiene by $PdCl_2$ in water gives a 34% yield of cro-

$$C_4H_6 + PdCl_2 \xrightarrow{H_2O} CH_3CH{=}CHCHO$$

$$C_4H_6 + PdCl_2 \xrightarrow{HOAc} CH_2{=}CHCH{=}CHOAc$$

$$C_4H_6 + HOAc \xrightarrow[CuCl_2]{PdCl_2} CH_3CH{=}CHCH_2OAc + CH_2{=}CHCH(OAc)CH_3$$

$$C_4H_6 + HOAc + O_2 \xrightarrow{PdCl_2,\ CuCl_2\ \text{LIQUID PHASE}} AcOCH_2CH{=}CHCH_2OAc$$

$$C_4H_6 + HOAc + O_2 \xrightarrow{Pd/SiO_2\ \text{VAPOR PHASE}} AcOCH_2CH{=}CHCH_2OAc$$

Figure 6.5 Oxidation of butadiene by palladium salts and catalysts.

tonaldehyde [11]. Similarly, oxidation by $PdCl_2$ in buffered acetic acid solutions gives $CH_2{=}CHCH{=}CHOAc$ [40]. These simple stoichiometric oxidations correspond nicely to the acetaldehyde and vinyl acetate syntheses from ethylene. In contrast, when $PdCl_2$ and $CuCl_2$ are used together in acetic acid, a mixture of allylic acetates forms [41] as shown in the figure. These reactions are simple 1,4- and 1,2-additions of acetic acid to butadiene. The 1,4 addition is especially interesting because it opens the way to an allylic oxidation to give the economically desirable 1,4-diacetoxy-2-butene.

Diacetoxylation of butadiene provides an attractive alternative to the conventional acetylene-based synthesis of 1,4-butanediol (Section 8.2). 1,4-Diacetoxy-2-butene is readily converted to the diol by successive hydrogenation and hydrolysis steps [25,37]:

$$AcOCH_2CH{=}CHCH_2OAc \xrightarrow[2.\ H_2O,\ H^+]{1.\ H_2,\ \text{Raney Ni}} HO(CH_2)_4OH + 2HOAc$$

The reaction of butadiene with acetic acid in the liquid phase yields a mixture of 1,4-diacetoxy-2-butene and 3,4-diacetoxy-1-butene [42]. As in glycol acetate synthesis the catalyst system comprises $PdCl_2$, $CuCl_2$, and lithium acetate. Most industrial attention has centered on vapor-phase processes, just as in synthesis of vinyl and allyl acetate. Reactions of butadiene with oxygen and acetic acid vapor over silica-supported palladium catalysts give moderately good yields of 1,4-diacetoxy-2-butene. Certain catalyst additives such as tellurium enhance the yield. Addition of CO to the reaction mixture is also said to be advantageous [43].

The mechanisms of diene oxidations are not well established. One useful working hypothesis is that the vinylic acetoxylation products arise from a simple oxypalladation reaction when butadiene is coordinated as a monodentate ligand:

$$(\eta^2\text{-}C_4H_6)\text{—}Pd^{2+} \xrightarrow{OAc^-} AcOCH_2\text{—}CH(CH{=}CH_2)\text{—}Pd^+ \xrightarrow{-H^+} Pd^0 + AcOCH{=}CHCH{=}CH_2$$

The complementary hypothesis (not proved) is that allylic products arise when butadiene acts as a bidentate ligand:

$$(C_4H_6)Pd^{2+} \xrightarrow{OAc^-} (AcOCH_2C_3H_4)Pd^{+} \xrightarrow{OAc^-} AcOCH_2CH{=}CHCH_2OAc + Pd^0$$

Allylic methoxide addition products analogous to the proposed acetoxybutenyl intermediate have been isolated. 1,3-Cyclooctadiene and Na_2PdCl_4 react in methanol to form a 4-methoxycyclooctenylpalladium chloride complex [44].

$$C_8H_{12} + [PdCl_4]^{2-} \xrightarrow{CH_3OH} [(CH_3OC_8H_{12})PdCl]_2 + HCl$$

6.3 OTHER PALLADIUM-CATALYZED OXIDATIONS

In some potentially useful organic syntheses palladium salts oxidize olefins to form C—Cl and C—C bonds. One general category is simple substitution of vinylic or allylic hydrogen by Cl or CN. Another is coupling of two molecules of olefin to form a diene. A closely related reaction, the oxidative coupling of olefins and arenes is discussed in Section 7.2.

Substitution

Superficially, it would appear attractive to synthesize vinyl chloride and acrylonitrile by oxidations of olefins (X = Cl, CN):

$$C_2H_4 + HX + \tfrac{1}{2}O_2 \rightarrow CH_2{=}CHX + H_2O$$

Both reactions occur with either homogeneous or heterogeneous palladium catalysts, but these processes are not economically attractive [45]. Vinyl chloride is easily available by dehydrochlorination of 1,2-dichloroethane. The cooxidation of propylene and ammonia is a superior synthesis of acrylonitrile.

The oxychlorination reaction has been applied to both ethylene and propylene. In high boiling solvents such as benzonitrile, a $PdCl_2$-$CuCl_2$ mixture catalyzes the reaction of ethylene, HCl, and oxygen to form vinyl chloride [46]. A similar reaction is observed in the vapor phase with supported palladium metal as the catalyst. The vapor-phase oxidation of propylene at 250° and 1 atmosphere pressure gives allyl chloride along with small amounts of 1- and 2-chloropropanes [47].

The stoichiometric oxidation of ethylene by $Pd(CN)_2$ in benzonitrile at 150° and 55 atmospheres pressure gives acrylonitrile in 50% yield [48]. Propionitrile, a simple hydrocyanation product, is also formed, along with small amounts of polyethylene. Propylene is cyanated under similar conditions to give methacrylonitrile but yields are low.

The simplest view of these substitution reactions is that they proceed by nucleophilic attack on a coordinated olefin. By analogy with vinyl acetate synthesis the following mechanism seems likely (X = Cl, CN):

$$X^- + CH_2{=}CH_2 \text{—} Pd^{2+} \longrightarrow X\text{—}CH_2\text{—}CH_2\text{—}Pd^+ \xrightarrow{-\beta\text{-H}} XCH{=}CH_2 + Pd^0 + H^+$$

Little evidence is available, but *exo* (or *trans*) attack on the olefin seems likely. Cyanoethyl complexes like the proposed intermediate have been isolated by addition of rhodium and ruthenium hydrides to acrylonitrile.

Oxidative Coupling of Olefins

The reaction of 1,1-diphenylethylene with palladium(II) salts in aqueous acetic acid does not form the expected vinylic acetate. Instead, a butadiene is produced by oxidative coupling (X = Cl, OAc) [49]:

$$2\,Ph_2C{=}CH_2 + PdX_2 \rightarrow Ph_2C{=}CH\text{—}CH{=}CPh_2 + Pd^0 + 2\,HX$$

Similar couplings occur with other styrenes [31] and with vinyl acetate [50]. These reactions differ from the nonoxidative olefin dimerization discussed in Section 4.3 in that the metal salt is reduced. A supplementary oxidizing agent such as $CuCl_2$ is needed to make the reaction catalytic in palladium.

The oxidative coupling is closely related to the coupling of olefins and arenes discussed in Section 7.2. Substantial evidence [49] suggests a mechanism based on metallation of the olefin:

$$Ph_2C{=}CH_2 + Pd(OAc)_2 \rightleftharpoons Ph_2C{=}CHPdOAc + HOAc$$

This reaction closely parallels the chemistry of arene-olefin coupling and seems in accord with the electrophilic character of the Pd^{2+} ion. Once the Pd—C bond is established, a conventional insertion and β-hydrogen elimination sequence can effect dimerization:

$$Ph_2C{=}C(H)\text{—}Pd(OAc)(H_2C{=}CPh_2) \longrightarrow Ph_2C{=}C(H)\text{—}CH_2\text{—}C(Ph)(Ph)\text{—}PdOAc$$

$$\xrightarrow{-\beta\text{-H}} Ph_2C{=}CH\text{—}CH{=}CPh_2 + [H\text{—}Pd\text{—}OAc]$$

An alternative mechanism [49] is based on simple coupling of two $Ph_2C{=}CH$- ligands in a binuclear complex. The exact nature of the catalytic species is unclear.

Kinetic studies of the reaction suggest that the trinuclear palladium acetate complex $Pd_3(OAc)_6$ is the most active catalyst. The dinuclear $[Pd_2(OAc)_6]^{2-}$ ion is less effective, and $[Pd(OAc)_4]^{2-}$ is unreactive in the dimerization of styrene to 1,4-diphenylbutadiene [51].

6.4 GLYCOL ACETATE SYNTHESES

Ethylene glycol is a major intermediate for synthesis of polyesters and polyurethanes (Chapter 11), in addition to its well-known use as an antifreeze component. It is made by oxidation of ethylene, as shown in Figure 6.6. The conventional process

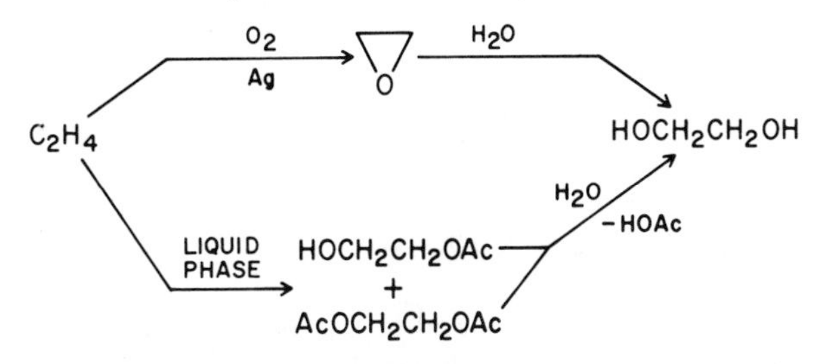

Figure 6.6 Alternative routes for synthesis of ethylene glycol.

involves reaction of an ethylene/oxygen mixture over a metallic silver catalyst to form ethylene oxide in 65–75% yield. The glycol is obtained by hydrolysis of the epoxide. Recently, however, Halcon Corporation and Teijin Limited have patented homogeneous catalytic processes for the oxidation of ethylene. The Teijin process [52] yields ethylene glycol directly. The Halcon process [53,54], indicated in the figure, produces a mixture of monoacetates and diacetates that is hydrolyzed cleanly to form the glycol. Both processes are based on nonprecious metals. Although glycol acetates can be made in high yields under mild conditions by the oxypalladation chemistry of Section 6.2, economic factors favor the use of semicatalytic processes based on cheap metals [25,53].

The Halcon glycol acetate process is based on a sequence of fairly conventional reactions that together amount to a catalytic oxidative acetoxylation of ethylene [53]:

$$C_2H_4 + Br_2 \rightarrow BrCH_2CH_2Br$$

$$BrCH_2CH_2Br \xrightarrow{H_2O/HOAc} HOCH_2CH_2OAc + 2\,HBr + AcOCH_2CH_2OAc$$

$$2\,HBr + Te^{6+} \rightarrow Br_2 + Te^{4+} + 2\,H^+$$

$$Te^{4+} + 2\,H^+ + \tfrac{1}{2}O_2 \rightarrow Te^{6+} + H_2O$$

$$C_2H_4 + \tfrac{1}{2}O_2 \xrightarrow{HOAc} HOCH_2CH_2OAc + AcOCH_2CH_2OAc$$

Philosophically, this development is analogous to the Wacker acetaldehyde synthesis in its creative use of known chemistry. Much less is known, however, about the chemistry in this instance. It is unlikely that the mechanism is as simple as would be suggested by the equations just shown. In particular, analogy to SeO_2 oxidation of organic compounds suggests the interaction of $Te(OH)_6$ with the olefin. Intermediates that contain the $Te—CH_2CH_2OAc$ moiety are likely.

The first commercial application of the Halcon technology is being made by Oxirane Corporation. In a continuous process [54] a reactor is fed with ethylene, oxygen, and an acetic acid solution of TeO_2, 48% HBr, and 2-bromoethyl acetate. Reaction occurs in about 1 hour at 160° and 20–30 atmospheres pressure to produce a liquid effluent that contains 28% $AcOCH_2CH_2OAc$, 16% monoacetate, and 2% ethylene glycol in acetic acid solution. The ratios of monoacetates and diacetates can be controlled by variation of reaction conditions, but careful distillation is necessary to separate them from brominated intermediates such as the $BrCH_2CH_2OAc$, which is recycled. Despite the complexity of the recycle operations and the corrosive nature of the reaction mixture, the process is economically attractive because it gives high yields. The overall yield from ethylene to glycol is more than 90%.*

Similar oxidations of propylene and higher olefins can also be effected with similar Te/Br systems [55]. In a stoichiometric oxidation with TeO_2 as the oxidant (rather than O_2), propylene is oxidized at 120° and 13 atmospheres pressure to give 1,2-diacetoxypropane in 56% yield. The reaction can also be effected with air as the oxidant and a little TeO_2 as the catalyst. It seems likely that yields are lower in the catalytic mode because allylic oxidation occurs as a side reaction. The propylene oxidation is interesting as a potential route to propylene glycol and propylene oxide [25]:

$$CH_3CH(OAc)—CH_2(OAc) \xrightarrow{H_2O} CH_3CH(OH)—CH_2(OH) + 2\ HOAc$$

$$CH_3CH(OAc)—CH_2(OAc) \xrightarrow{\Delta} CH_3\text{-}\overset{O}{\triangle} + Ac_2O$$

Many other oxidation couples can be used like the Te/Br system for olefin oxidation. Several different metals, including V, Mn, Co, and Cu, can be used as catalysts to oxidize HBr to Br_2 with oxygen as the ultimate oxidant. Other analogous systems are based on metal/iodine couples in which Fe(III) or Mn(IV) ions reoxidize HI to I_2. All these systems can be viewed as a series of coupled redox cycles, as illustrated for the Cu/Br system involved in the Teijin process [52].

* Largely because of unsolved corrosion problems, Oxirane has suspended development of this process. (*Chem. Eng. News*, 6 (Dec. 3, 1979))

$$\begin{array}{ccccccc} & 2\,H_2O & & & & 2\,H^+ & \\ C_2H_4 & \downarrow & Br_2 & \leftarrow\;\rightarrow & Cu_2^{2+} & \uparrow & \tfrac{1}{2}\,O_2 \\ C_2H_4(OH)_2 & \downarrow & 2\,Br^- & & 2Cu^{2+} & & H_2O \\ & 2\,H^+ & & & & & \end{array}$$

This series of redox couples illustrates nicely the many interlocking reactions required to achieve a nominally simple process. As with the Halcon chemistry, however, it understates the role of the metal ion. The copper(I) ions almost certainly serve to coordinate and activate the ethylene molecule.

The future of this acetoxylation technology depends heavily on progress in related fields. Production of ethylene glycol from synthesis gas (CO/H_2) could displace ethylene-based processes if the price of ethylene becomes too high (Section 5.9).

6.5 OLEFIN EPOXIDATION BY HYDROPEROXIDES

A major new application of homogeneous catalysis began in 1967 when Halcon Corporation announced a new synthesis of propylene oxide [56], based on a previously known [57] catalytic transfer of oxygen from an alkyl hydroperoxide to an olefin. The reaction with propylene is:

$$CH_3CH{=}CH_2 + ROOH \longrightarrow \underset{\diagdown\, O\, \diagup}{CH_3CH{-}CH_2} + ROH$$

This process appears likely to displace older technology based on dehydrochlorination of propylene chlorohydrin [58]. The impact will be substantial because propylene oxide (methyloxirane) is a major intermediate for propylene glycol, glycerine, and polyethers. As practiced by Oxirane Corporation propylene is reacted with the hydroperoxides obtained by oxidation of ethylbenzene or isobutane. The 1-phenylethanol obtained as a coproduct from the ethylbenzene reaction is dehydrated to produce styrene. *t*-Butyl alcohol, coproduced from *t*-butyl hydroperoxide, is blended with gasoline to inhibit engine knock. Additionally, it can be oxidized to methacrylic acid in a promising new process [59].

In a laboratory experiment [60] that may simulate commercial conditions [61] excess liquid propylene was reacted with a *t*-butyl hydroperoxide solution at 105° for 1 hour in an autoclave under autogeneous pressure. A small amount of $Mo(CO)_6$ was used as the catalyst. Under these conditions 92% of the hydroperoxide was converted and propylene oxide was formed in 86% yield. Similar results can be attained with other alkyl hydroperoxides [62], but the *t*-butyl derivative is especially desirable for the laboratory because it is moderately stable. An excess of the olefin is commonly used to ensure complete conversion of the hydroperoxide. In the presence of barium oxide, however, the hydroperoxide is stabilized against its normal radical decomposition process, and stoichiometric amounts of olefin can be used [63].

The epoxidation can be done with a great variety of olefins [60,62]. Under conditions similar to those described for propylene, 1-octene gives a 92% yield of 1,2-epoxyoctane. Internal olefins such as 2-butene, cyclohexene, and 2-methyl-2-pentene react faster than terminal olefins and give nearly quantitative yields of epoxide. For laboratory-scale preparations of epoxides, however, it is often more convenient to use the uncatalyzed reaction of a peroxy acid with the olefin.

The catalytic epoxidations have an advantage over the peroxy acid reactions in that they operate under neutral conditions. The epoxidation with *t*-BuOOH is very selective for allylic alcohols when a vanadium complex is used as the catalyst. With a cyclooctadienol, only the allylic double bond is epoxidized [64]:

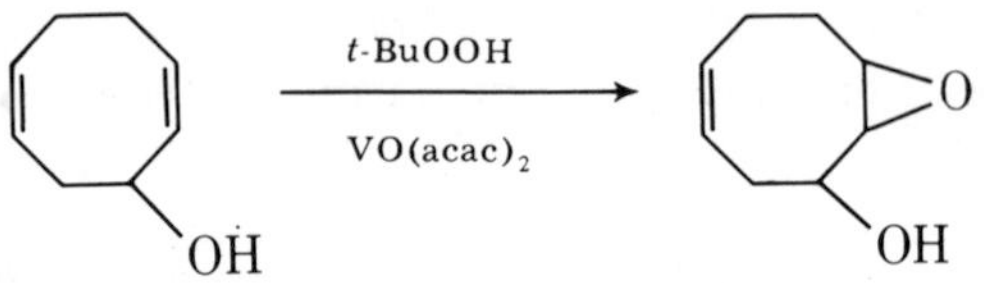

The most active catalysts for catalytic epoxidation are molybdenum salts and complexes [60,62]. Vanadium compounds are also quite effective, and many other metals such as Ti, Zr, and Cr have some activity. With the less effective epoxidation catalysts, both free radical and catalytic decompositions of the alkyl hydroperoxide become competitive reaction pathways. The formation of acetone in the reactions of *t*-butyl or cumene hydroperoxides is evidence of these undesired side reactions.

Most molybdenum compounds have some activity as catalysts for epoxidation. Curiously, zerovalent and hexavalent compounds, for example, $Mo(CO)_6$ and MoO_3, can have equivalent activity [60]. The most likely explanation is that the compounds added to the mixture are catalyst precursors rather than catalysts per se. Indeed there is an induction period before the reaction begins. This time is said to be consumed in formation of a complex with the olefin and the alkylhydroperoxide [65]. Formation of such a complex fits the kinetic pattern that is sometimes observed. The rate is often directly proportional to the concentrations of molybdenum, olefin, and hydroperoxide.

One appealing mechanism that has been proposed [66] involves formation of an olefin complex with a molybdenum(VI) peroxy compound, $[O{=}Mo(O_2)_2]$. Reaction of cyclohexene with MoO_5L_2 (L = hexamethylphosphoramide) gives 1,2-epoxycyclohexane in high yield. The proposal is that the olefin displaces the ligand L from molybdenum and that oxygen transfer occurs within the coordination sphere of the metal. Specific transfer of peroxy oxygen is indicated by the finding that $^{18}O{=}Mo(O_2)_2(HMPA)_2$ reacts with *trans*-cyclododecene to form epoxide that contains no ^{18}O [67]. Despite the strong evidence that $OMo(O_2)_2$ complexes oxidize olefins stoichiometrically, there is no evidence for their presence in epoxidations with alkyl hydroperoxides. Attempts to form $OMo(O_2)_2$ complexes from ROOH under typical epoxidation conditions have been unsuccessful.

The most likely mechanism for catalytic epoxidation is shown in Figure 6.7. In

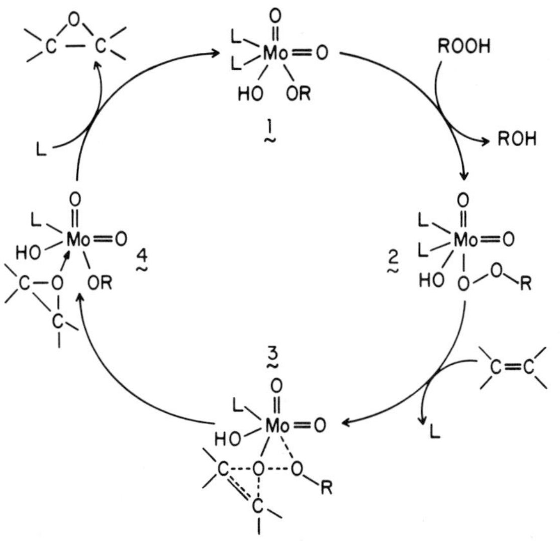

Figure 6.7 Molybdate-catalyzed epoxidation of an olefin.

this proposal by Sharpless [68], an alkyl molybdate (**1**) undergoes ester exchange with alkyl hydroperoxide to form an alkylperoxomolybdate (**2**). Reaction of this complex with an olefin probably occurs by displacement of a neutral ligand such as ROH. At any event, an intermediate or transition state (**3**) develops in which the Mo-bound peroxy oxygen transfers to olefin. This transfer forms the epoxide, which is still complexed to the metal in **4**. Displacement by alcohol or hydroperoxide completes the catalytic cycle.

The selectivity for oxidation of allylic alcohols is probably based on specific binding of the alcohol to the metal as an alkoxide. This complexation brings the allylic C—C bond close to the peroxo oxygen:

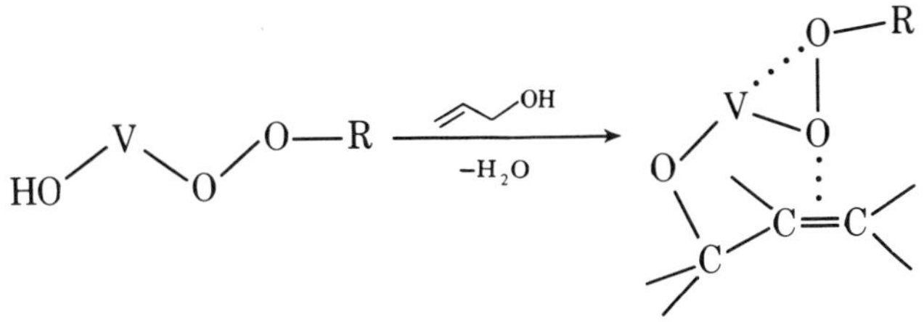

Asymmetric induction occurs during epoxidation of allylic alcohols with vanadium catalysts that bear chiral ligands [69].

SPECIFIC REFERENCES

1. P. M. Maitlis, *The Organic Chemistry of Palladium*, Academic Press, 1971, especially Vol. 2, pp. 77–108.

2. P. M. Henry, "Palladium-Catalyzed Organic Reactions," in *Adv. Organomet. Chem.*, **13,** 363 (1975); *Accounts Chem. Res.*, **6,** 16 (1973).
3. E. W. Stern, *Catal. Rev.*, **1,** 74 (1968).
4. R. Jira and W. Freiesleben, "Olefin Oxidation and Related Reactions with Group VIII Noble Metal Compounds," *Organomet. React.*, **3,** 1 (1972).
5. R. Jira, W. Blau, and D. Grimm, *Hydrocarbon Proc.*, 97 (Mar. 1976).
6. F. C. Phillips, *Am. Chem. J.*, **16,** 255 (1894).
7. R. Jira, "Manufacture of Acetaldehyde Directly from Ethylene," in S. A. Miller, Ed., *Ethylene and Its Industrial Derivatives,* Ernest Benn Ltd., 1969, p. 650.
8. K. I. Matveev, *Kinet. Katal.*, **18,** 862 (1977); British Patent 1,508,331 (1978).
9. L. F. Hatch and S. Matar, *Hydrocarbon Proc.*, 149 (June 1978).
10. *Ullmanns Encyklopädie der technischen Chemie,* Vol. 7, 4th ed., Verlag Chemie, 1977, p. 31.
11. J. Smidt et al., *Angew. Chem.* **71,** 176 (1959).
12. H. Stangl and R. Jira, *Tetrahedron Lett.*, 3589 (1970).
13. J.-E. Backvall, B. Akermark, and S. O. Ljunggren, *J. Am. Chem. Soc.*, **101,** 2411 (1979).
14. J. K. Stille and R. Divakaruni, *J. Organomet. Chem.*, **169,** 239 (1979).
15. J. Hillis, J. Francis, M. Ori, and M. Tsutsui, *J. Am. Chem. Soc.*, **96,** 4800 (1974).
16. I. I. Moiseev, M. N. Vargaftik, and Y. K. Syrkin, *Dokl. Akad. Nauk SSSR,* **133,** 377 (1960).
17. Ref. 4, p. 154.
18. W. Kronig and G. Scharfe, U.S. Patent 3,822,308 (1974).
19. A. Sabel, J. Smidt, R. Jira, and H. Prigge, *Chem. Ber.*, **102,** 2939 (1969).
20. R. N. Pandey and P. M. Henry, *Can. J. Chem.*, **53,** 2223 (1975).
21. R. L. Adelman, *J. Org. Chem.*, **14,** 1057 (1949).
22. S. Winstein, J. McCoskie, H. B. Lee and P. M. Henry, *J. Am. Chem. Soc.*, **98,** 6913 (1976).
23. P. M. Henry, *J. Org. Chem.*, **32,** 2575 (1967).
24. M. Tamura and T. Yasui, *Chem. Comm.*, 1209 (1968); A. Aguilo and A. W. Schnizer, U.S. Patent 3,859,336 (1975).
25. *Proc. Dev. Digest,* Chem Systems, Inc., Fairfield, N.J., **1** (1), 1 (1977).
26. M. Kosaki, M. Isemura, Y. Kitaura, S. Shinoda, and Y. Saito, *J. Mol. Catal.*, **2,** 351 (1977).
27. T. Majima and H. Kurosawa, *J.C.S. Chem. Comm.*, 610 (1977).
28. P. M. Henry, *J. Org. Chem.*, **39,** 3871 (1974).
29. J. E. Bäckvall, *Tetrahedron Lett.*, 467 (1977).
30. M. G. Volkhonskii, V. A. Likholobov, and Yu. I. Ermakov, *Kinet. Katal.*, **18,** 790 (1977).
31. R. Hüttel, J. Kratzer, and M. Bechter, *Chem. Ber.*, **94,** 766 (1961).
32. H. C. Volger, *Rec. Trav. Chim.*, **87,** 225 (1968).
33. H. J. Schmidt and G. Roscher, *Compend. Deut. Ges. Mineraloelwiss. Kohlechem.*, **1975,** pp. 75–76, 318–326; W. Kronig and G. Scharfe, British Patent 1,247,595 (1971).
34. W. E. Smith, British Patent 1,461,831 (1977).
35. P. Hayden, W. Featherstone, and J. E. Lloyd, Canadian Patent 900,510 (1972).
36. T. Shimizu, British Patent 1,493,154 (1977).
37. Y. Tsutsumi, *Chem. Econ. Eng. Rev.*, **8** (5), 45 (1976); A. M. Brownstein and H. L. List, *Hydrocarbon Proc.*, 159 (Sept. 1977).
38. R. Hüttel and H. Christ, *Chem. Ber.*, **97,** 1439 (1964).
39. J. Tsuji, *Accounts Chem. Res.*, **6,** 8 (1973).
40. E. W. Stern and M. L. Spector, *Proc. Chem. Soc.*, 370 (1961).
41. T. Inagaki, Y. Takahashi, S. Sakai, and Y. Ishii, *Bull. Jap. Petrol Inst.*, **13,** 73 (1971).

42. D. A. White, British Patent 1,138,366 (1969).
43. H. M. Weitz and J. Hartig, U.S. Patent 4,038,307 (1977).
44. S. D. Robinson and B. L. Shaw, *J. Chem. Soc.*, 5002 (1964).
45. F. R. Hartley, *J. Chem. Ed.*, **50**, 263 (1973).
46. C. W. Capp, G. W. Godin, R. F. Neale, J. B. Williamson, and B. W. Harris, U.S. Patent 3,194,847 (1965).
47. L. Hörnig, L. Hirsch, G. Mau, and T. Quadflieg, U.S. Patent 3,489,816 (1970).
48. Y. Odaira, T. Oishi, T. Yukawa, and S. Tsutsumi, *J. Am. Chem. Soc.*, **88**, 4105 (1966).
49. H. C. Volger, *Rec. Trav. Chim. Pays-Bas*, **86**, 677 (1967).
50. C. F. Kohll and R. van Helden, *Rec. Trav. Chim. Pays-Bas*, **86**, 193 (1967).
51. A. K. Yatsimirsky, A. D. Ryabov, and I. V. Berezin, *J. Mol. Catal.*, **4**, 151 (1978).
52. I. Hirose and H. Okitsu, U.S. Patent 4,008,286 (1977).
53. A. M. Brownstein, *Hydrocarbon Proc.*, **53**, 129 (June 1974).
54. J. Kollar, U.S. Patent 3,985,795 (1976); R. Hoch and J. Kollar, British Patent 1,351,243 (1974).
55. J. Kollar, British Patent 1,351,242 (1974).
56. *Hydrocarbon Proc.*, **46**, 141 (April 1967).
57. N. Indictor and W. F. Brill, *J. Org. Chem.*, **30**, 2074 (1965).
58. R. B. Stobaugh, V. A. Calarco, R. A. Morris, and L. W. Stroud, *Hydrocarbon Proc.*, **52**, 99 (Jan. 1973).
59. *Chem. Eng. News*, 17 (Oct. 17, 1977); 8 (Aug. 28, 1978).
60. M. N. Sheng and J. G. Zajacek, "Hydroperoxide Oxidations Catalyzed by Metals," in F. R. Mayo, Ed., *Oxidation of Organic Compounds*, Vol. 2, *Advances in Chemistry Series* **76**, American Chemical Society 1968, pp. 418–431.
61. J. Kollar, U.S. Patent 3,351,635 (1967).
62. R. Hiatt, "Epoxidation of Olefins by Hydroperoxides," in R. L. Augustine and D. J. Trecker, Eds., *Oxidation*, Vol. 2, Marcel Dekker, 1971, pp. 113–140.
63. C. Y. Wu and H. E. Swift, *J. Catal.*, **43**, 380 (1976).
64. T. Itoh, K. Jitsukawa, K. Kaneda, and S. Teranishi, *J. Am. Chem. Soc.*, **101**, 159 (1970).
65. V. N. Sapunov, I. Yu. Litvintsev, R. B. Svitych, and N. N. Rzhevskaya, *Kinet. Katal.*, **18**, 408 (1977).
66. H. Mimoun, I. Seree de Roch, and L. Sajus, *Tetrahedron*, **26**, 37 (1970).
67. A. O. Chong and K. B. Sharpless, *J. Org. Chem.*, **42**, 1587 (1977).
68. K. B. Sharpless, private communication, cited in G. W. Parshall, *J. Mol. Catal.*, **4**, 243 (1978).
69. K. B. Sharpless et al, *J. Am. Chem. Soc.*, **99**, 1990 (1977); S. Yamada et al., ibid, **99**, 1988 (1977); C. Dobler and E. Höft, *Z. Chem.*, **18**, 218 (1978).

7 ARENE REACTIONS

Benzene and its derivatives form a broad range of π-complexes and σ-aryl derivatives with transition metal ions. These arene and aryl complexes are intermediates in many catalytic reactions that have potential application in industry and in laboratory syntheses. Some palladium(II)-catalyzed reactions accomplish substitutions or couplings of arenes that are difficult to effect by conventional methods. Similarly copper complexes catalyze some unique reactions that are used on a moderate scale industrially.

One major limitation on use of the metal-catalyzed reactions is that classical organic methods for electrophilic aromatic substitution work so well. Many large-scale arene reactions (Figure 7.1) are catalyzed by simple Lewis or Bronsted acids. Although these acids are soluble catalysts in a broad sense, they fall outside the general context of this book. Almost universally, the primary interaction of the acid is with the attacking reagent rather than with aromatic substrate. In chlorination, for example, $FeCl_3$ reacts with chlorine to form a polarized complex that behaves as though free Cl^+ were formed:

$$Cl_2 + FeCl_3 \rightleftharpoons [Cl_2 \cdot FeCl_3] \rightleftharpoons Cl^+FeCl_4^-$$

This activated reagent attacks benzene to replace H^+ with Cl^+. Similar mechanisms generate incipient NO_2^+ from nitric acid or $C_2H_5^+$ from ethylene. These reactions are discussed thoroughly in standard organic texts and in reviews [1].

The palladium-catalyzed reactions discussed in Section 7.2 seem to involve electrophilic attack of Pd^{2+} on the aromatic ring. However, the arylpalladium compounds that result react by conventional organometallic mechanisms like those observed for olefins. Familiar mechanisms also apply to other reactions such as arene hydrogenation and H/D exchange, which are discussed because research interest in these topics is high. Heterogeneous catalysts are more effective for these reactions in practical terms but are difficult to study. Recent investigations of the soluble catalysts have provided mechanistic insights that should help explain the operation of the metallic catalysts.

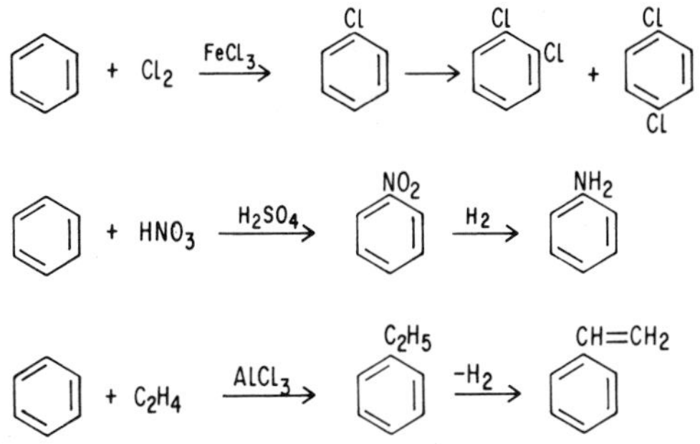

Cl Cl Cl

$+ Cl_2 \xrightarrow{FeCl_3}$ → Cl +

Cl

NO_2 NH_2

$+ HNO_3 \xrightarrow{H_2SO_4}$ $\xrightarrow{H_2}$

C_2H_5 CH=CH_2

$+ C_2H_4 \xrightarrow{AlCl_3}$ $\xrightarrow{-H_2}$

Figure 7.1 Some important industrial processes based on electrophilic substitution reactions of benzene.

7.1 BENZENE AS A LIGAND

The delocalized π orbitals that impart aromatic character to benzene have the right size and symmetry to bond to many transition metal ions. This π bonding interaction, rather like that in olefin π complexes, is present in a large family of mono- and bis(π-arene) complexes [2,3]. The classic example is bis(benzene)-chromium(0), in which a chromium atom is "sandwiched" between two benzene rings, as shown in Figure 7.2. Similar bis(benzene)metal complexes are formed by Ti^0, V^0, Mn^+, Fe^{2+}, and Ru^{2+} ions. Complexes in which only one arene is bonded to a metal ion are known for almost all the transition metals. A particularly important compound of this type is benzenetricarbonyl chromium, $Cr(CO)_3(C_6H_6)$, which catalyzes selective hydrogenation of dienes (Section 3.4).

In most of the stable Group VIII metal complexes the metal atom is centered over the face of the aromatic ring and bonds equally to the six arene carbon atoms. The benzene rings in $Cr(C_6H_6)_2$ are parallel and eclipsed in the crystalline state [4]. The normal sixfold symmetry of C_6H_6 is retained. In most molecular orbital descriptions the filled π orbitals of the benzene ring interact extensively with vacant chromium orbitals. The extent of back-bonding from the metal to the π^* orbitals

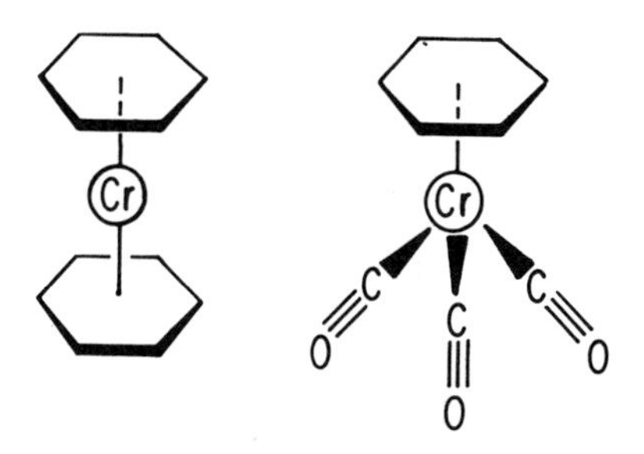

Figure 7.2 Structure of bis(benzene)chromium(O) and benzenetricarbonylchromium(O).

of the ring is variable. On the basis of ^{13}C nmr [5] and other spectra, it seems that more electron density is transferred from chromium to benzene in $Cr(C_6H_6)_2$ than in $Cr(CO)_3(C_6H_6)$. The ^{13}C spectra of substituted complexes such as $Cr(C_6H_5CH_3)_2$ indicate reduced transmission of substituent effects across the ring. In effect the aromatic character is reduced by complexation [5].

The presence of the metal ion on the face of the aromatic ring strongly affects the chemistry of benzene in these complexes. When the metal bears electron withdrawing substituents like carbon monoxide, as in $Cr(CO)_3(C_6H_6)$, the metal attracts electron density from the ring. The net effect is deactivation toward conventional electrophilic substitution but activation toward nucleophilic attack. Nucleophiles readily react with $Cr(CO)_3(C_6H_6)$ and $[Mn(CO)_3(C_6H_6)]^+$ to form cyclohexadienyl complexes. The cyanomethyl anion reacts with the chromium complex according to the following scheme [6]:

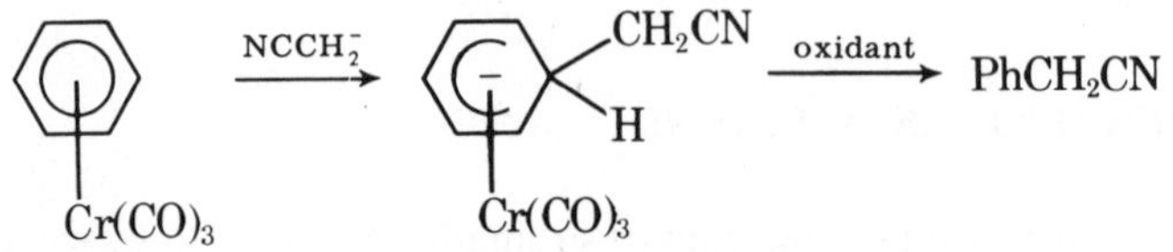

$C_6H_6Cr(CO)_3 \xrightarrow{NCCH_2^-} [C_6H_6(CH_2CN)]^-Cr(CO)_3 \xrightarrow{\text{oxidant}} PhCH_2CN$

Cyclohexadienyl complexes are also formed in electrophilic reactions of benzene complexes. In $Cr(CO)_2(C_6H_6)(PPh_3)$, the electron density on the metal and the ring is enhanced relative to $Cr(CO)_3(C_6H_6)$ by the phosphine ligand. Two reactions with proton are observed (Figure 7.3). Strong acids protonate the metal to form a "hydride" complex (**1**). In addition, protonation of the ring occurs. Even weak acids catalyze H/D exchange with D_2O, probably via cationic cyclohexadienyl complexes like **2** in the figure [7].

In the benzene complexes just discussed, the metal is bound symmetrically to all six carbons. Formally, the three double bonds in the valence tautomers of benzene donate six electrons to the metal ion. The 18-electron rule (Section 2.1) predicts the stability of $Cr(C_6H_6)_2$ and the π-cyclohexadienyl anion complexes if one considers C_6H_6 and $C_6H_6R^-$ as six-electron donors. In some benzene complexes,

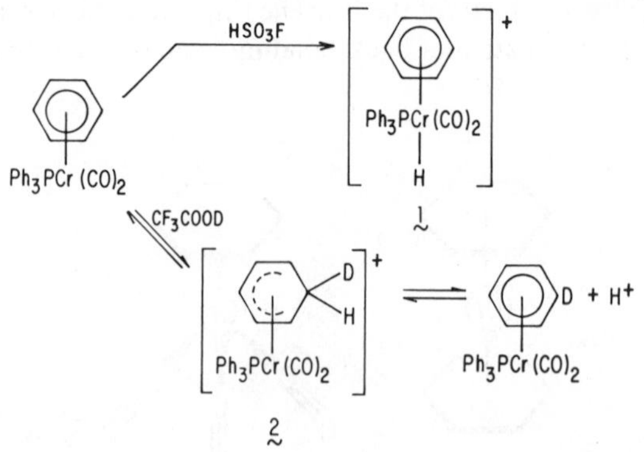

HSO3F
Ph3PCr(CO)2
Ph3PCr(CO)2–H (1)
CF3COOD
Ph3PCr(CO)2 (2)
D + H+
Ph3PCr(CO)2

Figure 7.3 Proton reactions of a π-benzene complex of chromium.

however, only one or two double bonds donate to the metal atom. In these compounds the metal is positioned over the face of the benzene ring but is displaced toward one side of the ring. Ruthenium complexes that involve only two double bonds may be intermediates in hydrogenation of C_6H_6 in solution and are discussed in Section 7.5.

Complexation to only one double bond is believed to be involved in many homogeneous catalytic reactions of benzene. The copper- and palladium-catalyzed reactions discussed later are in this category, although discrete π complexes are not isolated. The crystal structure of $[(\pi\text{-benzene})Cu]^+[AlCl_4]^-$ shows that the copper(I) ion is "almost directly above a C—C bond of the benzene ring [8]." This localization of the metal bonding suggests that the C═C bonds in the ring might be localized. The x-ray data suggest such localization with three C—C bonds of about 1.27 Å and three of about 1.4 Å, but the data are not sufficiently precise to confirm this proposal. As expected, the metal is centered over one of the short bonds.

In many catalytic reactions the π complex involving one double bond is believed to be transformed to a σ-aryl complex via electrophilic metallation or oxidative addition processes [9]:

$$C_6H_6 \cdot M^+ \longrightarrow C_6H_5\text{—}M + H^+$$

$$C_6H_6 \cdot M^+ \longrightarrow [C_6H_5\text{—}M\text{—}H]^+$$

The preliminary π complex formation facilitates attack on the aryl C—H bond. Alkanes are much less likely to react with metals than arenes are because the alkanes do not undergo π complexation.

The chemistry of the aryl-metal σ bonds formed by these processes is very similar to that of the alkyl-metal bonds generated in olefin reactions. Insertion of CO and olefins into Ar—M bonds is well known, and reactions with cleavage reagents such as HX and X_2 are similar to those of R—M bonds [10]. Interactions between the arene π bonds and the metal *d* orbitals are detectable by sensitive methods such as ^{13}C nmr [11] but do not substantially affect the chemistry of the Ar—M bond.

7.2 PALLADIUM-CATALYZED REACTIONS

Palladium(II) salts, especially the acetate, catalyze many oxidative substitution reactions of benzene and other aromatic hydrocarbons [9,12,13]. These reactions are not used commercially but have been studied as potential processes for manufacture of styrene, phenol, and substituted biphenyls. The reaction types closely parallel the Pd-catalyzed oxidations of olefins discussed in Chapter 6. For example, reaction of benzene with palladium acetate can give phenyl acetate, a reaction closely analogous to liquid-phase vinyl acetate synthesis. Like the olefin reactions, the arene reactions are very sensitive to reaction conditions. A change in acetate concentration in the benzene-$Pd(OAc)_2$ reaction makes biphenyl the major product.

These arene oxidations also resemble the olefin oxidations mechanistically, as illustrated in the oxidative coupling of olefins and arenes. This cross coupling may be regarded as a hybrid between olefin-olefin coupling (Section 6.3) and the arene-arene coupling discussed later in the present section.

Arene-Olefin Coupling

When styrene and $Pd(OAc)_2$ are heated in benzene that contains acetic acid, *trans*-stilbene is a major product [14]:

$$\left[(PhHC{=}CH_2)Pd(OAc)_2 \right] \xrightarrow{C_6H_6} trans\text{-}PhHC{=}CHPh + Pd^0 + 2\ HOAc$$

Labeling studies show that one of the phenyl groups in the stilbene comes from benzene. This cross coupling contrasts with the formation of 1,4-diphenylbutadiene when styrene and palladium(II) salts react in neat acetic acid (Section 6.3). Although this example uses the palladium salt as a stoichiometric oxidant, the reaction becomes catalytic when it is conducted under oxygen pressure [15]. Styrene, benzene, and palladium(II) acetate react at 80–100° under 10–20 atmospheres oxygen pressure to form stilbene in amounts corresponding to 2–11 catalyst cycles. Similarly ethylene can be phenylated catalytically to form styrene, stilbene, and higher derivatives [15]:

$$C_2H_4 \xrightarrow{C_6H_6} PhCH{=}CH_2 \rightarrow PhCH{=}CHPh \rightarrow Ph_2C{=}CHPh \rightarrow Ph_2C{=}CPh_2$$

The arene-olefin coupling reactions, whether stoichiometric or catalytic, generally do not give high yields of a single product. Arene-arene coupling is not observed in the presence of olefin [15], but addition of acetic acid to the C=C bond occurs. The styrene-$PdCl_2$ complex in benzene/acetic acid that contains sodium acetate gives α-phenylethyl acetate in addition to stilbene and a little β-acetoxystyrene [16]. This lack of selectivity has reduced the appeal of the benzene-ethylene reaction as a potential industrial process for styrene production.

Substituted benzenes and olefins couple under the conditions used for stilbene synthesis [16]. Toluene and styrene give β-(p-tolyl) styrene in 58% yield [14]. Benzene and methyl cinnamate react in the presence of $PdCl_2$ and sodium propionate to form the β-phenyl coupling product in 40–50% yield [17]:

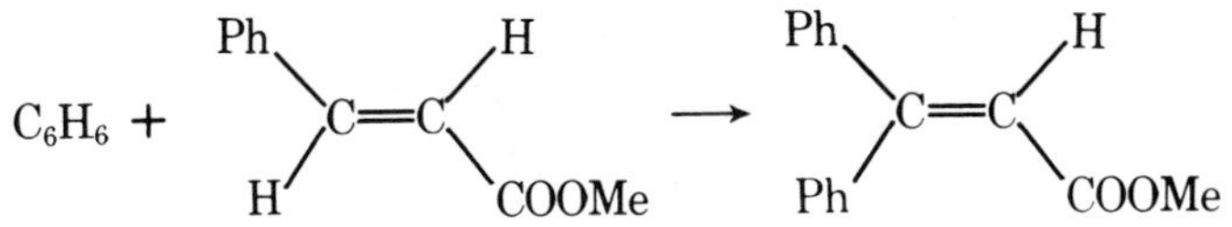

$$C_6H_6 + (Ph)(H)C{=}C(H)(COOMe) \longrightarrow (Ph)(Ph)C{=}C(H)(COOMe)$$

Similar yields of *β-para*-substituted diphenylacrylates are obtained from anisole, chlorobenzene, and nitrobenzene. This lack of substituent effect is curious because many Pd(II) oxidations of arenes are very sensitive to electronic effects. Cycloalkenes generally give low yields (7–28%) of phenylcycloalkanes in the reaction with benzene [16].

The mechanism of arene-olefin coupling is unclear. Two very plausible proposals appear in the literature [9]. Both involve metallation of the benzene ring by a Pd^{2+} electrophile as a key step. It seems likely that the electrophile coordinates to the face of the ring (perhaps off-center) and displaces a proton as illustrated for toluene:

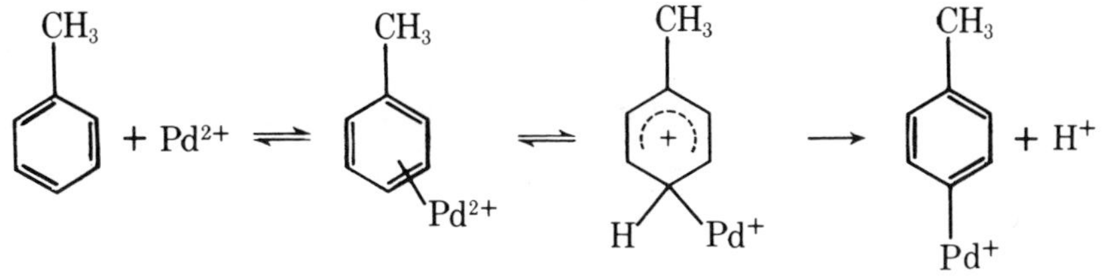

$$C_6H_5CH_3 + Pd^{2+} \rightleftharpoons [CH_3C_6H_5 \cdot Pd^{2+}] \rightleftharpoons [CH_3C_6H_5(H)(Pd^+)]^+ \longrightarrow p\text{-}CH_3C_6H_4Pd^+ + H^+$$

The formation of the *σ*-aryl derivative appears to be a common step in arene-olefin coupling, arene-arene coupling and arene acetoxylation. The *para* specificity observed with toluene in some of these reactions can be interpreted as evidence for attack by a bulky electrophile.

The coupling proceeds through coordination of the olefin to the Ar—Pd^+ compound to give a complex (**3**). Figure 7.4 shows two conventional mechanisms for formation of styrene from an ethylene complex. In the upper pathway [18], ethylene inserts into the Ar—Pd bond to form a *σ*-phenethyl complex (**4**). *β*-Hydrogen elimination by transfer to the metal gives styrene and an unstable Pd—H^+ species that decomposes to palladium metal. Alternatively (lower equation), the ethylene is metallated to form a *σ*-vinyl derivative (**5**). This compound then

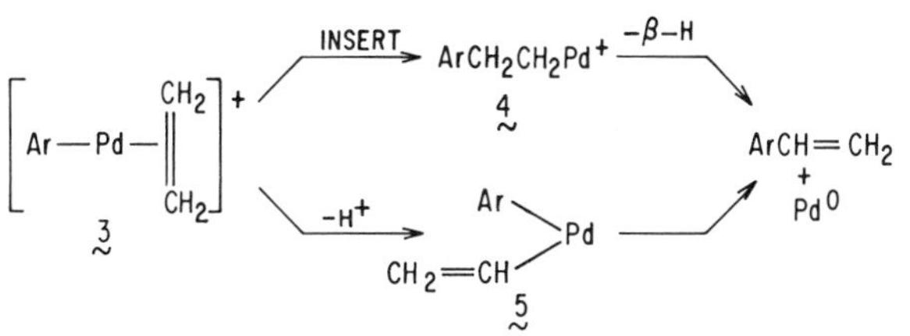

Figure 7.4 Two pathways for styrene synthesis from an ethylene complex of an aryl-palladium compound (other ligands omitted).

undergoes reductive elimination of the two organic ligands to form the styrene directly [19]. There is good precedent for both pathways. Preformed σ-vinyl complexes react with benzene to form styrenes as in the lower pathway. However, preformed σ-phenylpalladium compounds react with olefins to form similar products.

In a related reaction that may be useful in organic synthesis, aryl halides and olefins react to form styrenes. For example, *o*-bromotoluene and ethylene react at 125° and 9 atmospheres pressure to form *o*-methylstyrene in 86% yield [20]:

$$ArBr + C_2H_4 + Et_3N \rightarrow ArCH{=}CH_2 + Et_3NHBr$$

A palladium complex formed *in situ* from palladium acetate and triphenylphosphine is used as a catalyst rather than a reagent. It appears that the reaction proceeds via an arylpalladium complex formed by oxidative addition of aryl bromide to a palladium(O) complex.

$$PdX_2 + nPh_3P \xrightarrow{[H]} Pd(PPh_3)_n$$

$$ArBr + Pd(PPh_3)_n \rightarrow ArPdBr(PPh_3)_n$$

Once the aryl—Pd bond is formed, coordination, insertion, and coupling of the olefin proceed as shown in Figure 7.4.

Arene-Arene Coupling

Benzene, $PdCl_2$, and sodium acetate in acetic acid form biphenyl in high yield [21]:

$$2\,C_6H_6 + PdCl_2 \rightarrow C_6H_5{-}C_6H_5 + Pd^0 + 2\,HCl$$

This reaction uses the palladium salt as a stoichiometric oxidant, but the reaction becomes catalytic in palladium when it takes place under oxygen pressure [22,23]. The stoichiometric coupling to form biphenyl is ordinarily accompanied by some phenyl acetate formation, but acetoxylation is almost completely suppressed by 50 atmospheres oxygen pressure. Recently it has been found that heteropolymolybdate ions serve as catalysts to couple the palladium and O_2 redox systems [24]. With a $Pd(OAc)_2/Hg(OAc)_2/H_5[Mo_{10}V_2PO_{40}]$ catalyst, toluene is converted to dimethylbiphenyls with good rates and yields at 1.5 atmospheres pressure and 50–90°.

The oxidative coupling occurs with a variety of aromatic compounds [25]. Much of the industrial interest in this reaction lay in its potential use to couple toluene to 4,4′-dimethylbiphenyl and *o*-xylene to 3,4,3′,4′-tetramethylbiphenyl [23]. These compounds are possible precursors of biphenyldi- and tetracarboxylic acids, which yield condensation polymers with interesting physical properties. In practice, however, the reaction of toluene with palladium(II) salts is extraordinarily complex. In acetic acid, both benzyl acetate and tolyl acetates are formed in addition to isomeric bitolyls:

$$C_6H_5CH_3 + Pd^{2+} \xrightarrow[HOAc]{NaOAc}$$

$$C_6H_5CH_2OAc + CH_3C_6H_4OAc + CH_3C_6H_4{-}C_6H_4CH_3 + Pd^0$$

At high ratios of sodium acetate to palladium(II) acetate or chloride, benzyl acetate is formed in 92% yield based on palladium [26]. At low ratios, mixed bitolyls are formed in 75% yield [21]. All six possible bitolyl isomers are detected although only traces of the o,o′ isomer are formed. At 25° the o,p′ isomer is the largest single component (45% of the bitolyls) [27]. The m,p′ isomer (35%) is most abundant at 90–200°. The ratio of isomers also depends on the acid concentration in the reaction mixture. The sensitivity of the product distribution to reaction conditions suggests that there are several mechanistic pathways with little difference in activation energy.

In contrast to the arene-olefin coupling reaction, there is substantial evidence that arene-arene coupling proceeds via coupling of two σ-aryl ligands:

$$Ph{-}Pd^+ + C_6H_6 \xrightarrow{-H^+} Ph_2Pd \longrightarrow Ph{-}Ph + Pd^0$$

Tolylpalladium acetate prepared from mostly *para* TolHgOAc gives mostly *p,p′*-bitolyl at temperatures below 50° [27]. This result argues strongly for formation of a σ-aryl—Pd bond in the initial step. The major question in the mechanism seems to be whether the two phenyl groups are assembled at a single metal center as written or whether elimination occurs by interaction of two Ar—Pd species. The kinetics of the catalytic reaction under oxygen pressure are consistent with reaction at a single metal center [23].

The palladium-catalyzed coupling of arenes has been too unselective for industrial use, but a copper-catalyzed coupling of sterically hindered phenols is used commercially. The latter reaction is discussed in Section 7.3.

Oxidative Substitution

Arenes react with palladium(II) salts in the presence of anionic nucleophiles to form substitution products. The first and best defined reaction of this sort is acetoxylation [22]:

$$C_6H_6 + Pd^{2+} + OAc^- \rightarrow C_6H_5OAc + Pd^0 + H^+$$

The acetoxylation was studied as a potential phenol synthesis because phenyl acetate is easily hydrolyzed to phenol and acetic acid. The acetoxylation can be made catalytic in palladium by addition of inorganic oxidants such as $K_2Cr_2O_7$ [28], but

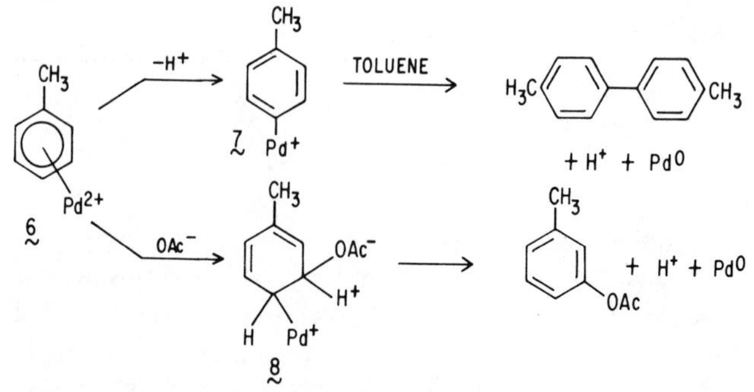

Figure 7.5 Competitive coupling and substitution pathways for a π-toluene palladium(II) complex (other ligands omitted).

it is repressed by oxygen. The best results in a catalytic sense come from use of a heterogeneous catalyst, just as in the closely analogous vinyl acetate synthesis (Section 6.2). Nearly quantitative yields of phenyl acetate and phenol are obtained by passing benzene and acetic acid vapors in a dilute oxygen stream over a supported palladium metal catalyst at 130–190° [29]. Despite the high yield the economics of this process have not been sufficiently attractive for large-scale commercial use.

Fairly high yields of acetoxylation products are obtained when $K_2S_2O_8$ is used as a cooxidant with $Pd(OAc)_2$ in acetic acid [30]. The acetoxylation is moderately fast and clean with a range of arene substituents from CH_3 to $COOCH_3$. Curiously, *meta* substitution products predominate even with toluene. If this reaction occurred by electrophilic attack by Pd^{2+}, *ortho* and *para* isomers might be expected to be most abundant. The best explanation for this phenomenon and for the delicate balance between substitution and coupling involves competitive reactions of an arene π-complex (**6**), as shown in Figure 7.5 [31]. Under the preferred conditions for arene-arene coupling, a σ-aryl palladium species (**7**) can form and react as shown in the upper equation. However, nucleophilic attack on the π-complex can give a product (**8**) in which a Pd—OAc bond has effectively added to an arene C═C bond. This reaction, which is closely analogous to oxypalladation of an olefin C═C bond, need not be *ortho-para* specific. Elimination of a proton and of palladium(0) forms the aryl acetate just as vinyl acetate results from ethylene.

Oxidative Carbonylation

The reaction of palladium(II) salts with CO and an arene is a potentially interesting synthesis of arylcarboxylic acid derivatives. Most attention has been given to carbonylation of arylmercury [32] and arylthallium [33] compounds in the presence of palladium acetate. Presumably, metal-metal exchange forms an arylpalladium complex that carbonylates readily [12,13]:

$$\mathrm{ArM + Pd^{2+} \rightleftharpoons \begin{matrix} Ar{-}Pd^{+} \\ + M^{+} \end{matrix} \xrightarrow{CO} ArC{-}Pd^{+}} \quad (\mathrm{C{=}O})$$

$$\mathrm{M = HgX, TlX_2}$$

The acylpalladium species thus formed reacts with alcohols to give esters. Particular attention has been given to carboxylation of toluene by this route because the metallation is highly *para*-specific. Both the mercury and thallium reactions are said to yield more than 90% methyl *p*-toluate, an intermediate in polyester manufacture (Chapters 10 and 11). Difficulties in reoxidizing the reduced metal salts have inhibited industrial use of this chemistry.

For laboratory syntheses of benzoic acid derivatives a closely related carbonylation of aryl halides may be useful [34]. Palladium complexes such as $PdBr_2(PPh_3)_2$ catalyze the reaction of bromobenzene with CO and butanol under mild conditions (100°, 1 atmosphere):

$$\mathrm{C_6H_5Br + CO + BuOH + Bu_3N \rightarrow C_6H_5COOBu + Bu_3NH^+Br^-}$$

With bromoarenes and iodoarenes, yields are high and many kinds of functional groups are tolerated by the catalyst. The mechanism of the reaction is not clearly established. It seems likely that the palladium(II) complex is reduced by CO or the alcohol. The resulting zerovalent complex reacts with the aryl halide to form an arylpalladium complex:

$$\mathrm{PdBr_2(PPh_3)_2 \xrightarrow[CO]{[H]} Pd(CO)(PPh_3)_2 \xrightarrow[-Ph_3P]{ArX} Ar{-}\overset{\displaystyle CO}{\underset{\displaystyle PPh_3}{Pd}}{-}X}$$

Carbonylation of the arene can then occur by mechanisms like those described in Section 5.4.

7.3 COPPER-CATALYZED OXIDATIONS

Copper(II) salts catalyze several synthetically useful oxidations of aromatic compounds. Two have been used industrially on a moderate scale. An oxidative decarboxylation of benzoic acid yields phenol. Oxidative coupling of 2,6-disubstituted phenols produces polymers or quinoid dimers. These reactions differ in mechanism from the analogous oxidative substitution and coupling reactions with palladium catalysts discussed earlier. The palladium-catalyzed oxidations clearly involve organometallic intermediates like the olefin oxidations of Chapter 6. The copper-catalyzed reactions are commonly described as free radical processes like those of Chapter 10, although organocopper intermediates may be present.

Decarboxylation

Copper salts catalyze both decarboxylation and oxidative decarboxylation of benzoic acid and its derivatives:

$$C_6H_5CO_2H \xrightarrow{Cu^I} C_6H_6 + CO_2$$

$$C_6H_5CO_2H \xrightarrow{O_2,\ Cu^{II}} C_6H_5OH + CO_2$$

The simple decarboxylation is often used in organic synthesis [35] and the oxidative decarboxylation is used commercially for manufacture of phenol [36]. In both reactions a copper(I) salt catalyzes CO_2 elimination, but in the oxidative process copper(II) also plays an important part.

The simple decarboxylation is very clean with arene carboxylic acids. When cuprous benzoate is heated above 200° in a high boiling solvent such as quinoline, benzene is formed in 99% yield [35]. The salt need not be preformed but can be prepared *in situ* by reaction with $CuOOCCH_3$, $CuOOCCF_3$, or an arylcopper(I) complex. The copper(I) compound may be used in catalytic quantities and is quite tolerant of other functional groups. For example, 0.1 equivalent of $[CuC_6F_5]_4$ catalyzes the decarboxylation of an indolecarboxylic acid in high yield:

$$\text{F}_3\text{CO-indole-3-}(CH_2)_2CO_2H\text{, 2-}CO_2H \longrightarrow \text{F}_3\text{CO-indole-3-}(CH_2)_2CO_2H + CO_2$$

The catalyst cycle for decarboxylation of benzoic acid is shown in Figure 7.6. Entry to the cycle occurs by acid cleavage of the C—Cu bond in the organometallic catalyst precursor. Migration of the phenyl group from C to Cu in copper(I) benzoate (**9**) forms a CO_2 complex of phenylcopper (**10**). Dissociation of CO_2 and acidolysis of the Ph—Cu bond completes the catalyst cycle. Support for this mechanism comes from the recently observed reversible carboxylation of a cyanomethylcopper(I) complex [37]:

$$CO_2 + (Bu_3P)_xCuCH_2CN \rightleftharpoons (Bu_3P)_xCuO_2CCH_2CN$$

Copper(II) benzoate, in contrast to the Cu(I) salt, undergoes oxidative decarboxylation. This reaction is the basis for phenol syntheses developed by Dow [36] and by Lummus [38]. Steam and air are blown through a solution of copper(II) and magnesium salts in molten benzoic acid at 230–240° [39]. Carbon dioxide evolves rapidly and phenol distills from the mixture in about 80% yield.

One particularly interesting feature of the oxidative decarboxylation is that the phenolic hydroxyl group occupies a position *ortho* to that of the original carboxyl group. *p*-Toluic acid yields *m*-cresol, for example. Similarly, 1-^{14}C-benzoic acid gives 2-^{14}C-phenol [40]. A clue to the origin of the *ortho* placement of the entering substituent comes from study of the stoichiometric pyrolysis of copper(II) benzoate

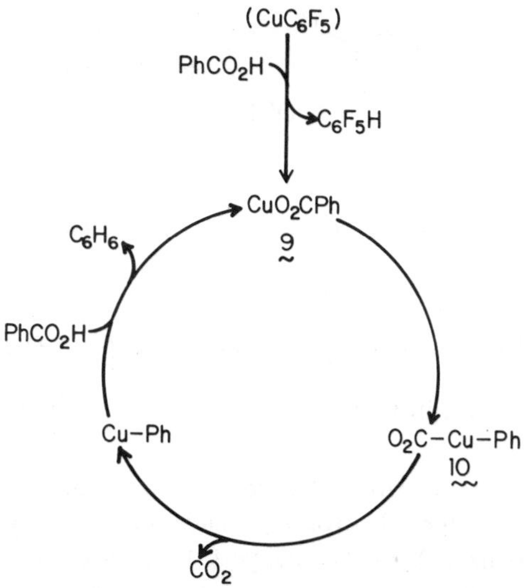

$$\text{Cu}(\text{C}_6\text{F}_5) \xrightarrow[-\text{C}_6\text{F}_5\text{H}]{\text{PhCO}_2\text{H}} \text{CuO}_2\text{CPh}\ (\mathbf{9}) \longrightarrow \text{O}_2\text{C}-\text{Cu}-\text{Ph}\ (\mathbf{10}) \xrightarrow{-\text{CO}_2} \text{Cu}-\text{Ph} \xrightarrow[-\text{C}_6\text{H}_6]{\text{PhCO}_2\text{H}} \text{CuO}_2\text{CPh}\ (\mathbf{9})$$

Figure 7.6 Catalysis of the decarboxylation of benzoic acid by pentafluorophenylcopper.

[41]. Heating this salt in mineral oil at 250° produces a mixture of copper(I) salts and some free benzoic acid:

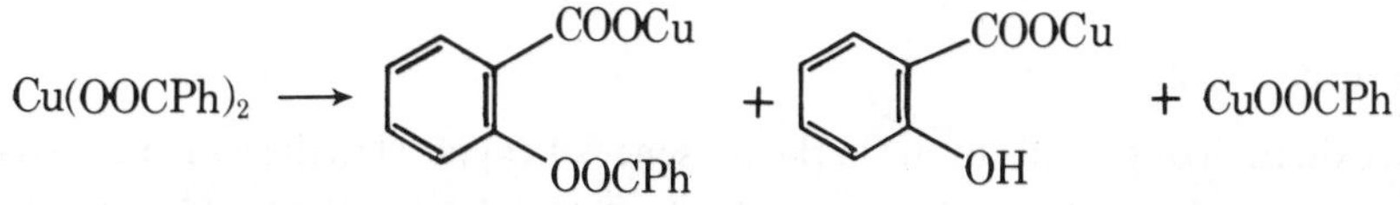

$$\text{Cu(OOCPh)}_2 \longrightarrow o\text{-C}_6\text{H}_4(\text{COOCu})(\text{OOCPh}) + o\text{-C}_6\text{H}_4(\text{COOCu})(\text{OH}) + \text{CuOOCPh}$$

The *ortho*-benzoatobenzoate and the salicylate salts almost certainly result from intramolecular attack on an *ortho* C—H of the benzene ring. The attack is usually described as the result of 2 one-electron transfers [42]:

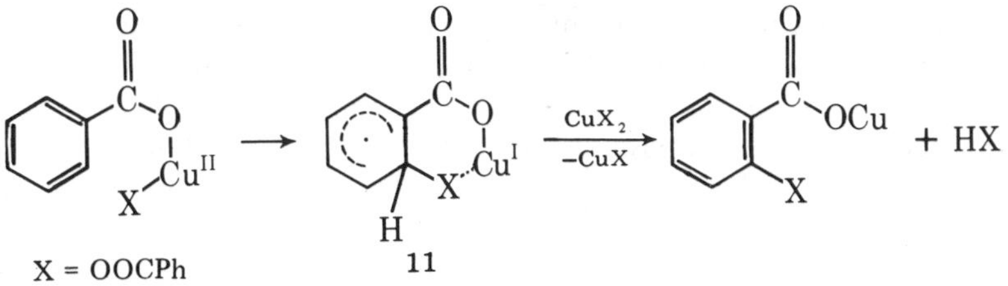

$$\text{PhC(=O)}-\text{O}-\text{Cu}^{\text{II}}-\text{X} \longrightarrow \mathbf{11}\ (\text{arenium radical, } \text{Cu}^{\text{I}}) \xrightarrow[-\text{CuX}]{\text{CuX}_2} o\text{-X-C}_6\text{H}_4\text{C(=O)OCu} + \text{HX}$$

X = OOCPh

This formulation of the mechanism is based on attack of the *ortho* position by an incipient benzoate radical. The arenium radical (**11**) thus formed is oxidized by a second copper(II) ion to produce the observed copper(I) *o*-benzoatobenzoate. No organocopper intermediates are involved in this description of the reaction.

The catalytic phenol synthesis is a combination of several reactions, the first of which is the pyrolysis of copper(II) benzoate just discussed.

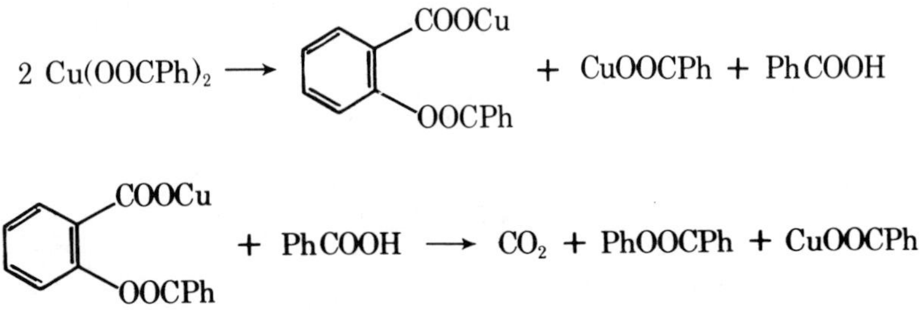

$$2\ Cu(OOCPh)_2 \longrightarrow o\text{-}C_6H_4(COOCu)(OOCPh) + CuOOCPh + PhCOOH$$

$$o\text{-}C_6H_4(COOCu)(OOCPh) + PhCOOH \longrightarrow CO_2 + PhOOCPh + CuOOCPh$$

$$PhOOCPh + H_2O \longrightarrow PhOH + PhCOOH$$

$$2\ CuOOCPh + 2\ PhCOOH + \tfrac{1}{2}\ O_2 \longrightarrow 2\ Cu(OOCPh)_2 + H_2O$$

The copper(I) *o*-benzoatobenzoate decarboxylates rapidly at 230–250° to form phenyl benzoate. This ester is hydrolyzed under the phenol synthesis conditions to give phenol and benzoic acid. The copper(I) salts are reoxidized by air to regenerate copper(II) benzoate.

The Dow process is not used as extensively for phenol production as are the hydrolysis of chlorobenzene and the oxidation of cumene. It has the virtue, however, of being based on toluene, which is often much cheaper than benzene. Hence, oxidation of toluene to benzoic acid (Chapter 10) and oxidative decarboxylation compete economically with benzene-based phenol syntheses.

Phenol Coupling

The oxidation of phenols by air in the presence of copper(I) salts can take several different pathways, as shown in Figure 7.7. Phenol itself is oxidized by oxygen to *p*-benzoquinone in 80% yield in a reaction catalyzed by copper(I) chloride in acetonitrile [43]. When a pyridine complex of copper(I) chloride is used in methanol solution, a major product is the monomethyl ester of *cis, cis*-muconic acid [44]. The same product is obtained from 1,2-dihydroxybenzene. Probably both oxidations involve *ortho*-benzoquinone as an intermediate prior to ring cleavage.

The copper(I)-catalyzed oxidations of phenols show considerable *ortho-para* specificity. When the *ortho* positions are blocked by alkyl or halo substituents, reaction occurs at the *para* position, even with the CuCl/pyridine complex as a catalyst. With very bulky *ortho* substituents such as *tert*-butyl, two phenol molecules couple to form the quinoid dimer shown in the figure.

From a technological viewpoint the most important oxidation in this class is that of 2,6-xylenol. This oxidation produces a *para*-phenylene oxide polymer by coupling an oxygen of one phenol molecule to the *para* carbon of another. This aromatic polyether is a high melting plastic that is very resistant to heat and to water. It has found wide use as an engineering thermoplastic under the trade name PPO [45]. An even more rigid and high melting material is obtained by the analogous oxidation of 2,6-diphenylphenol. The all-aromatic character of this material imparts outstanding thermal stability.

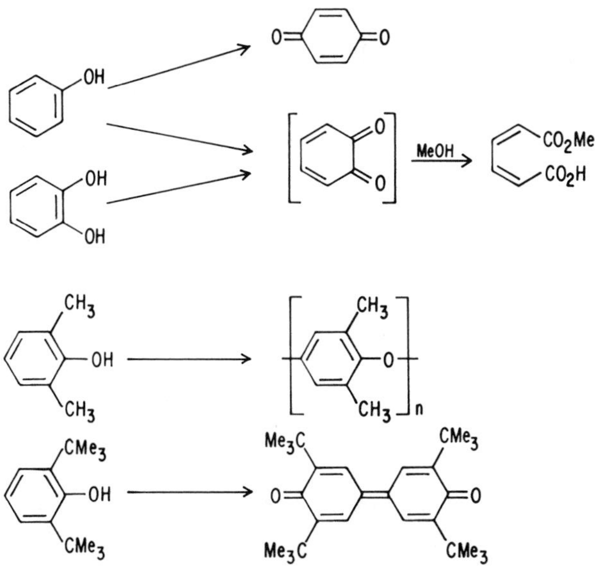

Figure 7.7 Oxidation of phenols by copper(I) chloride/amine/oxygen.

The oxidation of 2,6-xylenol is easy. In a semiworks experiment that may simulate commercial practice, 2,6-xylenol is reacted with oxygen in toluene at 40° in the presence of copper(II) chloride, dibutylamine, NaBr, and a quaternary ammonium dispersing agent [46]. A high-molecular-weight polymer is formed in about 80 minutes. The polymerization process is reversible in the presence of catalyst. The catalyst is deactivated with a chelating agent or with HCl in order to stabilize the polymer. The latter approach is used in a laboratory synthesis of poly(2,6-dimethyl-1,4-phenylene ether) [47].

The course of the oxidation is sensitive to the amine:copper ratio in the catalyst. High ratios of amine to copper produce the polyphenylene oxide polymer as described earlier. At low ratios, however, the major product is a quinoid dimer [48]:

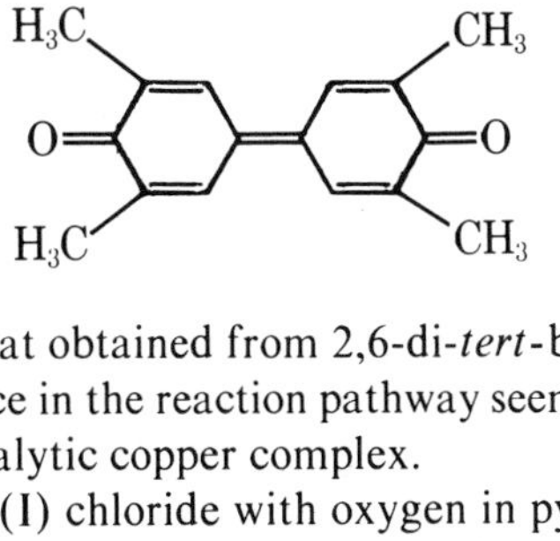

This product resembles that obtained from 2,6-di-*tert*-butylphenol (Figure 7.7). The basis for the divergence in the reaction pathway seems to lie in the interaction of the phenol with the catalytic copper complex.

The reaction of copper(I) chloride with oxygen in pyridine gives a polymeric copper(II) oxide, $[Cu(py)_xO]_n$ [49]. In methanol that contains pyridine the oxidation gives a copper(II) methoxide, [CuCl(OMe)(py)$_x$] [50]. This methoxide reacts with phenols to give phenolates that undergo the observed coupling processes.

$$\text{CuCl(OMe)} + \text{ArOH} \rightleftharpoons \text{CuCl(OAr)} + \text{MeOH}$$

Without excess pyridine the preformed complex gives mainly quinoid dimer. It seems likely that the copper(II) phenolate gives rise to a phenolate radical that is still associated with a copper(I) ion:

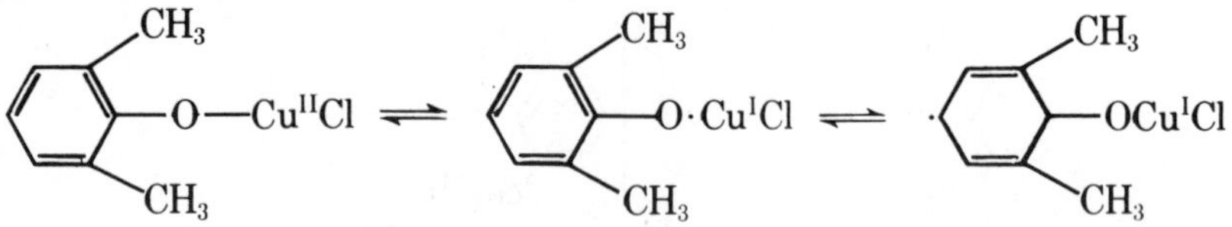
CH3
O—Cu^IICl
CH3
CH3
O·Cu^ICl
CH3
CH3
OCu^ICl
CH3

The O—Cu association inhibits O—C interactions between two radicals but leaves a clear path for C—C coupling of the *para*-quinoid form to give the observed dimer.

Heating the CuCl(phenolate) complex with excess pyridine gives mainly phenylene oxide polymer [48]. This reaction is most easily explained by assuming that the pyridine coordinates to copper and displaces the phenolate ligand. The displaced ligand has the properties of a free radical. Indeed, the polymerization process is most conveniently viewed as a coupling of phenolate radicals [51]. Both monomeric and polymeric radicals have been identified in the esr spectra of polymerization mixtures [52]. The simplest chain growth mechanism is coupling of a monomeric phenolate radical with a similar radical (**11**) generated from a polymer chain (Figure 7.8).

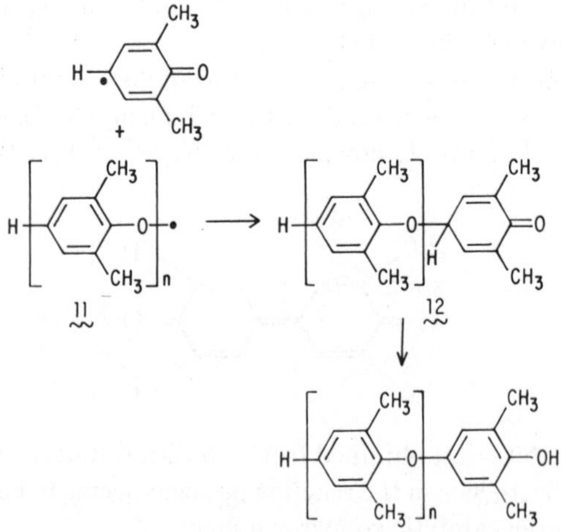
CH3
H
=O
CH3
+
CH3
H
O
CH3
n
11
CH3
H
O
H
CH3
CH3
=O
CH3
12
CH3
H
O
CH3
n
CH3
OH
CH3

Figure 7.8 A chain growth step in the copper-catalyzed oxidative polymerization of 2,6-xylenol.

Addition of the polymeric radical to the *para* position of a monomeric radical forms a coupling product (**12**) with a cyclohexadienone end group. Tautomerization of the end group to a phenol structure yields the enlarged polymer. The phenol end group can react with copper(II) again to form another polymeric radical that can undergo chain growth by a similar mechanism. Kinetic and product studies [51] indicate that many other processes including coupling of polymeric radicals are important parts of the polymerization process. The undesirable degradative processes require the catalyst deactivation step after polymerization.

In all the phenol oxidation processes shown in Figure 7.7 it seems likely that a copper(II) species is the actual oxidant, even though a copper(I) salt is often the preferred catalyst precursor. The role of copper(II) as a one-electron oxidant can also be filled by a cobalt or manganese complex that bears a tightly bound chelating ligand. Recent patents [53] suggest the use of oxime complexes of manganese or cobalt, for example:

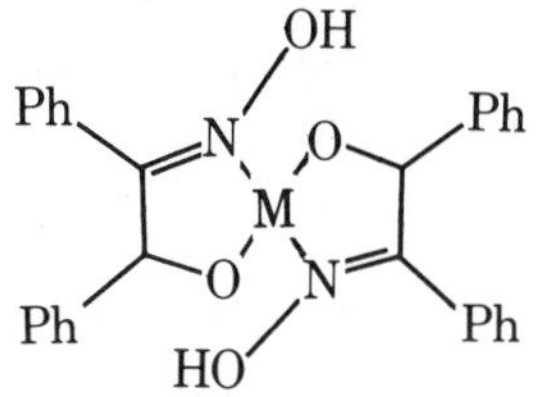

Closely related cobalt chelate complexes have received extensive study as "carriers" of molecular oxygen. The coordination of oxygen to the metal may be involved in the catalysis of phenol oxidation by these compounds [54].

7.4 COUPLING REACTIONS OF ARYL HALIDES

Many reactions of halobenzenes are catalyzed by soluble transition metal complexes. The Ar—X bond is notoriously sluggish toward direct substitution, even by powerful nucleophiles such as RS^-. Traditionally such nucleophilic substitutions have been catalyzed by copper salts. Recently, however, nickel and palladium complexes have been reported to catalyze halide displacement by $(CH_3)_2NH$ [55] and by CN^- [56]. Along with these substitution reactions there have been many reports of couplings between aryl halides and organometallic compounds such as Grignard reagents. The coupling catalysts or reagents are usually complexes of copper or nickel, although many other metals are active. These coupling reactions have been reviewed extensively [57,58] because they have many applications in organic synthesis.

Stoichiometric Coupling

The Ullmann reaction is a classic synthesis of substituted biphenyls [59]. Metallic copper reductively couples some of the more reactive halobenzenes to form the corresponding biaryls. For example, treatment of *o*-chloronitrobenzene with copper bronze at 215–225° produces 2,2′-dinitrobiphenyl in about 60% yield [60]:

$$2\ o\text{-}ClC_6H_4NO_2 \xrightarrow{Cu} 2,2'\text{-}(O_2N)_2C_6H_4\text{—}C_6H_4 + Cu_2Cl_2$$

Lower yields are obtained in the coupling when *o*-bromonitrobenzene is used as the starting material.

Recently it has been found that copper(I) salts bring about similar reductive couplings in high yields under mild conditions. An ammonia complex, $[Cu(NH_3)_2](O_3SCF_3)$, reduces *o*-bromonitrobenzene to 2,2′-dinitrobiphenyl in 90% yield [61]. The reaction occurs smoothly at 30° and has been amenable to mechanistic study. It was proposed that copper(I) reacts with the halobenzene to form an arylcopper(III) compound that couples with more halobenzene:

$$ArBr + Cu^{I} \rightarrow ArCu^{III}Br$$

$$ArCu^{III}Br + ArBr \rightarrow Ar\text{—}Ar + Cu^{III}Br_2^{\bullet}$$

$$Cu^{I} + Cu^{III}Br_2 \rightarrow 2Cu^{II}Br$$

It seems likely that similar processes occur on the surface of copper metal during the Ullman coupling. The metal serves as a reservoir of reducing potential to regenerate copper(I) continuously.

Nickel(0) complexes, which are isoelectronic with copper(I), effect similar couplings of halobenzenes in homogeneous solution. Bis(1,5-cyclooctadiene)-nickel(0) reduces bromobenzene in DMF to form biphenyl in 82% yield [62]. A triphenylphosphine complex, $Ni(PPh_3)_3$, gives similarly high yields in the reductive coupling of bromobenzenes [63]. With both reagents the initial step is reaction with the aryl halide to form a σ-arylnickel compound. This type of intermediate is isolable when trialkylphosphine ligands are present [64]:

$$ArX + Ni(PEt_3)_4 \longrightarrow Ar\text{—}\overset{PEt_3}{\underset{PEt_3}{Ni}}\text{—}X + 2\ Et_3P$$

The simplest course for the actual coupling step would involve aryl/halide exchange between two nickel centers. This process would form a diarylnickel species that can reductively eliminate a biaryl molecule:

$$2\ Ar\text{—}Ni\text{—}X \rightleftharpoons \underset{NiX_2}{NiAr_2 + } \rightarrow Ni^{0} + Ar\text{—}Ar$$

This sequence of events has much in common with catalytic coupling processes discussed later and with the palladium-catalyzed oxidative coupling discussed earlier (Section 7.2).

Catalytic Coupling

Salts and complexes of Fe, Co, Ni, Pd, and Cu catalyze the reactions of halobenzenes with Grignard reagentsto form alkylbenzenes and biphenyls [65–67]:

$$C_6H_5X + RMgX \rightarrow C_6H_5{-}R + MgX_2$$

This reaction does not proceed well in the absence of the transition metal compound, but it becomes rapid when the proper catalyst is present. Alkyllithium, zinc, and aluminum compounds can often be substituted for the Grignard reagent.

Nickel complexes have received the broadest study in this reaction. The most effective catalysts are $NiCl_2(PR_3)_2$ complexes, although the yields vary with the nature of the phosphine and of the Grignard reagent [65]. Generally, highest yields are obtained with complexes of chelating phosphines such as $Ph_2P(CH_2)_3PPh_2$(DPPP). For example, *o*-dichlorobenzene reacts with *n*-butylmagnesium bromide in the presence of $NiCl_2$(DPPP) to form *o*-dibutylbenzene in about 80% yield [66]:

$$o\text{-}C_6H_4Cl_2 + 2\ BuMgBr \longrightarrow o\text{-}C_6H_4Bu_2 + 2\ MgBrCl$$

This product would be difficult to prepare by conventional organic syntheses. Similar yields are obtained with a wide range of *n*-alkyl and aryl Grignard reagents. However, with sterically hindered aryl Grignard reagents such as 2,4,6-trimethylphenyl magnesium bromide, the nonchelate complex $NiCl_2(PPh_3)_2$ is most effective.

Although the coupling reactions are usually straightforward with the catalysts just described, seemingly minor changes in the ligands can change the course of the reaction drastically. This effect is illustrated in the coupling of $(CH_3)_2CHMgCl$ and chlorobenzene with various $NiCl_2L_2$ complexes [65].

	Product Distribution		
Ligand	*i*-PrC_6H_5	*n*-PrC_6H_5	C_6H_6
$Ph_2P(CH_2)_3PPh_2$	96	4	0
$Me_2P(CH_2)_2PMe_2$	9	84	7
$(PPh_3)_2$	16	30	54

The chelating phosphine ligand DPPP gives almost entirely the simple coupling product. However, the methyl chelate ligand leads to extensive isomerization. The triphenylphosphine complex produces extensive reduction of the chlorobenzene to benzene.

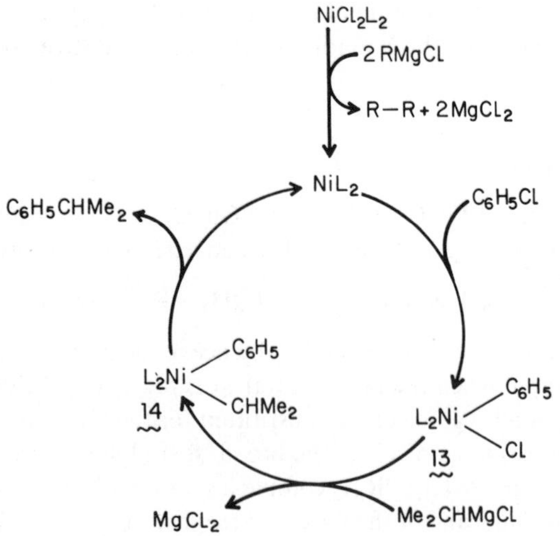

$NiCl_2L_2$

$2\,RMgCl$

$R{-}R + 2\,MgCl_2$

NiL_2

C_6H_5Cl

$C_6H_5CHMe_2$

$L_2Ni(C_6H_5)(CHMe_2)$ **14**

$L_2Ni(C_6H_5)(Cl)$ **13**

$MgCl_2$

$Me_2CHMgCl$

Figure 7.9 A simplified mechanism for coupling chlorobenzene and isopropylmagnesium chloride ($L = R_3P$).

These seemingly anomalous results can be accommodated by a mechanism proposed for the coupling reaction [64]. The mechanism is illustrated in Figure 7.9. As in the stoichiometric coupling described earlier, a nickel(0) complex is a key intermediate and provides entry to the catalyst cycle. Oxidative addition of chlorobenzene to NiL_2 gives a σ-phenyl complex (**13**). A metathetical reaction of the Grignard reagent gives a complex (**14**) containing both σ-aryl and σ-alkyl ligands. In the normal coupling process reductive elimination of the two organic ligands from nickel gives isopropylbenzene and NiL_2 to complete the catalytic cycle. However, when the phosphine ligands are small or weakly bound, other reactions can occur. β-Hydrogen elimination from the isopropyl ligand in **14** forms a hydrido olefin complex:

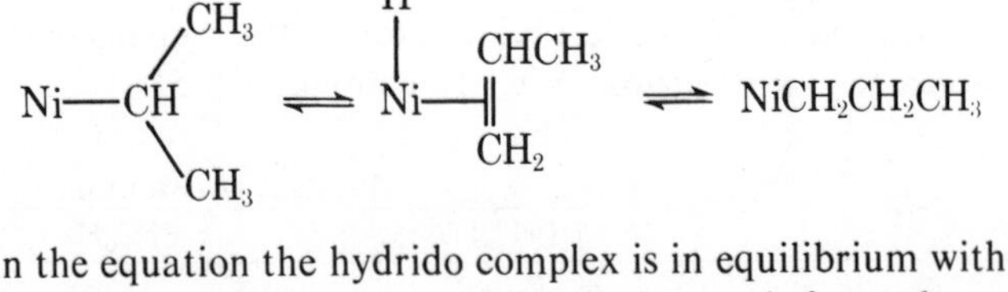

$$Ni{-}CH(CH_3)_2 \rightleftharpoons H{-}Ni{\leftarrow}\left(CH_2{=}CHCH_3\right) \rightleftharpoons NiCH_2CH_2CH_3$$

As indicated in the equation the hydrido complex is in equilibrium with both *n*- and *i*-propyl compounds. If isomerization of the alkyl group is faster than reductive elimination of the alkyl and aryl ligands, *n*-propylbenzene is the major product. Reductive elimination of the Ni—H and Ni—Ph bonds in the hydrido complex gives benzene.

The reaction pathways just described account for the observed products simply. They do not, however, represent a detailed mechanism, because they fail to account for the observations that oxidizing agents such as O_2 and ArBr accelerate the coupling reaction [68]. Part of this effect may be due to depletion of the phosphine concentration by oxidation of the ligand. However, the major rate increase probably results from acceleration of the reductive elimination process, which generates product. Reductive elimination processes have not been studied thoroughly, but it seems likely that some involve one-electron pathways as in oxidative addition (Section 2.3). Mechanisms that involve free radicals or one-electron transfers have been suggested for catalysts based on Fe, Co, and Cu complexes.

As with nickel, phosphine complexes of palladium catalyze coupling of halobenzenes with organomagnesium and lithium reagents. The high yields and mild conditions of these reactions often justify the use of the precious metal catalysts. The reaction pathways seem similar to those of nickel. Zerovalent complexes such as $Pd(PPh_3)_4$ are quite effective, but the more stable $PdCl_2(PPh_3)_2$ or Pd(Ar)-(Cl)$(PPh_3)_2$ complexes are more convenient for most purposes [67].

7.5 ARENE HYDROGENATION

Hydrogenation of aromatic compounds with soluble catalysts has received concerted scientific scrutiny only in recent years. Homogeneous catalysts for the hydrogenation of arenes have been known since the early 1950s [69], but they have received little attention because heterogeneous catalysts are extraordinarily effective for this reaction. For the organic laboratory Adams' catalyst (brown PtO_2) hydrogenates aromatics at 25° and 3 atmospheres pressure [70]. In commercial practice palladium-on-carbon or high-surface-area nickel catalysts are used for hydrogenation of benzene to cyclohexane on a very large scale.

The soluble catalysts for arene hydrogenation are largely based on cobalt and ruthenium. One of the soluble cobalt catalysts may be useful for synthesis of selectively deuterated cyclohexanes because it produces all-*cis* cyclohexane-d_6 derivatives [71].

Cobalt Carbonyl Systems

In early studies of the hydroformylation of olefins in the presence of aromatics it was found that hydrogenation of the aromatic ring was sometimes a major side reaction [69]. Benzene itself resisted hydrogenation but fused ring systems underwent extensive reduction. Naphthalene is reduced to tetrahydronaphthalene by a H_2/CO mixture at 200° and 200 atmospheres pressure with $Co_2(CO)_8$ as the catalyst [72]:

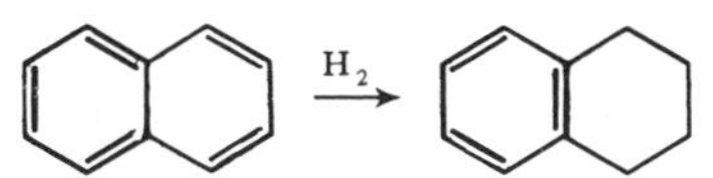

This selectivity for polynuclear aromatic compounds is particularly interesting in the context of converting coal to liquid fuels. Hydrogenation of the polynuclear segments of the coal structure should produce soluble derivatives. The cobalt carbonyl catalyst is attractive because it does not involve a precious metal and because cheap synthesis gas (H_2/CO) serves as the reducing agent.

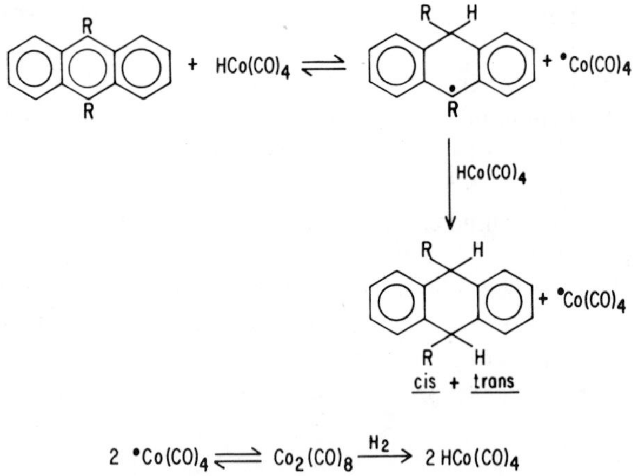

$$2\ {}^{\bullet}Co(CO)_4 \rightleftharpoons Co_2(CO)_8 \xrightarrow{H_2} 2\ HCo(CO)_4$$

Figure 7.10 Proposed mechanism for hydrogenation of anthracene (R = H or CH_3).

The active catalyst in a cobalt carbonyl system appears to be $HCo(CO)_4$. As in other reactions of this "hydride" (Section 2.3) a radical mechanism has been implicated [69,73]. The hydrogenation of anthracene to the 9,10-dihydro derivative was proposed to occur as shown in Figure 7.10. A key step is addition of a hydrogen atom to the central ring to form a resonance-stabilized aromatic radical. Hydrogen transfer from a second mole of $HCo(CO)_4$ produces *cis* and *trans* 9,10-dihydroanthracenes. The observed nearly equimolar mixture of *cis* and *trans* products is consistent with this proposal [73].

Ziegler Systems

The combinations of transition metal salts with alkylaluminum compounds discovered by Karl Ziegler and his coworkers catalyze many reactions in addition to the well-known olefin polymerizations (Chapter 4). Complexes prepared from triethylaluminum and a cobalt or nickel salt catalyze the hydrogenation of benzene and its derivatives. For example, benzene is reduced to cyclohexane rapidly and quantitatively at 150–190° and about 75 atmospheres pressure with $Al(C_2H_5)_3$ and Ni(2-ethylhexanoate)$_2$ as the catalyst [74]. Similarly, a combination of Co(2-ethylhexanoate)$_2$ and excess alkylaluminum compound reduces the xylenes to dimethylcyclohexanes [75]. The *cis*-dimethylcyclohexanes are favored over *trans* by about 2:1, consistent with a predominant *cis* addition of hydrogen.

These catalysts show some selectivity with mixtures of aromatic substrates. The cobalt catalyst hydrogenates benzene in preference to xylenes. It can also be used to hydrogenate naphthalene to either tetrahydronaphthalene or decahydronaphthalene by appropriate choice of reaction conditions [75].

The nature of the Ziegler-type catalysts is poorly defined. The reaction of $Al(C_2H_5)_3$ with either a cobalt or nickel salt gives a dark brown or black solution. In the case of nickel the mixture is neither pyrophoric nor paramagnetic and does not yield solids on ultracentrifugation [74]. It seems likely that the solutions contain metal hydride species that are stabilized by coordination to aluminum. The cobalt complexes may arise by the reaction sequence:

$$CoX_2 \xrightarrow{Et_3Al} CoEt_2 \longrightarrow CoH_2 + C_2H_4$$

$$CoH_2 + Et_3Al \longrightarrow Co\begin{smallmatrix} H \\ \\ H \end{smallmatrix}AlEt_3$$

A similar interaction of triethylaluminum with Cp_2TaH_3 is well characterized [76].

Relatively stable catalysts are obtained by the reaction of triethylaluminum with cobalt(II) acetylacetonate in the presence of tributylphosphine [77]. This catalyst system effects cohydrogenation of olefins and arenes under mild conditions. At 30° and 1.5 atmospheres hydrogen pressure, styrene and benzene are hydrogenated to form primarily ethylbenzene and cyclohexane. Hydrogenation of the olefin is much faster than reduction of the arenes. Although paramagnetic cobalt(0) complexes are detected in the reaction mixture by esr, it seems likely that these catalysts function much like the allylcobalt catalysts discussed next.

Allylcobalt Catalysts

Much of the recent interest in arene hydrogenation has derived from the discovery of discrete soluble catalysts that act under mild conditions. For example, the π-allyl complex $Co(C_3H_5)(P(OMe)_3)_3$ hydrogenates benzene to cyclohexane at room temperature and atmospheric pressure [71]. The hydrogenation is slow but very stereoselective. When D_2 is the reducing agent, all-*cis* cyclohexane-d_6 is formed in more than 95% yield. Similarly, naphthalene and anthracene give high yields of the *cis* perhydro derivatives. In contrast to the $HCo(CO)_4$ and Ziegler systems, benzene is hydrogenated more rapidly than naphthalene and anthracene with the allyl catalyst. With alkylbenzenes the rates fall in this order: benzene > toluene > xylenes > mesitylene > durene. Little or no cyclohexene is produced from benzene under ordinary conditions.

The major drawbacks of this catalyst are the low hydrogenation rate and the limited catalyst life. Higher rates can be attained with $Co(C_3H_5)(P(OR)_3)_3$ complexes in which R is ethyl or isopropyl [78]. Evidently the steric bulk of the higher alkyl phosphite accelerates ligand dissociation to open a coordination site for complexation of an arene or olefin. Unfortunately, ligand lability also accelerates decomposition of the catalyst. The isopropyl phosphite complex has a very short life in the presence of hydrogen. It appears that the decomposition involves cleavage of the allyl—Co bond to give a hydride with much lower catalytic activity:

$$Co(C_3H_5)L_3 + H_2 \rightarrow H_2C{=}CHCH_3 + HCoL_3$$

The allyl ligand seems essential to catalytic activity in this reaction. Its unique characteristic is its ability to occupy either one or two coordination sites on the metal:

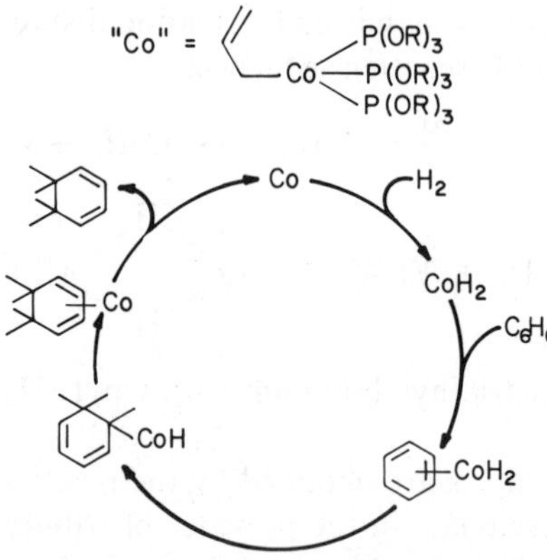

Figure 7.11 Catalytic cycle for hydrogenation of one double bond in benzene. The allyl and phosphite ligands have been omitted for clarity.

In the bidentate form the ligand fills the coordination sphere of the cobalt and stabilizes the complex. In the monodentate form it exposes an orbital for coordination of arene or hydrogen.

The probable catalytic cycle is sketched in Figure 7.11. The σ-allyl complex designated "Co" reacts with hydrogen to form an allyl dihydride complex, "CoH_2." This species has been observed spectroscopically. It is probably the species that decomposes by elimination of propylene. The dihydride complex dissociates a phosphite ligand in a step not shown in the catalytic cycle to vacate a site for coordination of benzene. The benzene ligand is probably complexed through a single double bond (Section 7.1). Addition of a Co—H bond to this C=C bond gives a cyclohexadienyl complex. Transfer of a second Co-H gives 1,3-cyclohexadiene, which is still coordinated to the cobalt. In the normal operation of the catalyst it probably remains coordinated to the metal for two more H_2 addition cycles that ultimately yield cyclohexane. The overall process is very similar to the hydrogenation of an olefin by Wilkinson's catalyst (Section 3.4).

A clever solution to the lability of the allylcobalt catalysts is based on continuous regeneration of the allyl ligand. Cohydrogenation of butadiene and benzene with $CoH_3(PBu_3)_3$ occurs under mild conditions [77]. The diene is hydrogenated faster than the arene, but the catalyst is supposed to be relatively stable. It is likely that a crotyl ligand is generated by reaction of the diene with the hydride:

$$CoH_3L_3 + C_4H_6 \longrightarrow \text{(crotyl)}CoH_2L_3 \rightleftharpoons \pi\text{-}C_4H_7CoH_2L_2 + L$$

In this way the presence of the diene preserves an allylic catalyst species like that present in the phosphite system.

Ruthenium Complexes

Ruthenium metal has received a great deal of attention as a catalyst for benzene hydrogenation. One of its characteristics is a tendency to form cyclohexene as a product in addition to cyclohexane. When used in the presence of aqueous NaOH, ruthenium-on-magnesia gives about 50% cyclohexene at moderate conversion of benzene [79]. This result is especially interesting because one might hope to oxidize cyclohexene cleanly to adipic acid, a nylon precursor.

Recently soluble ruthenium complexes have been found to catalyze the hydrogenation of benzene under mild conditions. Bis(hexamethylbenzene)ruthenium(0) hydrogenates benzene rapidly at 90° and 2–3 atmospheres pressure [80]. The reaction resembles that catalyzed by ruthenium metal in that substantial amounts of cyclohexene are formed (40–55% dimethylcyclohexenes from the xylenes). A second catalyst, $Ru(C_6Me_6)(H)(Cl)(PPh_3)$, also bears a hexamethylbenzene ligand. It catalyzes hydrogenation of benzene at 50° and 50 atmospheres pressure to form cyclohexane exclusively [81]. It also catalyzes transfer of hydrogen from 1-phenylethanol to benzene and olefins.

The bis(hexamethylbenzene)ruthenium catalyst differs in several respects from the allylcobalt catalysts discussed earlier but is said to operate by a similar mechanism. It differs in giving cyclohexene as a substantial product and in producing extensive H/D exchange when D_2 is the reducing agent. When xylene is treated with D_2, deuterium appears in the methyl groups of the unreduced xylene. In addition, the hexamethylbenzene ligands of recovered catalyst undergo methyl H/D exchange [80]. It was proposed that this exchange occurred via a π-benzyl intermediate.

The bis(hexamethylbenzene)ruthenium(0) is interesting in that one ligand is symmetrically complexed (η^6) but the other is coordinated through only two C=C bonds (η^4) [82]. As shown in structure **15** of Figure 7.12, the latter is folded so that the uncoordinated arene carbons bend away from the metal atom. The uncoordinated carbons form a C=C bond with a normal 1.33 Å length. It has been suggested [80] that, in solution, this structure is in equilibrium with a π-benzylic hydride complex (**16**). This benzyl ruthenium complex can then catalyze arene hydrogenation by a mechanism like that shown for the allylcobalt complex in Figure 7.11.

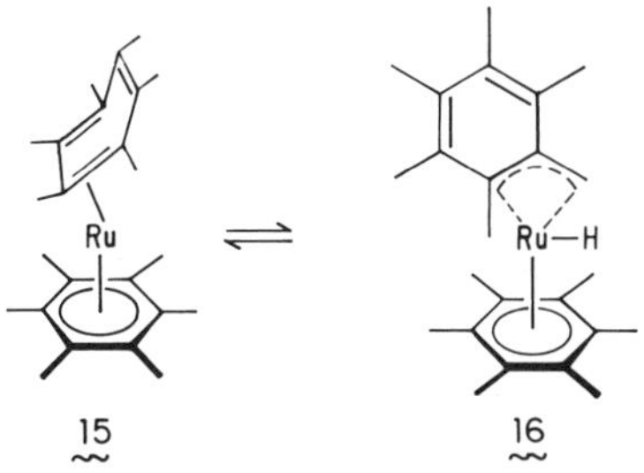

Ru ⇌ Ru—H

15 16

Figure 7.12 A proposed equilibrium between an η_4 arene complex and an η_3 benzyl hydride.

The H/D exchange that is observed occurs because of the presence of the hydride ligand not found in the cobalt system. The addition of D_2 to the ruthenium hydride provides an H/D scrambling mechanism:

$$\mathrm{Ru{-}H} + \mathrm{D_2} \rightleftharpoons \mathrm{Ru(H)(D){-}D} \rightleftharpoons \mathrm{Ru{-}D} + \mathrm{HD}$$

The similarity in chemistry between the homogeneous and heterogeneous ruthenium catalysts provides a new opportunity to understand the processes that occur on a metallic surface.

SPECIFIC REFERENCES

1. G. A. Olah, *Friedel Crafts Chemistry,* Wiley Interscience, 1973.
2. W. E. Silverthorn, *Adv. Organomet. Chem.,* **13,** 47 (1975).
3. M. A. Bennett, "Aromatic Compounds of the Transition Metals," in S. Coffey, Ed., *Rodd's Chemistry of Carbon Compounds,* Vol. 3B, 2nd ed., Elsevier, 1974, pp. 357–457.
4. F. A. Cotton, W. A. Dollase, and J. S. Wood, *J. Am. Chem. Soc.,* **85,** 1543 (1963); M. F. Bailey and L. F. Dahl, *Inorg. Chem.,* **4,** 1314 (1965).
5. V. Graves and J. J. Lagowski, *Inorg. Chem.,* **15,** 577 (1976).
6. M. F. Semmelhack, H. T. Hall, M. Yoshifuji, and G. Clark, *J. Am. Chem. Soc.,* **97,** 1247 (1975).
7. D. N. Kursanov, V. N. Setkina, N. K. Baranetskaya, V. I. Zdanovich, et al., *Izvest. Akad. Nauk SSSR, Ser. Khim.* 1359 (1973); *J. Organomet. Chem.,* **37,** 339 (1972).
8. R. W. Turner and E. L. Amma, *J. Am. Chem. Soc.,* **88,** 1877 (1966).
9. G. W. Parshall, *Catalysis,* **1,** 334 (1977).
10. R. R. Schrock and G. W. Parshall, *Chem. Rev.,* **76,** 243 (1976); G. W. Parshall and J. J. Mrowca, *Adv. Organomet. Chem.,* **7,** 157 (1968).
11. D. R. Coulson, *J. Am. Chem. Soc.,* **98,** 3111 (1976).
12. R. F. Heck, *Organotransition Metal Chemistry,* Academic Press, 1974, pp. 98–103.
13. P. M. Maitlis, *The Organic Chemistry of Palladium,* Vol. 2, Academic Press, 1971, pp. 8–15, 60–70, 115–118.
14. Y. Fujiwara, I. Moritani, S. Danno, R. Asano, and S. Teranishi, *J. Am. Chem. Soc.,* **91,** 7166 (1969).
15. R. S. Shue, *J. Catal.,* **26,** 112 (1972).
16. I. Moritani and Y. Fujiwara, *Synthesis,* 524 (1973).
17. T. Sakakibara, S. Nishimura, K. Kimura, I. Minato, and Y. Odaira, *J. Org. Chem.,* **35,** 3884 (1970).
18. R. F. Heck, *J. Am. Chem. Soc.,* **91,** 6707 (1969).
19. I. Moritani, Y. Fujiwara, and S. Danno, *J. Organomet. Chem.,* **27,** 279 (1971).
20. J. E. Plevyak and R. F. Heck, *J. Org. Chem.,* **43,** 2454 (1978).
21. R. van Helden and G. Verberg, *Rec. Trav. Chim. Pays-bas,* **84,** 1263 (1965).
22. J. M. Davidson and C. Triggs, *Chem. Ind.,* 1361 (1967).

23. H. Itatani, H. Yoshimoto, et al., *J. Org. Chem.*, **38**, 76 (1973); *Bull. Chem. Soc. Japan*, **46**, 2490 (1973); *J. Catal.*, **29**, 92 (1973).
24. A. I. Rudenkov, G. U. Mennenga, L. N. Rachkovskaya, K. I. Matveev, and I. V. Kozhevnikov, *Kinet. Katal.*, **18**, 915 (1977).
25. H. W. Krause, R. Selke, and H. Pracejus, *Z. Chem.*, **16**, 465 (1976).
26. D. R. Bryant, J. E. McKeon, and B. C. Ream, *Tetrahedron Lett.*, 3371 (1968).
27. M. O. Unger and R. A. Fouty, *J. Org. Chem.*, **34**, 18 (1969).
28. P. M. Henry, *J. Org. Chem.*, **36**, 1886 (1971).
29. L. Hörnig and T. Quadflieg, U.S. Patent 3,642,873 (1972).
30. L. Eberson and L. Jonsson, *Acta Chem. Scand.*, **B30**, 361 (1976).
31. L. Eberson and L. Gomez-Gonzales, *Acta Chem. Scand.*, **27**, 1255 (1973).
32. P. M. Henry, *Tetrahedron Lett.*, 2285 (1968); W. C. Baird, R. L. Hartgerink, and J. H. Surridge, Ger. Offenleg. 2,310,629 (1973).
33. J. J. van Venrooy, U.S. Patent 4,093,647 (1978).
34. A. Schoenberg, R. F. Heck, et al., *J. Org. Chem.*, **39**, 3318, 3327 (1974).
35. A. Cairncross, J. R. Roland, R. M. Henderson, and W. A. Sheppard, *J. Am. Chem. Soc.*, **92**, 3187 (1970).
36. W. W. Kaeding, *Hydrocarbon Proc.*, **43**, 173 (Nov. 1964).
37. T. Tsuda, Y. Chujo, and T. Saegusa, *J. Am. Chem. Soc.*, **100**, 630 (1978).
38. A. Gelbein and A. S. Nislick, *Hydrocarbon Proc.*, 125 (Nov. 1978).
39. W. W. Kaeding, R. O. Lindblom, and R. G. Temple, U.S. Reissue Patent 24,848 (1960).
40. W. Schoo, J. U. Veenland, J. A. Bigot, and F. L. J. Sixma, *Rec. Trav. Chim. Pays-Bas*, **80**, 134 (1961).
41. W. W. Kaeding and G. R. Collins, *J. Org. Chem.*, **30**, 3750 (1965).
42. W. W. Kaeding, H. O. Kerlinger, and G. R. Collins, *J. Org. Chem.*, **30**, 3754 (1965).
43. E. L. Reilly, U.S. Patent 3,987,068 (1976).
44. M. M. Rogic and T. R. Demmin, *J. Am. Chem. Soc.*, **100**, 5472 (1978).
45. A. S. Hay, *Poly. Eng. Sci.*, **16**, 1 (1976); H. L. Finkbeiner, A. S. Hay, and D. M. White, "Polymerizations by Oxidative Coupling," in C. E. Schildknecht and I. Skeist, Eds., *Polymerization Processes*, Wiley-Interscience, 1977, pp. 537–581.
46. G. D. Cooper and D. E. Floryan, Ger. Offenleg. 2,754,887 (1978).
47. A. S. Hay, H. S. Blanchard, G. F. Endres, and J. W. Eustance, *Macromol. Syn., Coll.*, **1**, 81 (1977).
48. G. F. Endres, A. S. Hay, and J. W. Eustance, *J. Org. Chem.*, **28**, 1300 (1963).
49. I. Bodek and G. Davies, *Inorg. Chem.*, **17**, 1814 (1978).
50. H. Finkbeiner, A. S. Hay, H. S. Blanchard, and G. F. Endres, *J. Org. Chem.*, **31**, 549 (1966).
51. G. D. Cooper and A. Katchman, in N. A. J. Platzer, Ed., *Addition and Condensation Polymerization Processes, Advances in Chemistry Series* **91**, American Chemical Society 1969, p. 660.
52. W. G. B. Huysmans and W. A. Waters, *J. Chem. Soc.* B, 1163 (1967).
53. T. F. Rutledge, U.S. Patent 4,100,203 (1978); W. K. Olander, Ger. Offenleg. 2,755,829 (1978).
54. P. Hudec, *J. Catal.*, **53**, 228 (1978).
55. R. Cramer and D. R. Coulson, *J. Org. Chem.*, **40**, 2267 (1975).
56. K. Takagi, T. Okamoto, Y. Sakakibara, and S. Oka, *Chem. Lett.*, 471 (1973); L. Cassar, *J. Organomet. Chem.*, **54**, C57 (1973).
57. R. Noyori, "Coupling Reactions via Transition Metal Complexes," in H. Alper, Ed., *Transition Metal Organometallics in Organic Synthesis*, Academic Press, 1976, pp. 83–187.

58. J. K. Kochi, "Coupling of Alkyl Groups Using Transition Metal Catalysts," in L. F. Albright and A. R. Goldsby, Eds., *Industrial and Laboratory Alkylations,* American Chemical Society Symposium Series, **55,** 167 (1977).
59. P. E. Fanta, *Chem. Rev.,* **64,** 613 (1964); *Synthesis,* 9 (1974).
60. R. C. Fuson and E. A. Cleveland, *Org. Syn., Coll.,* **3,** 339 (1955).
61. T. Cohen and I. Cristea, *J. Am. Chem. Soc.,* **98,** 748 (1976).
62. M. F. Semmelhack, P. M. Helquist, and L. D. Jones, *J. Am. Chem. Soc.,* **93,** 5908 (1971).
63. A. S. Kende, L. S. Liebeskind, and D. M. Braitsch, *Tetrahedron Lett.,* 3375 (1975).
64. G. W. Parshall, *J. Am. Chem. Soc.,* **96,** 2360 (1974).
65. K. Tamao, K. Sumitani, Y. Kiso, M. Zembayashi, A. Fujioka, S. Kodama, I. Nakajima, A. Minato, and M. Kumada, *Bull. Chem. Soc. Japan,* **49,** 1958 (1976); also *J. Am. Chem. Soc.,* **94,** 9268 (1972).
66. M. Kumada, K. Tamao, and K. Sumitani, *Org. Syn.,* **58,** 127 (1978).
67. A. Sekiya and N. Ishikawa, *J. Organomet. Chem.,* **118,** 349 (1976).
68. D. G. Morrell and J. K. Kochi, *J. Am. Chem. Soc.,* **97,** 7262 (1975).
69. I. Wender, R. Levine, and M. Orchin, *J. Am. Chem. Soc.,* **72,** 4375 (1950).
70. R. Adams and J. R. Marshall, *J. Am. Chem. Soc.,* **50,** 1970 (1928).
71. L. S. Stuhl, M. Rakowski Du Bois, F. J. Hirsekorn, J. R. Bleeke, A. E. Stevens, and E. L. Muetterties, *J. Am. Chem. Soc.,* **100,** 2405 (1978).
72. S. Friedman, S. Metlin, A. Svedi, and I. Wender, *J. Org. Chem.,* **24,** 1287 (1959).
73. H. M. Feder and J. Halpern, *J. Am. Chem. Soc.,* **97,** 7186 (1975).
74. S. J. Lapporte and W. R. Schuett, *J. Org. Chem.,* **28,** 1947 (1963).
75. G. Bressan and R. Broggi, *Chim. Ind. (Milan),* **50,** 1194 (1968).
76. F. N. Tebbe, *J. Am. Chem. Soc.,* **95,** 5412 (1973).
77. F. K. Shmidt, Y. S. Levkovskii, V. V. Saraev, N. M. Ryutina, O. L. Kosinski, and T. I. Bakunina, *React. Kinet. Catal. Lett.,* **7,** 445 (1977).
78. M. C. Rakowski, F. J. Hirsekorn, L. S. Stuhl, and E. L. Muetterties, *Inorg. Chem.,* **15,** 2379 (1976).
79. W. C. Drinkard, U.S. Patent 3,767,720 (1973).
80. J. W. Johnson and E. L. Muetterties, *J. Am. Chem. Soc.,* **99,** 7395 (1977).
81. M. A. Bennett, T-N Huang, A. K. Smith, and T. W. Turney, *J.C.S. Chem. Comm.,* 582 (1978).
82. G. Huttner and S. Lange, *Acta Cryst.,* **B28,** 2049 (1972).

8 ACETYLENE REACTIONS

Acetylene was once a major starting material for the organic chemical industry. Commercial processes for vinyl acetate, vinyl chloride, acetaldehyde, acrylonitrile, acrylates, and chloroprene were largely based on acetylene in the years following World War II. However, development of technology for synthesis of these materials from ethylene, propylene, or butadiene gradually made acetylene-based processes obsolete. Its major advantage chemically, the large amount of energy stored in the C≡C function, became a disadvantage economically. The 1973 "energy crisis" made high-energy materials such as acetylene ($\Delta F^0 = 50.84$ kcal/mole) extravagantly expensive as feedstocks. Consequently, most acetylene-based processes have been replaced by olefin-based technology. One of the few remaining large-scale uses is the synthesis of butyne-1,4-diol, a precursor of 1,4-butanediol and tetrahydrofuran. There has been speculation that acetylene might again become economically attractive because it is produced as a byproduct in the cracking of heavy petroleum fractions to produce olefins. Although such feedstocks will be used more extensively in the future, the availability of large quantities of acetylene seems questionable.

Nearly all the major acetylene processes were catalytic. The first major industrial application of homogeneous catalysis appears to be the $FeCl_3$-catalyzed chlorination of acetylene to tetrachloroethane, which was commercialized about 1910 [1]. This innovation was followed by development of practical catalysts for several other reactions of acetylene. Both soluble and heterogeneous catalysts were used. In practice it is sometimes difficult to determine if a relatively insoluble catalyst such as copper acetylide acts in solution. When it is used as a slurry in a liquid reaction mixture, it is quite possible that catalysis occurs on the surface of the solid.

Generally, reactions of the acetylenic C—H bond are catalyzed by copper salts, which presumably form copper acetylides, $Cu(C{\equiv}CR)_n$. Additions to the C≡C function are catalyzed by copper or mercury salts. Group VIII metal compounds are used to catalyze carbonylation of C_2H_2 to acrylate esters. The Group VIII metals also yield some remarkable dimers, trimers, and tetramers of acetylenes. These four classes of reactions are discussed separately after a brief introduction to the coordination chemistry of acetylene.

8.1 COORDINATION CHEMISTRY OF ACETYLENES

Acetylenes interact with transition metal ions in many different ways [2,3]. The π orbitals of the C≡C bond can donate electrons to vacant metal orbitals and the π^* orbitals can accept electron density from filled metal *d* orbitals. The simplest situation, π bonding to a single metal atom, is shown in structure **1** in Figure 8.1 [4]. The bonding in this diphenylacetylene complex is like that discussed in detail for "side-on" bonded complexes of molecular nitrogen in Chapter 2. The presence of two mutually orthogonal π systems in the acetylene bond also permits simultaneous π bonding to two metal atoms. This situation, which is quite common, is illustrated by a cobalt carbonyl complex of acetylene (**2**). The two cobalt atoms are sufficiently close to permit formation of a Co—Co bond [5]. In fact the compound may be regarded as a derivative of a pseudotetrahedral C_2Co_2 cluster.

The bonding situation can be more complex for acetylenes that bear hydrogen substituents. The acetylenic hydrogens are modestly acidic and form acetylide complexes with many metals [6]. The acetylide ion, C_2^{2-}, is isoelectronic with N_2 and CN^-. Like these two ligands, acetylides often prefer to bond "end-on" to a transition metal ion. The bonding scheme is like that shown for N_2 complexes in Figure 2.1, but the anionic acetylide ligands seem to be better *sigma* donors than N_2. In this respect acetylides resemble cyanide ion. Many anionic polyacetylide

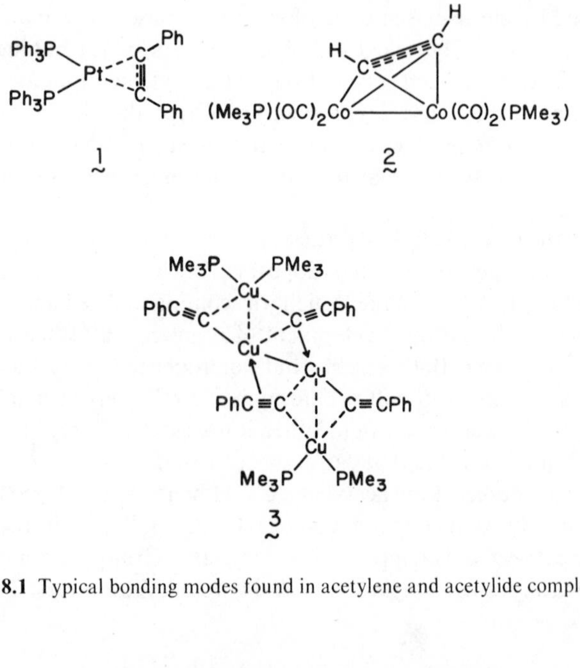

Ph3P
Ph3P
Pt
C
C
Ph
Ph
1

H
C
C
H
(Me3P)(OC)2Co
Co(CO)2(PMe3)
2

Me3P
PMe3
Cu
PhC≡C
C≡CPh
Cu
Cu
PhC≡C
C≡CPh
Cu
Me3P
PMe3
3

Figure 8.1 Typical bonding modes found in acetylene and acetylide complexes.

complexes, analogous to polycyano complexes, have been prepared. For example, copper(I) forms a tris(phenylacetylide) complex, $[Cu(C{\equiv}CPh)_3]^{2-}$ [7].

Many acetylide complexes, especially the catalytically important copper compounds, are polymeric [6,8]. The acetylide ligands in these complexes form *sigma* bonds to one metal atom and form *pi* bonds between the C≡C function and a second metal atom. The crystal structure (**3**) of the seemingly simple complex, $[Cu(C{\equiv}CPh)(PMe_3)]_4$, shows that two phenylacetylides are *sigma* bonded to two coppers, and two other acetylide ligands are *pi*-bonded to the same two metal atoms [9]. Such complexity is typical of copper and silver acetylide complexes and is involved in catalysis of acetylene reactions by these compounds.

The simple acetylene complexes are usually stable in the absence of oxygen, but the acetylide complexes are often violently explosive [8]. This danger is prevalent for the copper(I) acetylide complexes used as catalysts for many commercial reactions of acetylene. Gaseous acetylene reacts with ammoniacal aqueous solutions of copper(I) chloride to form yellow, red, or brown precipitates of Cu_2C_2. These precipitates detonate on heating (120–123° in air) or mechanical disturbance. Explosive deposits form when oxidized copper surfaces are exposed to acetylene. Consequently, copper fittings should never be used to handle acetylene. (Acetylene itself is treacherously explosive, especially in the liquid state. Consequently, acetylene is usually shipped in cylinders as a solution in acetone. A scrubbing process is necessary to obtain acetone-free acetylene.)

The copper(I) acetylides and the even more dangerous silver acetylides appear to have complex structures analogous to that of the copper(I) phenylacetylide shown in Figure 8.1. Although the insolubility of Cu_2C_2 and Ag_2C_2 have prevented growth of crystals for x-ray diffraction studies, it is likely that both compounds are polymers [8]. Presumably, at least one metal atom is *sigma*-bonded to each end of the acetylide ion and the structure is cross-linked by π bonds from the C≡C function.

The reactions of mercury salts with acetylenes are also important catalytically and just as complex as those of copper salts. Discrete complexes of mercury(II) with alkylacetylides are known. These presumably have complex structures analogous to those of the copper acetylides. However, the presence of positive charge on an ion such as

$$\left[\begin{array}{c} XHg{-}C{\equiv}CR \\ \quad\quad\quad | \\ \quad\quad\quad Hg^{2+} \end{array}\right]$$

should render the C≡C function susceptible to nucleophilic attack as was noted for olefin complexes earlier. The situation is complicated, however, by the tendency of mercuric salts to add to the acetylenic triple bond. For example, $HgCl_2$ adds to acetylene to form *cis* or *trans* chlorovinylmercury derivatives, depending on reaction conditions [8]:

$$HgCl_2 + C_2H_2 \rightarrow ClCH{=}CHHgCl$$

Similar reactions probably occur with copper(II) chloride and are important in catalysis of chlorination and other additions to acetylenes.

8.2 ACETYLIDE-CATALYZED REACTIONS

A major family of catalytic reactions of acetylene is based on reactions of copper acetylides. These reactions include oxidative coupling to diynes and the addition of the Cu—C bond to aldehydes and ketones. The addition to C≡C was once an especially important process for manufacture of vinylacetylene as an intermediate in neoprene production.

Oxidative Coupling

The oxidation of acetylenes to give diacetylenes:

$$2RC{\equiv}CH \rightarrow RC{\equiv}C{-}C{\equiv}CR$$

is a useful synthesis procedure [10]. It is not used commercially, but it illustrates some principles of copper acetylide catalysis. The oxidative coupling discovered by Glaser in 1869 involves reaction of terminal acetylenes with copper(I) salts to give copper(I) acetylides that are oxidized by air to give diacetylenes. For example, 1-ethynylcyclohexanol is coupled in 93% yield by bubbling oxygen through an acetone solution of the acetylene in the presence of a copper(I) chloride-tetramethylethylenediamine complex [11]:

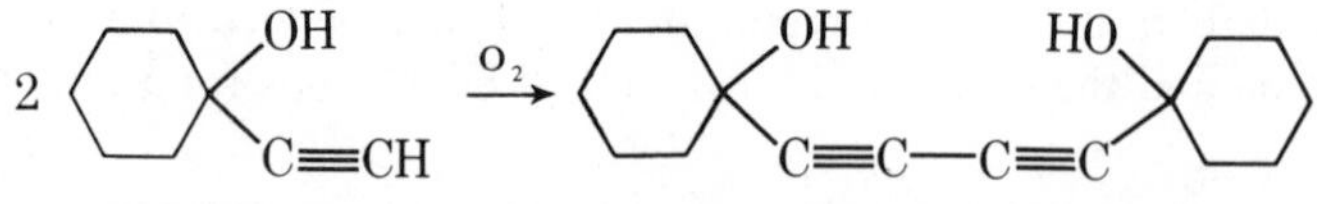

2 (1-ethynylcyclohexanol: cyclohexane bearing OH and C≡CH) $\xrightarrow{O_2}$ (cyclohexane bearing OH)–C≡C—C≡C–(cyclohexane bearing HO)

In a variant of this procedure that is sometimes more convenient, an amine solution of a copper(II) salt is used as the oxidant.

Both the catalytic Glaser coupling and the stoichiometric oxidation with copper(II) salts seem to involve the same mechanism [10,12]. As illustrated in Figure 8.2, the catalytic cycle begins with the formation of a copper(I) acetylide from a copper(I) salt and a terminal acetylene. This acetylide is presumed to have a dimeric structure (**4**) like that found in the crystal structure of $[Me_3PCuC{\equiv}CPh]_4$. Oxidation of **4** by a copper(II) salt is proposed to give an unstable dimeric copper(II) acetylide (**5**). Decomposition of **5** forms the observed diacetylene product and completes the catalytic cycle by regeneration of copper(I) ions.

In the catalytic Glaser coupling the copper(II) oxidant is supplied by oxidation of copper(I) by air or oxygen (lower cycle in Figure 8.2). The reaction of copper(I) chloride with oxygen in pyridine has been shown to form $CuCl_2(Py)_2$ and a soluble copper(II) oxide polymer [13]. The latter species is proposed to be the reagent for oxidative coupling of phenols (Chapter 7), anilines, and acetylenes.

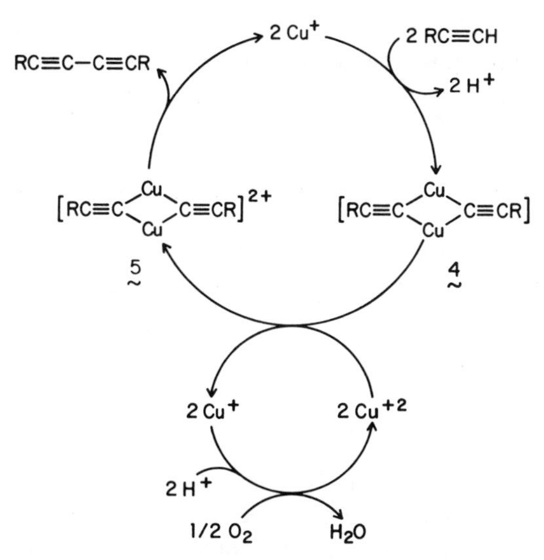

Figure 8.2 A proposed mechanism for catalytic oxidative coupling of acetylenes (amine ligands omitted for simplicity).

Addition to Aldehydes and Ketones

The largest remaining use of acetylene as a chemical intermediate is in the synthesis of tetrahydrofuran via the sequence:

$$C_2H_2 + 2HCHO \longrightarrow HOCH_2C{\equiv}CCH_2OH \xrightarrow{H_2} HO(CH_2)_4OH \xrightarrow{H^+} \text{(tetrahydrofuran)}$$

Similar additions of the C—H bonds of acetylene to higher aldehydes and ketones are also performed commercially. For example, acetone is condensed with acetylene to give

$$\underset{\displaystyle OH}{\underset{|}{Me_2C}}—C{\equiv}C—\underset{\displaystyle OH}{\underset{|}{CMe_2}}$$

Such condensations of acetylene with carbonyl compounds can be performed by reaction of sodium acetylide (NaC≡CH) with the aldehyde or ketone [14]. These stoichiometric reactions are the most convenient way to effect condensation on a laboratory scale. However, industrial practice employs copper acetylide to catalyze the reaction of acetylene itself with the carbonyl compound.

The condensation of acetylene and formaldehyde is usually effected as a heterogeneous catalytic reaction [14]. A supported form of copper acetylide may be prepared by adding gaseous acetylene to a suspension of copper oxide on magnesium silicate powder in a 10% solution of formaldehyde in water [15]. Other minerals such as alumina and kieselguhr may also be used. The acetylene reacts with the copper oxide and formaldehyde to give a complex acetylide:

$$CuO \xrightarrow{HCHO} Cu_2O \xrightarrow{C_2H_2} Cu_2C_2 \xrightarrow{HC\equiv CH} Cu_2C_2 \cdot nC_2H_2$$

Typical conditions for the acetylene-formaldehyde condensation involve reaction of 29–37% aqueous formaldehyde with acetylene at 2 atmospheres pressure and 95° [15]. Acetylene and formaldehyde are fed continuously as required to a slurry of the catalyst. The reaction proceeds in high conversion to give more than 90% 1,4-butynediol with a little $HC\equiv CCH_2OH$ as a byproduct.

The reaction of formaldehyde with acetylene may be directed to form propargyl alcohol:

$$HC\equiv CH + HCHO \rightarrow HC\equiv CCH_2OH$$

if the amount of acetylene is carefully controlled. This procedure, which uses a mixture of copper acetylide and Fuller's earth as the catalyst, has been recommended as a laboratory synthesis of the alcohol [14,16]. Some butynediol is formed as a byproduct because formaldehyde reacts rapidly with the acetylenic C—H in propargyl alcohol. The reactions of higher aldehydes and ketones with acetylenic C—H bonds proceed much more slowly. Usually polar organic solvents such as dimethylformamide are used to attain good solubility of both reactants in laboratory-scale preparations [14].

The catalytic addition of acetylene to a carbonyl function appears to be a straightforward organometallic reaction. One can visualize the following steps:

$$RC\equiv C{-}Cu + HCHO \rightarrow RC\equiv C{-}CH_2O{-}Cu$$

$$\xrightarrow{RC\equiv CH} RC\equiv CCH_2OH + RC\equiv C{-}Cu$$

The addition of the C—Cu bond to the carbonyl group resembles the addition of a Grignard reagent to an aldehyde or ketone. In this instance the alkoxide is protonated by the modestly acidic acetylenic C—H ($K_a \sim 10^{-22}$). Although such a weak acid will not ordinarily protonate an alkoxide, precipitation of the insoluble copper(I) acetylide provides a driving force for this step. Overall the addition of C—H to C=O is thermodynamically favorable.

Chloroprene Synthesis

The dimerization of acetylene was a key step in an obsolete synthesis of chloroprene, the monomer for neoprene rubber [17]:

$$2C_2H_2 \rightarrow CH_2{=}CHC\equiv CH \xrightarrow{HCl} CH_2{=}CHCCl{=}CH_2$$

Both acetylene dimerization and the addition of HCl to vinylacetylene are catalyzed by copper(I) chloride. In contrast to butynediol synthesis, the catalysts in these processes are soluble in the reaction mixtures. Even though the normally insoluble copper(I) acetylides probably form in these solutions, the copper remains in solution. Coordination to chloride ions and to excess acetylene, both of which are good ligands for copper(I) ions, inhibits formation of acetylide polymers.

The dimerization of acetylene is a formal addition of an acetylenic C—H bond to the C≡C function of a second acetylene molecule. The process probably involves nucleophilic attack of an acetylide ion on a C≡C bond that is activated by coordination to a copper ion. In commercial operation acetylene is fed continuously to a dilute aqueous HCl solution of copper(I) and potassium chlorides. At 55–75° dimerization is rapid and vinylacetylene is swept out of the mixture into a drying system. Distillation removes the major byproduct, divinylacetylene, an acetylene trimer that is an isomer of benzene. The divinylacetylene is a treacherously explosive material, a major hazard in the operation of this process.

The addition of hydrogen chloride to vinylacetylene is conceptually similar to the Cu_2Cl_2-catalyzed addition of HCl to acetylene itself (Section 8.3). However, in contrast to vinyl chloride synthesis, it occurs in two steps, both of which are catalyzed by copper(I) chloride:

$$CH_2{=}CHC{\equiv}CH \xrightarrow{HCl} [ClCH_2CH{=}C{=}CH_2] \rightarrow CH_2{=}CH\overset{\overset{Cl}{|}}{C}{=}CH_2$$

The initial addition of HCl occurs in a 1,4-manner to form 4-chloro-1,2-butadiene. This compound is isomerized in the reaction mixture to give chloroprene. This 1,3 shift of chlorine resembles that in the Cu_2Cl_2-catalyzed isomerization of dichlorobutenes (Section 11.3). The overall HCl addition to vinylacetylene to give chloroprene occurs under mild conditions that can be achieved easily in the laboratory [18]. Reaction of vinylacetylene with concentrated aqueous HCl, Cu_2Cl_2 and NH_4Cl at room temperature gives chloroprene in 97% yield at 94% conversion of vinylacetylene.

Monosubstituted acetylenes can also be dimerized by CuCl or by Wilkinson's catalyst, $RhCl(PPh_3)_3$. The latter dimerizes a substituted propargyl alcohol in yields up to 73% under mild conditions [19]:

$$2\ \underset{\underset{OH}{|}}{Me_2C}C{\equiv}CH \longrightarrow \underset{\underset{OH}{|}}{Me_2C}C{\equiv}CCH{=}CH\underset{\underset{OH}{|}}{C}Me_2$$

Both the copper-catalyzed dimerization of acetylene and the HCl addition appear to involve addition to a C≡C bond activated by coordination to a copper(I) ion [20–22]. In the dimerization the metal ion also serves to activate the C—H bond. The latter aspect resembles its role in butynediol synthesis. These two functions are illustrated in Figure 8.3. Acetylene coordinates to the metal ion to form a complex (**6**) that dissociates a proton to form the acetylide complex (**7**). Although acetylene is a very weak acid, dissociation is assisted by coordination to the metal ion. In the presence of excess acetylene the copper acetylide coordinates another molecule of C_2H_2. Insertion of the coordinated C_2H_2 into the Cu—acetylide bond in **8** forms the dimer as a copper complex (**9**). Protonation yields vinylacetylene.

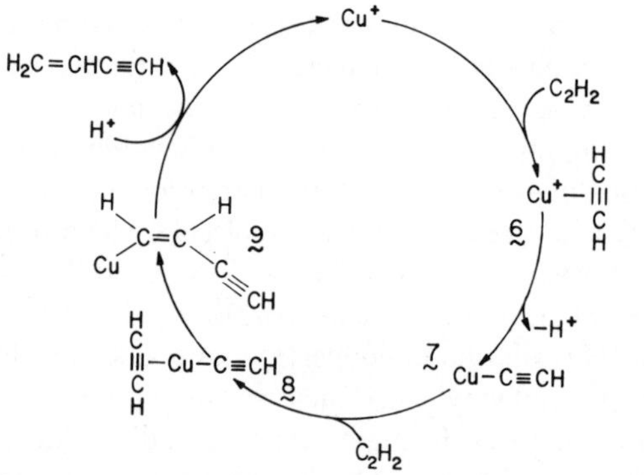

Figure 8.3 The role of copper(I) ion in dimerization of acetylene.

It has been proposed [22] that all these reactions take place in a cluster complex, $[Cu_4Cl_4C{\equiv}CH]^-$. Yellow chlorocopper(I) acetylide complexes of this composition were delineated by spectroscopic studies of simulated reaction mixtures.

A simple model for the acetylide addition is the reaction of lithium dialkylcuprates with acetylenes [23], for example:

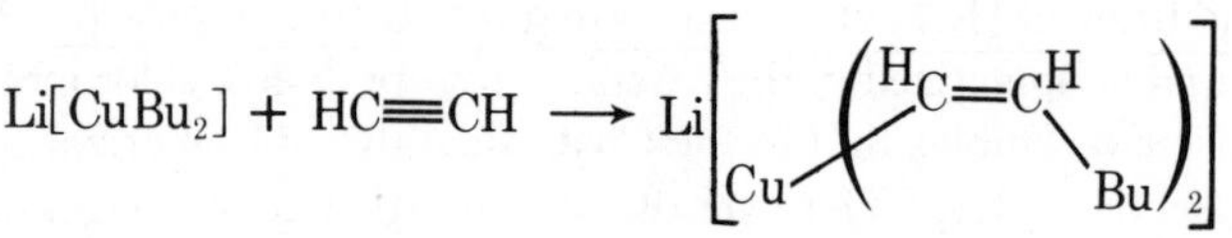

$$Li[CuBu_2] + HC{\equiv}CH \longrightarrow Li[Cu(CH{=}CHBu)_2]$$

In this reaction the acetylene must coordinate to the cuprate ion before insertion into the Cu—C bond. In a competitive experiment with a $[C_7H_{15}{-}Cu{-}C{\equiv}CBu]^-$ salt, acetylene is inserted into the Cu—alkyl bond in preference to the Cu—acetylide function.

A similar coupling of a nucleophile and an acetylene is proposed to account for the addition of HCl to vinylacetylene [24]. Coordination of the C≡C bond to a dichlorocuprate (−1) ion sets the stage for intramolecular attack by chloride ion:

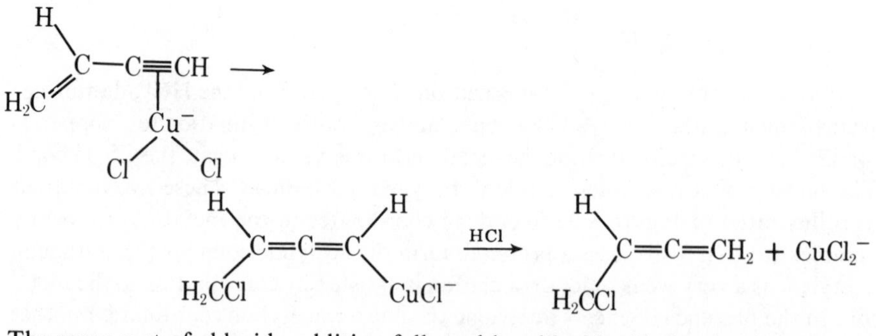

$$H_2C{=}CH{-}C{\equiv}CH\cdot CuCl_2^- \longrightarrow$$

$$H_2CCl(H)C{=}C{=}C(H)CuCl^- \xrightarrow{HCl} H_2CCl(H)C{=}C{=}CH_2 + CuCl_2^-$$

The same sort of chloride addition followed by chloride elimination is proposed to account for the isomerization of the allenic compound to chloroprene [25]:

$$\text{H(H}_2\text{C)(Cl)C}{=}\text{C}{=}\text{CH}_2\cdot\text{CuCl}_2^- \longrightarrow \text{H(H}_2\text{C(Cl))C}{=}\text{C(CuCl}^-)\text{-C(Cl)}{=}\text{CH}_2 \longrightarrow \text{H(H}_2\text{C}{=})\text{C}{-}\text{C(Cl)}{=}\text{CH}_2 + \text{CuCl}_2^-$$

This pattern of attack of a coordinated nucleophile on a complexed acetylene is seen repeatedly in the following section.

8.3 ADDITIONS TO ACETYLENES

In the 1950s the synthesis of vinyl monomers was based largely on reactions in which HX molecules (X = Cl, OAc, CN) were added to the triple bond of acetylene [1]. Similarly acetaldehyde production was accomplished by addition of water to acetylene, presumably via vinyl alcohol as a transient intermediate. These processes are now obsolete, but the chemistry of these additions continues to be of interest.

Most of the additions were catalyzed by soluble salts of copper(I) or mercury(II). A general mechanistic picture of these reactions has emerged. Acetylenes form *pi* complexes with these salts, which activate the C≡C bond toward nucleophilic attack to give *sigma* vinyl complexes:

$$X^- + HC{\equiv}CH\cdot M^+ \longrightarrow (X)(H)C{=}C(H)(M) \xrightarrow{H^+} (X)(H)C{=}C(H)(H) + M^+$$

This catalysis of nucleophilic attack by a metal cation closely resembles that described for cationic olefin complexes. Electron density is transferred from the acetylene to the metal. This depletion of electron density on the acetylenic carbons makes them susceptible to nucleophilic attack. In contrast to the situation with the alkylmetal complexes arising from olefins, the vinylmercury and vinylcopper compounds protonate cleanly to give $CH_2{=}CHX$ compounds and regenerate the catalytic metal ion.

Acetaldehyde Synthesis

Before the development of the Wacker process (Section 6.1), most acetaldehyde was made by hydration of acetylene. Typically, gaseous acetylene was passed through a sulfuric acid solution of $HgSO_4$ and $FeSO_4$ at 95° and about 2 atmospheres pressure [1,26]. About 55% of the acetylene was converted in the reactor. The unchanged acetylene swept out the acetaldehyde before it underwent deleterious side reactions. Fractionation of the volatile products gave acetaldehyde in about 95% yield. Copper(I) chloride also catalyzes the hydration of acetylene to acetaldehyde, but this process was not used commercially.

Similar hydrations of higher acetylenes have been studied extensively. The addition of water typically occurs in Markovnikov fashion. For example, terminal acetylenes give largely methyl ketones rather than aldehydes. The hydration of 1-heptyne gives a 94% yield of 2-heptanone with a catalyst obtained by reacting mercury(II) oxide with a perfluoroalkylsulfonic acid resin [27].

Substantial evidence has accumulated for the formation of acetylene complexes with mercury(II) ions in the early stages of the addition process [28]. When a substituted acetylene is reacted with a mercury salt that contains no coordinating anions, a discrete 2:1 complex is formed. With phenylacetylene and $Hg(ClO_4)_2$ this complex can be observed to form and subsequently decay as the acetylene is hydrated to acetophenone. Presumably, the hydration occurs by nucleophilic attack of water on the coordinated acetylene:

$$2\ PhC{\equiv}CH + Hg^{2+} \rightleftharpoons \left[\begin{matrix} Ph & & Ph \\ C & & C \\ \| & -Hg- & \| \\ C & & C \\ H & & H \end{matrix}\right]^{2+} \xrightarrow{H_2O} Hg^{2+} + 2\ Ph\overset{\overset{\displaystyle O}{\|}}{C}CH_3$$

Strongly coordinating anions such as chloride compete with the acetylene for sites on the metal ion and prevent spectroscopic detection of the acetylene complex. An acetylene complex evidently forms with $HgCl_2$, but it reacts rapidly with chloride ion to give a chlorovinyl complex.

Vinyl Acetate Synthesis

The early syntheses of vinyl acetate were based on addition of acetic acid to acetylene with a homogeneous catalyst [1]. This technology was displaced by a vapor-phase process that used a carbon-supported zinc acetate catalyst. This process, in turn, has been largely displaced by a heterogeneous catalytic acetoxylation of ethylene (Section 6.2).

The liquid-phase HOAc addition closely resembles other additions to the acetylenic triple bond. Typically, acetylene is swept through an acetic acid solution of $Hg(OSO_2OAc)_2$ at 35°. Rapid reaction takes place to give a mixture of vinyl acetate and ethylidene diacetate [1]:

$$C_2H_2 + HOAc \rightarrow CH_2{=}CHOAc + CH_3CH(OAc)_2$$

Rapid gas passage and low conversions of acetylene are used to minimize formation of the latter product. Yields of vinyl acetate up to about 70% are obtained. Similar technology could also be used to prepare other volatile vinyl esters. Nonvolatile esters are best prepared by a mechanistically related transvinylation [29]:

$$CH_2{=}CHOAc + RCOOH \xrightleftharpoons{HgSO_4} CH_2{=}CHOOCR + HOAc$$

The mechanism of acetic acid addition to acetylenes is similar to that for hydration. It appears that a *pi* complex forms from the C≡C bond and a mercury(II) ion. The *pi* complex then isomerizes to β-acetoxyvinylmercury acetate, which reacts with a proton to give vinyl acetate. Diphenylacetylene, which is known to form a complex with mercury(II) ion [28], readily undergoes the π-σ transformation [30]:

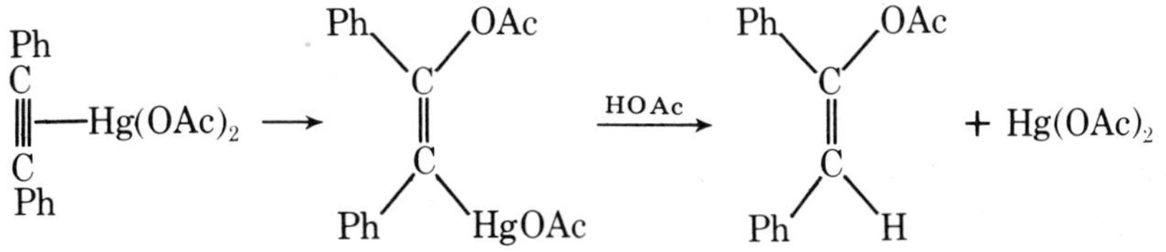

$$\text{PhC}{\equiv}\text{CPh}\cdot\text{Hg(OAc)}_2 \longrightarrow \text{Ph(AcO)C}{=}\text{C(Ph)HgOAc} \xrightarrow{\text{HOAc}} \text{Ph(AcO)C}{=}\text{C(Ph)H} + \text{Hg(OAc)}_2$$

In contrast to the hydration process the direction of acetic acid addition is often anti-Markovnikov. Methylacetylene gives

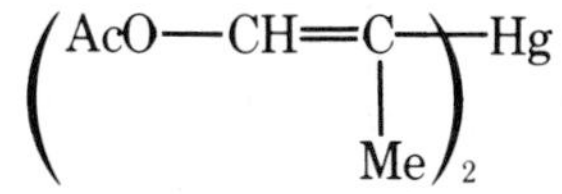

$$\left(\text{AcO—CH}{=}\underset{\text{Me}}{\underset{|}{\text{C}}}\right)_2\text{Hg}$$

Hydrolysis gives propionaldehyde rather than acetone, which is formed by direct hydration of methylacetylene.

Vinyl Chloride Synthesis

The addition of hydrogen chloride to acetylene is less facile than addition of water or acetic acid. In the industrial process used before oxychlorination of ethylene was developed HCl was added to acetylene over a carbon-supported $HgCl_2$ catalyst at 100–200° [1]. Homogeneous catalysts were relatively ineffective, but the mechanism of hydrochlorination has been studied extensively in solution with both $HgCl_2$ and Cu_2Cl_2 as catalysts [21]. As in the additions of water and acetic acid it seems likely that a *pi* complex of acetylene rearranges to 2-chlorovinyl derivative, which is protonated to give vinyl chloride. The $HgCl_2$-catalyzed addition of HCl to ethylene resembles the CuCl-catalyzed addition of vinylacetylene discussed earlier.

Hydrocyanation

Until about 1960 the major industrial synthesis of acrylonitrile was the addition of hydrogen cyanide to acetylene: $HC{\equiv}CH + HCN \rightarrow CH_2{=}CHCN$. Like most acetylene-based processes it was replaced by an olefin reaction, the Sohio process for cooxidation of propylene and ammonia.

The acrylonitrile synthesis from acetylene was done in aqueous solution [1].

Both the process and the catalyst closely resembled those used for synthesis of vinylacetylene. Dilute acetylene and hydrogen cyanide streams were fed to a solution of copper(I) and ammonium chlorides at 70–90° and 1.3 atmospheres pressure. The products were swept out of the reactor by unconverted acetylene (HCN conversion is quantitative). A complex purification scheme was required to give polymer-grade acrylonitrile, but the overall yield was 80–90%. As might be expected from the similarity of the catalyst to those used in other processes, major byproducts were vinylacetylene and acetaldehyde. The catalyst for hydrocyanation of acetylene is also similar to one described [31] for the hydrocyanation of butadiene (Chapter 4).

The general similarity between dimerization and hydrocyanation of acetylene also extends to mechanism [32]. The similarity between $[C{\equiv}CH]^-$ and CN^- as ligands suggests that these two nucleophiles add to acetylene similarly. The steps in hydrocyanation shown in the following equations parallel those of the dimerization mechanism of Figure 8.3.

$$Cu^+ + HCN \rightleftharpoons CuCN + H^+$$

$$CuCN + C_2H_2 \rightleftharpoons (HC{\equiv}CH)\text{—}CuCN \longrightarrow \underset{Cu}{\overset{H}{}}\!\!>C{=}C<\!\!\underset{CN}{\overset{H}{}}$$

$$\underset{Cu}{\overset{H}{}}\!\!>C{=}C<\!\!\underset{CN}{\overset{H}{}} \xrightarrow{H^+} CH_2{=}CHCN + Cu^+$$

Indeed copper(I) ions form an extensive series of cyanocuprate complexes. In the hydrocyanation mixture the presence of chloride ions further complicates the structure of the catalytic species. The presence of a strong acid like HCl is required for catalytic activity, probably to cleave the Cu—C bond in the cyanovinylcopper σ complex.

Chlorination

As noted earlier the liquid-phase chlorination of acetylene may have been the first large-scale homogeneous catalytic process. The chlorination gives 1,1,2,2-tetrachloroethane, an intermediate in the production of trichloroethylene, a widely used solvent.

$$C_2H_2 + 2\,Cl_2 \rightarrow CHCl_2CHCl_2 \xrightarrow{-HCl} Cl_2C{=}CHCl$$

Although mixtures of acetylene and chlorine explode in the presence of air or light, the pure gases do not react significantly in the dark. A catalyst is necessary to produce controllable reaction to form tetrachloroethane. In the commercial process used from 1910 until its obsolescence in the 1960s the catalyst was anhydrous iron(III) chloride [1]. Acetylene was fed to a solution of chlorine and $FeCl_3$ in tetrachloroethane at 80–90° under reduced pressure. Rapid chlorination occurred and tetrachloroethane was continuously distilled out of the system. Yields exceeded 95%.

The uncontrollable air- or light-catalyzed reaction of chlorine with acetylene is almost certainly a radical chain process. It seems likely that the liquid-phase chlorination involves electrophilic catalysis because the best catalysts are Lewis acids like $SbCl_5$, $AlCl_3$ and $FeCl_3$ [1]. Although $SbCl_5$ forms a 1 : 1 complex with acetylene, a major factor in the catalysis is activation of chlorine. Lewis acids assist heterolytic dissociation of molecular chlorine, for example,

$$Cl_2 + SbCl_5 \rightleftharpoons Cl^+SbCl_6^-$$

Incipient Cl^+ ion is a potent electrophile capable of attack on arenes, olefins, or, in this case, acetylene.

8.4 ACETYLENE-CO REACTIONS

Carbon monoxide and acetylenes react with transition metal complexes in an extraordinary number of ways [33]. Some of these reactions are useful in synthesis. In this section we cover three reactions that had considerable industrial potential when acetylene was a viable feedstock. These are the nickel-catalyzed synthesis of acrylate esters, the cobalt-catalyzed synthesis of "bifurandione," and an iron- or ruthenium-catalyzed synthesis of hydroquinone.

Acrylate Synthesis

Until recently, most of the acrylic acid and acrylate esters used in polymers were made by carbonylation of acetylene:

$$HC{\equiv}CH + CO + ROH \rightarrow CH_2{=}CHCOOR$$

This process has been displaced by syntheses based on propylene oxidation in the United States but is used extensively elsewhere. Two major variants of this process are practiced. One is a true catalytic process that uses a nickel salt as a catalyst. The other is semicatalytic in that about 20% of the acrylate is formed in a stoichiometric reaction of nickel tetracarbonyl with acetylene. This reaction provides a catalyst for the other 80% of the carbonylation that occurs in the same vessel.

In the catalytic process for acrylic acid an acetylene solution in tetrahydrofuran reacts with CO, water, and $NiBr_2$ at about 200° and 60 atmospheres pressure [34]. Acrylic acid is formed in about 90% yield based on acetylene. The major byproducts are acrylic acid oligomers. The semicatalytic process [35] is performed very much as originally developed by Reppe in the 1940s [36]. In the synthesis of ethyl acrylate, acetylene, CO, ethanol, $Ni(CO)_4$, and HCl are fed to a reactor at 30–50°. Acrylate yields of about 85% based on acetylene are attained. Careful control of temperature and reactant ratios is necessary to keep the rather unstable catalyst system effective.

In both processes it seems likely that carbonyl nickel halides are the active catalytic species.

The mechanisms of these processes are still controversial [37]. One plausible reaction sequence for the semicatalytic process closely parallels the chemistry of olefin carboxylation (Section 4.4). Catalyst formation was proposed to occur via protonation of $Ni(CO)_4$ to give an unstable hydride [38]:

$$Ni(CO)_4 + HCl \rightarrow HNi(CO)_2Cl + 2CO$$

Similar species may be generated from NiX_2 in the reducing environment of the catalytic process. (The shift reaction, $CO + H_2O \rightleftharpoons CO_2 + H_2$, provides a source of hydrogen.) Once the hydride is formed, the analogy to the olefin reaction seems straightforward:

$$HC{\equiv}CH + H{-}\overset{|}{\underset{|}{Ni}}{-}CO \longrightarrow CH_2{=}CH{-}\overset{|}{\underset{|}{Ni}}{-}CO \longrightarrow CH_2{=}CH\underset{\underset{O}{\|}}{C}{-}Ni\langle$$

Cleavage of the acyl—Ni bond by water gives acrylic acid. Presumably HCl gives acrylyl chloride, and alcohols give the acrylate esters. It is reported [39] that ester synthesis does not occur under anhydrous conditions, but this limitation may reflect failure to form the necessary catalytic species.

Similar conditions also bring about the carbonylation of higher alkynes [37,39], but yields tend to be much lower than for acetylene. Markovnikov addition seems to be the rule for alkylacetylenes. For example, methylacetylene gives a 50% yield of methyl methacrylate in a stoichiometric reaction with $Ni(CO)_4$, HCl, and methanol. Disubstituted acetylenes give exclusively *cis* addition of the H and COOR moieties to the triple bond. This result is consistent with the Ni—H addition to the C≡C bond that was suggested in the mechanistic scheme earlier.

Both iron and cobalt carbonyls can be used as catalysts for acrylate synthesis, but yields are low [37]. These metals tend to direct the reaction of acetylenes with CO to form more complex products as described next.

Bifurandione Synthesis

Cobalt carbonyls catalyze the reaction of acetylene and CO to form two isomeric bifurandiones [40,41].

$$2C_2H_2 + 4\ CO \longrightarrow \text{(bifurandione)} + \text{(isomeric bifurandione)}$$

This remarkable reaction was investigated extensively because the products could be hydrogenated to give useful polymer intermediates. Hydrogenation with a copper chromite catalyst at 160–190° gives octanedioic acid and at 250–300° gives 1,8-octanediol [42].

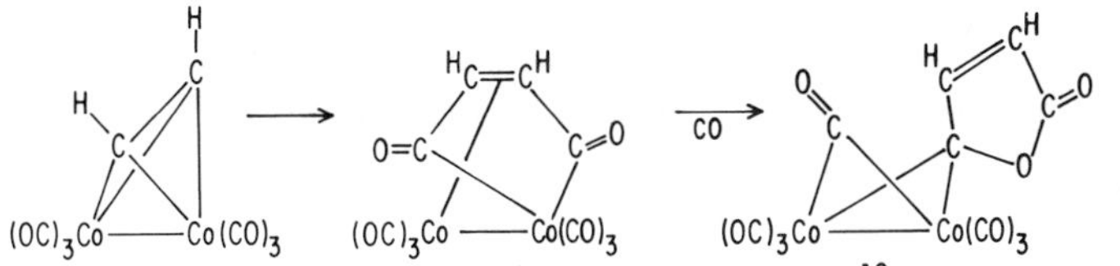

Figure 8.4 Steps in lactone synthesis from acetylene and CO.

The synthesis of bifurandiones is typically achieved by adding acetylene and carbon monoxide to a solution of $Co_2(CO)_8$ and heating the mixture to 90–100°. High pressures (more than 200 atmospheres) are generally used, but yields of bifurandione can be fairly good (50–70%) if CO is in excess [40]. The *trans* isomer usually predominates, but the *cis* isomer is the only product when tetramethylurea is the solvent. Monoalkylacetylenes undergo the same sort of reaction to give complex mixtures of isomeric dialkylbifurandiones.

The mechanism of formation of bifurandione is unclear, but some possible intermediates in the process have been characterized. Both acetylene and disubstituted acetylenes react with cobalt carbonyls to form acetylene-bridged dicobalt complexes like **10** in Figure 8.4, some of which have been characterized crystallographically. It has been suggested [37] that insertion of CO into two of the acetylene-cobalt bonds in **10** gives a product such as **11.** Isomerization of **11** and further CO absorption gives **12,** which has also been characterized crystallographically [43]. The lactone unit in **12** may be regarded as a carbenoid ligand analogous to CO or C≡NR. Such pseudocarbenes are known to bridge metal atoms in cluster compounds. Coupling of two carbene-like lactone ligands from **12** would form the observed bifurandiones, although the mechanism of coupling is not known.

In support of the double CO insertion that converts **10** to **11,** the reaction of acetylene, CO, and water in the presence of cobalt salts gives good yields of succinic acid [44]. Hydrolysis of **11** directly would give maleic acid, but cobalt carbonyls are good catalysts for hydrogenation of olefinic double bonds. Hydrogen is readily available in this system from the water gas shift reaction, $CO + H_2O \rightleftharpoons CO_2 + H_2$. Carbon dioxide is observed in the gaseous byproducts.

Hydroquinone Synthesis

Hydroquinone is formed in iron- and ruthenium-catalyzed reactions of acetylene, CO, and water [37,45]:

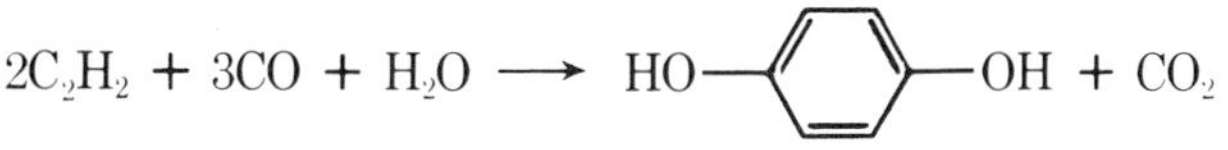

$$2C_2H_2 + 3CO + H_2O \longrightarrow HO{-}C_6H_4{-}OH + CO_2$$

Other alkynes react under similar conditions to form substituted quinone derivatives. These reactions have created much interest in industry because hydroquinone is widely used as a reducing agent, especially in photography.

Perhaps the best results are obtained when $Ru_3(CO)_{12}$ is the catalyst and moist 2-butanone is the solvent [37]. Acetylene and excess carbon monoxide react at 250° and 200 atmospheres pressure to give hydroquinone in 73% yield. Roughly 1 mole of carbon dioxide forms per mole of hydroquinone, as predicted from the equation just given. Tetrahydrofuran and other ether and ketone solvents give the best yields and rates. Careful control of temperature and reactant concentrations is necessary for good yields and to avoid acetylene decomposition.

For a potential commercial process a nonnoble metal catalyst is much more attractive than $Ru_3(CO)_{12}$ or a ruthenium salt. Iron-based systems have received a great deal of attention. As in acrylate synthesis one of the best procedures is semicatalytic [45]. Acetylene, $Fe(CO)_5$, and water react stoichiometrically at 50–80° to form hydroquinone in fair yields. However, under high CO pressure (600–700 atmospheres), the reaction becomes partially catalytic, and yields up to 70% are obtained. Substituted acetylenes react under similar conditions to form the corresponding alkylhydroquinones. A mixture of methylacetylene and dimethylacetylene produces trimethylhydroquinone, a vitamin E precursor:

$$CH_3C{\equiv}CCH_3 + CH_3C{\equiv}CH + CO \longrightarrow \text{HO–}C_6H(CH_3)_3\text{–OH}$$

The desired trimethyl derivative forms 55% of the hydroquinone products along with 22% tetramethylhydroquinone and 23% mixed dimethyl derivatives.

The mechanism of hydroquinone synthesis is not clear, but several possible intermediates have been isolated and characterized. Some of these acetylene-CO complexes are shown in Figure 8.5, which illustrates two possible pathways for the overall process.

The reactions of acetylene with iron carbonyls are extremely complex [33]. The logical initial product, $Fe(C_2H_2)(CO)_4$, is not observed in the reaction of acetylene with either $Fe(CO)_5$ or $Fe_3(CO)_{12}$. Instead, reaction with a second iron carbonyl unit probably forms **13** [46], an acetylene-bridged $Fe_2(CO)_6$ unit analogous to the product from acetylene and $Co_2(CO)_8$ or its derivatives (Figure 8.4). Insertion of CO into two Fe—acetylene bonds in **13** would give **14**, which is analogous to the second intermediate in the cobalt-catalyzed reaction. If this postulate is correct, the C_4Fe ring in **14** can undergo several reactions. One pathway forms a maleyliron carbonyl complex (**15**) that seems well established [46,47]. Insertion of acetylene into the acyl—Fe bonds in **15** would give a quinone complex (**16**). Several analogous alkyl-substituted quinone-$Fe(CO)_3$ complexes are known. Reduction of the quinone ligand by $[HFe(CO)_4]^-$, a powerful reducing agent present in alkaline aqueous solutions of $Fe(CO)_5$, would give hydroquinone, possibly as an iron carbonyl complex (**18**). Several analogous (arene)$Fe(CO)_2$ complexes have been characterized.

Another equally plausible pathway converts complex **14** to **17** by reduction with $[HFe(CO)_4]^-$. Compound **17** has been isolated from reactions of acetylene in aqueous alkaline iron carbonyl solution and has been characterized crystallo-

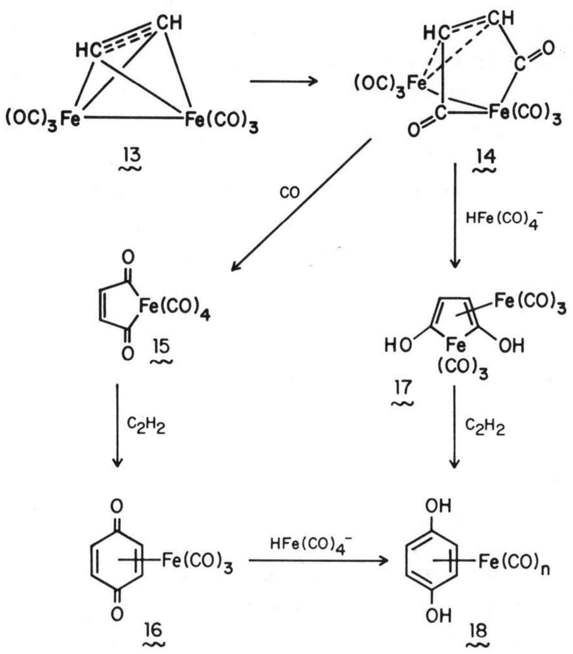

Figure 8.5 Two possible schemes for $Fe(CO)_5$-catalyzed synthesis of hydroquinone (as an iron carbonyl complex).

graphically [47]. If acetylene were to add to the diene system in place of face-bound $Fe(CO)_3$ unit, a hydroquinone complex (**18**) would form directly. This addition can be viewed as a Diels-Alder reaction.

Many analogous intermediates are found in the reaction of higher alkynes with iron carbonyls [33]. The reaction pathways have not been shown conclusively, but the results support mechanisms like those shown in Figure 8.5.

8.5 OLIGOMERIZATION

In the absence of carbon monoxide, Group VIII metal complexes catalyze some remarkable dimerizations, trimerizations, and tetramerizations of acetylenes. Other unsaturated substrates such as olefins and dienes can also be incorporated into the products. Even nitriles react to give aromatic heterocycles. None of these processes are used commercially, but several are useful in synthesis of cyclic compounds that would be difficult to prepare in any other way.

The mechanism of alkyne cyclization has been controversial for 20 years [48,49]. One early proposal, a "π-complex multicenter mechanism" [50], was that all the alkyne molecules that are to be assembled into a ring gather on a single metal atom. The alkynes then link to form a cyclobutadiene, a benzene, or a cyclooctatetraene ligand. The number of alkyne molecules that are present on the metal determines the size of the ring. This theory was appealing because it related several processes very simply. Empirically, nickel complexes that bear no strongly bonded ligands produce cyclooctatetraene from acetylene. This result fits the hypothesis that a "naked nickel" [49] should have four empty coordination sites and should assemble four acetylene molecules. Similarly, a nickel complex that bears one strongly bound ligand such as triphenylphosphine yields benzene, the expected product from assembly of three acetylene molecules. The formation of cyclobutadiene ligands by assembly of two acetylene molecules also fits this pattern.

A more commonly accepted proposal is that two π-bonded acetylene ligands join together with the metal to form a metallocyclopentadiene ring. This ring can rearrange to a cyclobutadiene ligand or can expand by addition of a third or fourth alkyne to form benzene or cyclooctatetraene. This mechanism is discussed in more detail in connection with trimerization. Interestingly, one version of this scheme is indistinguishable from a stepwise application of the π-complex multicenter mechanism.

Cyclooctatetraene Synthesis

The preparation of this acetylene tetramer was studied intensively as a potential commercial process [45,48]. Reppe originally observed that nickel salts catalyze cyclotetramer formation [51]:

$$4\ HC{\equiv}CH \longrightarrow \text{(cyclooctatetraene)}$$

For example, a nickel(II) cyanide slurry in anhydrous tetrahydrofuran slowly absorbs acetylene at 60–70° and 15–20 atmospheres. The product is about 70% 1,3,5,7-cyclooctatetraene, along with some benzene and an acetylene polymer known as cuprene.

Functionally substituted acetylenes also take part in this reaction. Conditions like those discovered by Reppe are used to form cotetramers from acetylene and acetylenic alcohols [52]. Propargyl alcohol ($HC{\equiv}CCH_2OH$) reacts with excess acetylene to form hydroxymethylcyclooctatetraene in about 20% yield. Zerovalent nickel catalysts such as $Ni(CO)_4$ or Ni(1,5-cyclooctadiene)$_2$ are also effective catalysts [53]. These systems convert $HC{\equiv}CCMe_2OH$ into mixtures of isomeric tetrakis(2-hydroxy-2-propyl)cyclooctatetraenes in yields of 85–90%. Bis-(acrylonitrile)nickel(0) is less effective as a catalyst and gives only stoichiometric yields (based on Ni) of cyclooctatetraene and benzene from acetylene [50].

In contrast to the trimerization reaction discussed next the cyclooctatetraene synthesis seems to work well only with nickel catalysts. This observation fits the proposal that four coordination sites disposed in a tetrahedral array are needed to make the cyclic tetramer. It does not distinguish between concerted and stepwise processes for joining four acetylene molecules into a ring.

Trimerization

The joining of three molecules of acetylene to give one of benzene is very favored thermodynamically but requires a catalyst to proceed well. The uncatalyzed trimerization discovered more than 100 years ago requires high temperatures (300–400°) and gives poor yields. In the 1940s Reppe discovered that some nickel complexes produce benzene from acetylene in good yield under mild conditions. More recently the trimerization has been turned into a valuable tool for synthesis of complex organic molecules [48,54], especially when cobalt complexes are used as catalysts.

In the original Reppe synthesis [55] acetylene was heated at 60–70° and 15 atmospheres pressure with an acetonitrile solution of $Ni(CO)_2(PPh_3)_2$. Benzene formed in 88% yield, along with 12% styrene. The latter product probably arose from cotrimerization of acetylene with vinylacetylene, which is formed under reaction conditions. Under similar conditions, propargyl alcohol ($HC{\equiv}CCH_2OH$) trimerizes to a mixture of 1,2,4- and 1,3,5-tris(hydroxymethyl)benzenes in high yield. The $Ni(CO)_2(PPh_3)_2$ catalyst trimerizes a great variety of monosubstituted acetylenes to trisubstituted benzenes [56]. The catalyst tolerates many functional groups such as ester, ketone, hydroxyl, ether, olefin, and nitro. Disubstituted acetylenes generally do not react, but 1,4-butynediol gives hexakis(hydroxymethyl)benzene in good yield. Generally, phosphine- or phosphite-substituted nickel(0) complexes give benzenoid products preferentially.

For synthesis of benzene, in contrast to cyclooctatetraene, complexes of cobalt and of palladium are also very useful. With these metals [54,57] trimerization seems to occur by a stepwise mechanism that can be used to good advantage in synthesis of unsymmetrical benzenes. Cobalt catalysts are extraordinarily tolerant of steric strain. For example, 1,5-hexadiyne reacts with monoalkylacetylenes in the presence of $Co(CO)_2(C_5H_5)$ to form monoalkylbenzocyclobutenes [58].

$$RC{\equiv}CH + HC{\equiv}C{-}CH_2{-}CH_2{-}C{\equiv}CH \longrightarrow \text{R-benzocyclobutene}$$

Unsymmetrical cotrimerization can be favored by using an acetylene with two bulky substituents as one reactant. An elegant application in the synthesis of polycyclic compounds involves several C—C bond-making and bond-breaking steps [54,59]:

$$Me_3SiC{\equiv}CSiMe_3 + (HC{\equiv}C)_2\text{-diyne-ene} \longrightarrow \text{bis}(Me_3Si)\text{ tricyclic product (H, H)}$$

80%

Palladium chloride complexes catalyze oligomerization of diphenylacetylene [60]

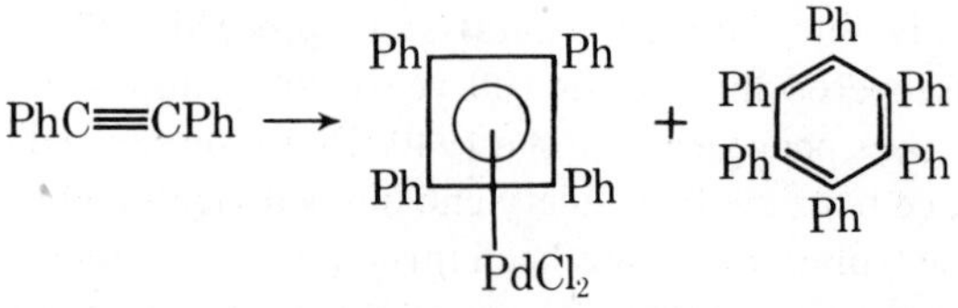
PhC≡CPh
Ph
Ph
Ph
Ph
PdCl2
+
Ph
Ph
Ph
Ph
Ph
Ph

In nonhydroxylic solvents hexaphenylbenzene is formed in 80–85% yield. Similarly mono-*tert*-butylacetylene gives largely 1,3,5-tri-*tert*-butylbenzene. A heterogeneous catalyst, palladium-on-carbon, is also effective. Acetylenedicarboxylate esters trimerize to benzenehexacarboxylates in more than 90% yield [61].

A metallocyclopentadiene mechanism seems well established for cobalt [62], rhodium [19], and iridium [63] catalysts and is also likely to operate with zerovalent nickel and palladium catalysts. The distinctive feature is replacement of two labile ligands to form a bis(alkyne) complex that rearranges to a metallocyclopentadiene. With $C_5H_5Co(PPh_3)_2$, the metallocycle (**21**) and a monoalkyne complex (**19**) are isolable [62,64], even though the bis(alkyne) complex (**20**) is not (R=CH_3, C_6H_5).

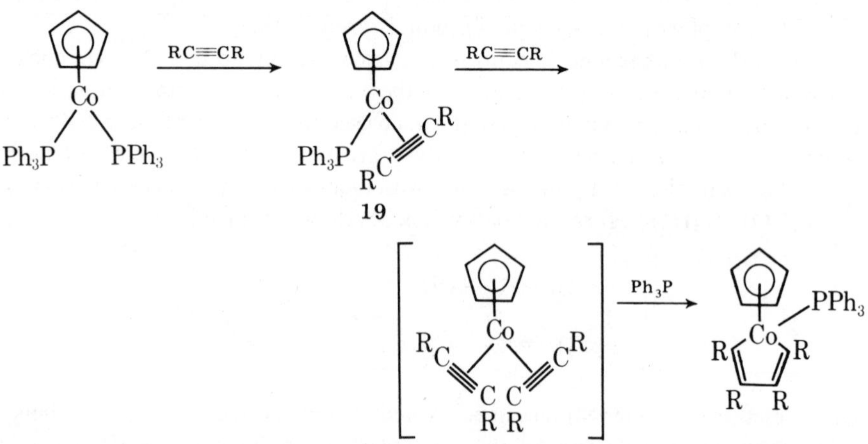
RC≡CR
RC≡CR
Co
Ph3P
PPh3
Co
Ph3P
CR
RC
Co
RC
CR
C C
R R
Ph3P
PPh3
Co
R
R
R R

19 **20** **21**

The subsequent addition of a third alkyne molecule to form the cyclic trimer could occur by several mechanisms. Three likely paths are shown in Figure 8.6. The simplest is reaction of the metallocyclopentadiene (**22**) with an acetylenic dienophile in Diels-Alder fashion to form **23,** which unfolds to a π-bonded benzene complex (**24**). This pathway seems to operate when a good dienophile such as dimethyl acetylenedicarboxylate is present [64].

With less electrophilic acetylenes precoordination of the third acetylene to the metallocycle is essential for trimerization to occur. In the cobalt system it is necessary to replace the phosphine ligand in **21** to give the alkyne complex equivalent

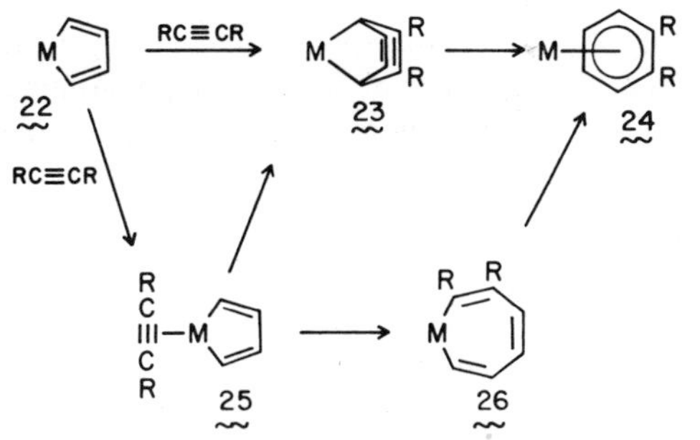

Figure 8.6 Three pathways for conversion of a metallocyclopentadiene to a coordinated benzene. Unreactive ligands are omitted for simplicity.

to **25** in the figure. At this point two paths are available. A pseudo-Diels-Alder reaction within the coordination sphere of the metal can give **23** and **24,** as in the process discussed earlier. However, insertion of the coordinated alkyne into an M—C bond of the metallocyclopentadiene seems more likely. The resulting metallocycloheptatriene (**26**) can undergo reductive elimination by coupling the M—C bonds to form the benzene complex (**24**).

The relative stability of the metallocyclopentadienes such as **22** facilitates selective syntheses of unsymmetrical benzenes as noted earlier. The versatility also extends to incorporation of substrates other than alkynes. The cobaltacyclopentadiene (**21**) reacts with nitriles to form pyridines [58] and with olefins to yield 1,3-cyclohexadienes [62].

The palladium chloride-catalyzed trimerization of acetylenes follows a different mechanism [57,65], which is characterized by a series of acetylene insertion steps. As in the copper- and mercury-catalyzed reactions discussed earlier, an acetylene complex of $PdCl_2$ readily rearranges to a chlorovinyl derivative by insertion.

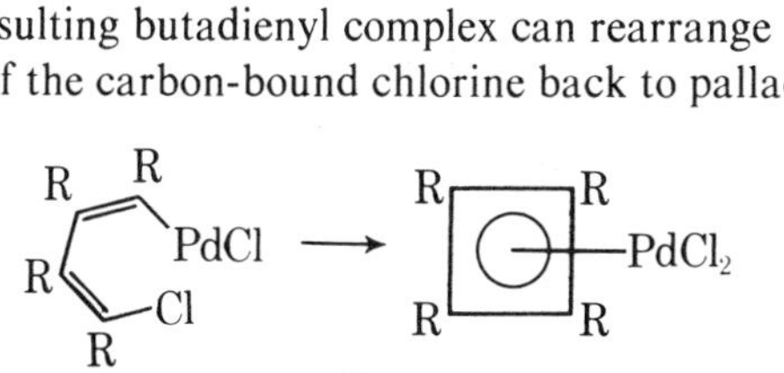

In contrast to the Cu and Hg systems further insertion into a Pd—C bond occurs readily [66]. The resulting butadienyl complex can rearrange to a cyclobutadiene ligand by transfer of the carbon-bound chlorine back to palladium:

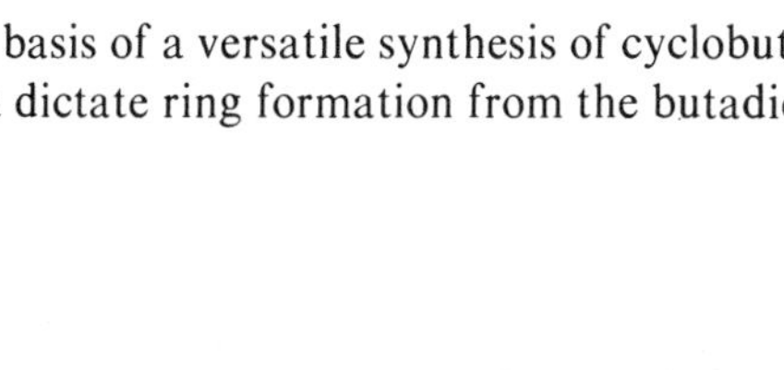

This reaction is the basis of a versatile synthesis of cyclobutadiene complexes. If steric factors do not dictate ring formation from the butadienyl ligand, insertion of a third alkyne molecule may form a 6-chlorohexatrienyl ligand. Rearrangement of such ligands to benzenes by chlorine transfer to palladium should be easy, but there is strong evidence that a less direct path involving a cyclopentadienylmethyl ligand is more favorable [66].

Cotrimerization Reactions

The catalytic trimerization of alkynes can also be modified by addition of other substrates. In the cobalt-catalyzed trimerization of acetylene the presence of acetonitrile gives 2-methylpyridine as a major product [67–69].

$$2HC{\equiv}CH + CH_3C{\equiv}N \longrightarrow \text{2-methylpyridine } (C_5H_4N\text{-}CH_3)$$

Higher nitriles react even better. The chelating agent, 2,2′-bipyridine can be prepared in 95% yield by reaction of acetylene with 2-cyanopyridine with the 1,5-cyclooctadiene complex $Co(C_5H_5)(COD)$ as the catalyst. α,ω-Dinitriles give similarly high yields of alkylene-bridged bipyridines. 1-Alkynes can also be used. Acetonitrile and 1-butyne give an 84% yield of a mixture of 3,6- and 4,6-diethyl-2-methylpyridines. The main byproducts are triethylbenzenes.

GENERAL REFERENCES

Viehe, H. G., Ed., *Chemistry of Acetylenes,* Marcel Dekker, 1969.

Patai, S., Ed., *The Chemistry of the Carbon-Carbon Triple Bond,* Wiley, 1978.

SPECIFIC REFERENCES

1. S. A. Miller, *Acetylene,* Vol. 2, Academic Press, 1966.
2. A. C. Hopkinson, "Acidity, Hydrogen Bonding and Complex Formation," in S. Patai, Ed., *The Chemistry of the Carbon-Carbon Triple Bond,* Part 1, 1978, pp. 75–136.
3. R. Mason and K. M. Thomas, *Ann. N.Y. Acad. Sci.,* **239,** 225 (1974); S. Otsuka and A. Nakamura, *Adv. Organomet. Chem.,* **14,** 245 (1976).
4. J. O. Glanville, J. M. Stewart, and S. O. Grim, *J. Organomet. Chem.,* **7,** P9 (1967).
5. J. J. Bonnet and R. Mathieu, *Inorg. Chem.,* **17,** 1973 (1978).
6. A. M. Sladkov and L. Yu. Ukhin, *Russ. Chem. Rev.,* **37,** 748 (1968).
7. R. Nast and W. Pfab, *Chem. Ber.,* **89,** 415 (1956).
8. G. E. Coates, *Organometallic Compounds,* 2nd ed., Wiley, 1960; 3rd ed., Methuen, 1967.
9. P. W. R. Corfield and H. M. M. Shearer, *Acta Cryst.,* **21,** 957 (1966).
10. P. Cadiot and W. Chodkiewicz, "Couplings of Acetylenes," in H. G. Viehe, Ed., *Chemistry of Acetylenes,* Marcel Dekker, 1969, p. 597.
11. A. S. Hay, *J. Org. Chem.,* **27,** 3320 (1962).
12. F. Bohlmann, H. Schönowsky, E. Inhoffen, and G. Grau, *Chem. Ber.,* **97,** 794 (1964).
13. I. Bodek and G. Davies, *Inorg. Chem.,* **17,** 1814 (1978).

14. W. Ziegenbein, "Synthesis of Acetylenes and Polyacetylenes by Substitution Reactions," in H. G. Viehe, Ed., *Chemistry of Acetylenes,* Marcel Dekker, 1969, p. 169, especially pp. 237–241.
15. E. V. Hort, U.S. Patents 3,920,759 (1975) and 4,002,694 (1977).
16. W. Reppe et al., *Liebig's Ann. Chem.,* **596,** 1 (1955).
17. C. A. Hargreaves and D. C. Thompson, "2-Chlorobutadiene Polymers," in *Encyc. Poly. Sci. Tech.,* **3,** 705 (1965).
18. W. H. Carothers, G. J. Berchet, and A. M. Collins, *J. Am. Chem. Soc.,* **54,** 4066 (1932).
19. H. J. Schmitt and H. Singer, *J. Organomet. Chem.,* **153,** 165 (1978); L. Carlton and G. Read, *J. Chem. Soc. Perkin* **1,** 1631 (1978).
20. O. A. Chaltykyan, *Copper-Catalytic Reactions,* Consultants Bureau, 1966, pp. 39–57.
21. O. N. Temkin and R. M. Flid, *Catalytic Conversions of Acetylenic Compounds in Solutions of Metal Complexes,* Nauka, 1968.
22. O. N. Temkin et al., *Kinet. Katal.,* **10,** 1004, 1230 (1969).
23. A. Alexakis, J. Normant, and J. Villieras, *Tetrahedron Lett.,* 3461 (1976).
24. G. K. Shestakov, F. I. Belškii, S. M. Airyan, and O. N. Temkin, *Kinet. Katal.,* **19,** 334 (1978).
25. G. K. Shestakov, S. M. Airyan, F. I. Belškii, and O. N. Temkin, *Zhur. Org. Khim.,* **12,** 2053 (1976).
26. D. F. Othmer, K. Kon, and T. Igarashi, *Ind. Eng. Chem.,* **48,** 1258 (1956).
27. G. A. Olah and D. Meidar, *Synthesis,* 671 (1978).
28. W. L. Budde and R. E. Dessy, *J. Am. Chem. Soc.,* **85,** 3964 (1963).
29. D. Swern and E. F. Jordan, *Org. Syn., Coll.,* **4,** 977 (1963).
30. A. N. Nesmeyanov, A. E. Borisov, I. S. Savelėva, and M. A. Osipova, *Izvest. Akad. Nauk, SSSR, Otdel. Khim. Nauk* 1249 (1961).
31. D. Y. Waddan, U.S. Patent 3,869,501 (1975).
32. O. N. Temkin, R. M. Flid, G. F. Tikhonov, L. V. Melnikova, T. G. Sukhova, G. K. Shestakov, A. A. Khorkin, and L. N. Reshetova, *Katal. Reakts. Zhidk. Faze,* Tr. 2nd Vsesouizhii. Konferentsia, Alma-Ata, 1966, pp. 49–57; *Chem. Abs.,* **69,** 76328 (1968).
33. W. Hübel, "Organometallic Derivatives from Metal Carbonyls and Acetylene Compounds," in I. Wender and P. Pino, Eds., *Organic Syntheses via Metal Carbonyls,* Vol. 1, Wiley-Interscience, 1968, pp. 273–342.
34. T. Toepel, *Chem. Ind.* (*Paris*), **91,** 139 (1964).
35. F. T. Maher and W. Bauer, "Acrylic Acid and Its Derivatives," in J. J. McKetta and W. A. Cunningham, Eds., *Encyclopedia of Chemical Processing and Design,* Marcel Dekker, **1,** 401 (1976).
36. W. Reppe, *Liebigs Ann. Chem.,* **582,** 1 (1953).
37. P. Pino and G. Braca, "Carbon Monoxide Addition to Acetylenic Substrates," in I. Wender and P. Pino, Eds., *Organic Syntheses via Metal Carbonyls,* Vol. 2, Wiley-Interscience, 1977, p. 419.
38. R. F. Heck, *J. Am. Chem. Soc.,* **85,** 2013 (1963).
39. J. Falbe, *Carbon Monoxide in Organic Synthesis,* Springer-Verlag, 1970, pp. 87–96.
40. J. C. Sauer, R. D. Cramer, V. A. Engelhardt, T. A. Ford, H. E. Holmquist, and B. W. Howk, *J. Am. Chem. Soc.,* **81,** 3677 (1959).
41. G. Albanesi and M. Tovaglieri, *Chim. Ind.* (*Milan*), **41,** 189 (1959); G. Albanesi and E. Gavezzotti, ibid, **47,** 1322 (1965).
42. H. E. Holmquist, F. D. Marsh, J. C. Sauer, and V. A. Engelhardt, *J. Am. Chem. Soc.,* **81,** 3681 (1959).
43. O. S. Mills and G. Robinson, *Proc. Chem. Soc.,* 156 (1959).

44. G. Natta and G. Albanesi, *Chim. Ind. (Milan)*, **48,** 1157 (1966).
45. W. Reppe, N. von Kutepow, and A. Magin, *Angew. Chem. Int. Ed.*, **8,** 727 (1969); W. Reppe and H. Vetter, *Liebig's Ann., Chem.*, **582,** 133 (1953).
46. E. Weiss, W. Hübel, and R. Merenyi, *Chem. Ber.*, **95,** 1155, (1962).
47. R. Clarkson, E. R. H. Jones, P. C. Wailes, and M. C. Whiting, *J. Am. Chem. Soc.*, **78,** 6206 (1956); J. R. Case, R. Clarkson, E. R. H. Jones, and M. C. Whiting, *Proc. Chem. Soc.*, 150 (1959).
48. C. Hoogzand and W. Hübel, "Cyclic Polymerization of Acetylenes by Metal Carbonyl Compounds," in I. Wender and P. Pino, Eds., *Organic Syntheses via Metal Carbonyls,* Vol. 1, Wiley-Interscience, 1968, p. 343.
49. P. W. Jolly and G. Wilke, *The Organic Chemistry of Nickel,* Vol. 2, Academic Press, 1975, p. 94.
50. G. N. Schrauzer, *Chem. Ber.*, **94,** 1403 (1961); *Adv. Organomet. Chem.*, **2,** 1 (1964).
51. W. Reppe, O. Schlichting, K. Klager, and T. Toepel, *Liebig's Ann. Chem.*, **560,** 104 (1948).
52. A. C. Cope et al., *J. Am. Chem. Soc.*, **75,** 3215, 3220 (1953).
53. P. Chini, N. Palladino, and A. Santambrogio, *J. Chem. Soc.*, **C,** 836 (1967).
54. K. P. C. Vollhardt, *Accounts Chem. Res.*, **10,** 1 (1977); *Nachr. Chem., Tech. Lab.*, **25,** 584 (1977).
55. W. Reppe and W. J. Schweckendiek, *Liebig's Ann. Chem.*, **560,** 104 (1948).
56. L. S. Meriwether, E. C. Colthup, G. W. Kennerly, and R. N. Reusch, *J. Org. Chem.*, **26,** 5155 (1961).
57. P. M. Maitlis, *Accounts Chem. Res.*, **9,** 93 (1976).
58. K. P. C. Vollhardt and R. G. Bergman, *J. Am. Chem. Soc.*, **96,** 4996 (1974).
59. R. L. Funk and K. P. C. Vollhardt, *J. Am. Chem. Soc.*, **101,** 215 (1979).
60. A. T. Blomquist and P. M. Maitlis, *J. Am. Chem. Soc.*, **84,** 2329 (1962).
61. D. Bryce-Smith, *Chem. Ind.*, 239 (1964).
62. H. Yamazaki and Y. Wakatsuki, *J. Organomet. Chem.*, **139,** 157, 169 (1977).
63. J. P. Collman, J. W. Kang, W. F. Little, and M. F. Sullivan, *Inorg. Chem.*, **7,** 1298 (1968).
64. D. R. McAlister, J. E. Bercaw, and R. G. Bergman, *J. Am. Chem. Soc.*, **99,** 1666 (1977).
65. G. M. Whitesides and W. J. Ehmann, *J. Am. Chem. Soc.*, **91,** 3800 (1969).
66. B. E. Mann, P. M. Bailey, and P. M. Maitlis, *J. Am. Chem. Soc.*, **97,** 1275 (1975).
67. R. A. Clement, U.S. Patent 3,829,429 (1974).
68. H. Bönnemann, *Angew. Chem. Int. Ed.*, **17,** 505 (1978).
69. Y. Wakatsuki and H. Yamazaki, *Synthesis,* 26 (1976); *J. Chem. Soc. Dalton,* 1278 (1978).

9 CARBENE COMPLEXES IN OLEFIN METATHESIS AND ALKANE REACTIONS

"Carbene" ligands in transition metal complexes have been known for about 15 years, but their role in catalysis has been recognized only recently. About 1960 it became apparent that coordination to a metal ion might stabilize reactive organic groups such as cyclobutadiene and "methylene." The first well-characterized complexes with carbenoid ligands were synthesized in E. O. Fischer's laboratory [1]. The ligands usually bore electron-withdrawing substituents that polarized the M═C bond as shown in **1b**:

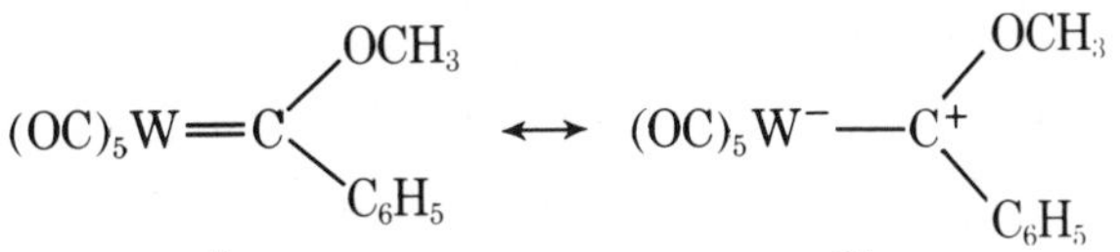

$$(OC)_5W{=}C(OCH_3)(C_6H_5) \longleftrightarrow (OC)_5W^-{-}C^+(OCH_3)(C_6H_5)$$

1a **1b**

Another major advance came in the mid 1970s with synthesis of the first simple alkylidene complexes and a methylene complex **2** [2]:

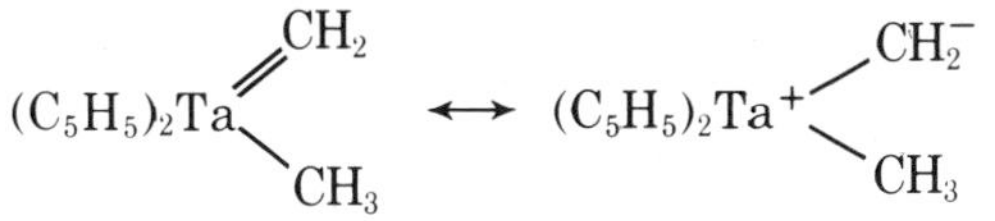

$$(C_5H_5)_2Ta(=CH_2)(CH_3) \longleftrightarrow (C_5H_5)_2Ta^+(CH_2^-)(CH_3)$$

2a **2b**

The M═C bond in the simple alkylidene complexes is polarized in the opposite sense from those of the Fischer compounds. The CH_2 ligand readily coordinates to Lewis acids such as $(CH_3)_3Al$ [3]. This property has permitted isolation of very labile methylene complexes of titanium [4].

The isolation of methylene and simple alkylidene complexes was especially significant for two reasons.

- The synthesis of these compounds confirmed the importance of α-hydrogen elimination as a decomposition process for alkyl metal compounds [5].

- Alkylidene ligand complexes were being proposed as major intermediates in catalytic processes such as olefin metathesis [6] and alkane activation.

The mechanism of olefin metathesis has been one of the greatest intellectual challenges in homogeneous catalysis. It is now generally accepted that alkylidene complexes are intermediates.

The following sections discuss the synthesis and properties of alkylidene complexes and their role in olefin metathesis and other catalytic reactions. There seems to be a good reason to suspect the intermediacy of alkylidene surface species in alkane reactions such as isomerization or hydrogenolysis with metallic catalysts. In practice all these processes employ heterogeneous catalysts, but the reactions are easily explained in the terminology of organometallic chemistry.

9.1 ALPHA-HYDROGEN ELIMINATION

The transfer of an α-hydrogen from an alkyl ligand to a metal atom or another alkyl ligand is a common reaction. It has usually been overlooked because β-hydrogen transfer (Section 2.3) is even easier. The α-hydrogen transfer is most easily observed in ligands such as CH_3, CH_2Ph, and CH_2CMe_3 in which no β-hydrogen is present. The transfer of α-hydrogen to metal has not been observed directly, but chemical evidence suggests the equilibrium [7]:

$$[(C_5H_5)_2W{-}CH_3]^+ \rightleftharpoons \left[(C_5H_5)_2W\genfrac{}{}{0pt}{}{=CH_2}{-H}\right]^+$$

The transfer of α-hydrogen to an adjacent alkyl ligand is more thoroughly documented [5]. Most simple polymethyl complexes such as $TiMe_4$, $TaMe_5$, and WMe_6 decompose at room temperature or below. The decomposition is autocatalytic, exothermic, and sometimes violent. Usually the only organic product is methane. The decomposition may be intermolecular, but it does not involve the solvent. Decomposition of $Nb(CD_3)_5$ in ether gives 96% CD_4 [8]. Nearly all the deuterium present in the methyl groups appears in the methane. The decomposition of some higher alkyl compounds proceeds similarly. With $Ta(CH_2C_6H_5)_5$, the process is cleanly intramolecular [9]. The first step may be visualized as α-hydrogen abstraction by an adjacent benzyl carbon:

$$Ta\genfrac{}{}{0pt}{}{\diagup CHPh \diagdown}{\diagdown CH_2Ph \cdots}H \longrightarrow Ta{=}CHPh + CH_3Ph$$

The hydrogen atom appears to be most easily transferred as a proton. Both the transfer to metal and the transfer to an adjacent carbon are favored in cationic complexes. This characteristic was used in synthesis of the first isolable methylene complex [3]. Treatment of a cationic dimethyl complex with base removes a proton to give a CH_2 ligand:

$$\left[(C_5H_5)_2Ta\begin{matrix}CH_3\\CH_3\end{matrix}\right]^+ \xrightarrow{-H^+} (C_5H_5)_2Ta\begin{matrix}{=}CH_2\\CH_3\end{matrix}$$

Many of the organometallic catalysts for olefin metathesis are prepared by reaction of a metal chloride with an alkylaluminum compound. The most commonly used is a mixture of WCl_6, $C_2H_5AlCl_2$ and ethanol [10]. This mixture may generate an ethylidene complex stabilized by coordination to aluminum [11]. A similar stabilizing effect has been used in the isolation of a highly reactive methylene complex of titanium [4]:

$$(C_5H_5)_2TiCl_2 \xrightarrow{AlMe_3} CH_4 + (C_5H_5)_2Ti\begin{matrix}CH_2\\Cl\end{matrix}AlMe_2 \quad \mathbf{3}$$

Complex **3** reacts as though it were $(C_5H_5)_2Ti{=}CH_2$, apparently because the alkylaluminum entity dissociates to some extent in solution. It catalyzes a form of olefin metathesis and has provided convincing evidence for the role of alkylidene complexes in metathesis [12].

The nucleophilic character of the alkylidene complexes of the early transition metals is apparent in many of their reactions. As noted earlier, the methylene complexes coordinate to Lewis acids. The alkylidene complexes also behave like Wittig reagents in their reactions with carbonyl compounds. The neopentylidene complex (**4**) reacts with aldehydes, ketones, esters, and amides to replace O with $CHCMe_3$ [13].

$$\underset{\mathbf{4}}{(Me_3CCH_2)_3Ta{=}CHCMe_3} \xrightarrow{C_6H_{10}{=}O} C_6H_{10}{=}C\begin{matrix}H\\CMe_3\end{matrix}$$

The methylenetitanium complex (**3**) reacts similarly. It converts cyclohexanone to methylenecyclohexane. It seems likely that the reaction involves nucleophilic attack on the somewhat electrophilic carbonyl carbon atom:

$$\begin{matrix}M^{\delta+}{=}CH_2^{\delta-}\\O^{\delta-}{=}C^{\delta+}\end{matrix} \longrightarrow \begin{matrix}M^+{-}CH_2\\ |\\O^-{-}C\end{matrix} \longrightarrow \begin{matrix}M\\ \|\\O\end{matrix} + \begin{matrix}CH_2\\ \|\\C\end{matrix}$$

9.2 OLEFIN METATHESIS PROCESSES

The nature of olefin metathesis is nicely illustrated by the reaction of ethylene with stilbene to form styrene:

$$H_2C{=}CH_2 + PhHC{=}CHPh \rightleftharpoons \begin{matrix} H_2C \\ \| \\ PhHC \end{matrix} + \begin{matrix} CH_2 \\ \| \\ CHPh \end{matrix}$$

In simplest terms the methylene and benzylidene groups of the two olefins recombine to produce an equilibrium mixture of the three olefins. The other C—C bonds and the C—H bonds in the olefins are unaffected. This remarkable reaction has stimulated much research in both industrial and academic laboratories. Some good reviews of olefin metathesis are listed at the end of the chapter.

Most simple olefins undergo metathesis, although some catalysts are ineffective with terminal olefins. Like most catalytic reactions, metathesis gives an equilibrium mixture of products. For example, in the "Phillips Triolefin Process" [14], propylene is converted to ethylene and butene:

$$2\,CH_3CH{=}CH_2 \rightleftharpoons H_2C{=}CH_2 + CH_3CH{=}CHCH_3$$

The initial product is an equilibrium mixture of the three olefins, but the volatility of ethylene permits its selective removal from the process. Like the volatility of ethylene the stability of the six-membered ring can be a strong driving force. Cyclohexene is a poor substrate for metathesis, but its formation can be a strong driving force for the reaction, for example:

$$\text{(1,7-octadiene)} \longrightarrow \text{(cyclohexene)} + C_2H_4$$

Metathesis catalysts have been prepared from almost all the transition metals, but best results are obtained with Mo, W, and Re catalysts. The soluble catalysts, which have received extensive study in both academic and industrial laboratories, are of two major types. One family might be designated Ziegler catalysts because they are based on combinations of alkyl aluminum, magnesium, or lithium reagents with transition metal salts. The best seem to be obtained from WCl_6 or $WOCl_4$. The other family is derived from metal carbonyls, especially $Mo(CO)_6$ and $W(CO)_6$ and ligand-substituted derivatives such as $[W(CO)_5Cl]^-$. The metal carbonyl catalysts usually require activation by photolysis or by addition of a Lewis acid. The metal carbonyls can also be used to prepare heterogeneous catalysts. The reaction of $Mo(CO)_6$ with silica, alumina, or magnesia gives surface species such as

$$\begin{matrix} \text{—O} \\ \text{—O} \\ \text{—O} \end{matrix} \!\!>\! Mo(CO)_n,$$

which are good catalysts for metathesis.

Most industrial applications of olefin metathesis use heterogeneous catalysts. These generally take the form of molybdenum, tungsten, or rhenium oxides that have been deposited on silica or alumina. Although the oxide is usually applied in its maximum oxidation state such as MoO_3, reaction with the olefin at elevated temperature brings about reduction. It seems likely that Mo at the catalyst site has an oxidation state in the range +2 to +4 whether the molybdenum is deposited as MoO_3 or $Mo(CO)_6$.

Many industrial applications of metathesis have been seriously considered. The styrene synthesis from stilbene listed earlier has been considered for commercialization by Monsanto [15]. (The stilbene would be produced by oxidative coupling of the methyl groups of toluene in a heterogeneous catalytic process.) The "Phillips Triolefin Process" mentioned earlier was operated when propylene prices were low and butene was desired as a feedstock for butadiene production [14]. Two metathesis processes are used commercially at present. Neohexene, a monomer for specialty plastics, is made by metathesis of ethylene and diisobutylene:

$$(CH_3)_2C{=}CHC(CH_3)_3 + H_2C{=}CH_2 \rightleftharpoons (CH_3)_2C{=}CH_2 + CHC(CH_3)_3{=}CH_2$$

The catalyst, a mixture of magnesia and WO_3 on silica, is maintained at about 370°, while a mixture of diisobutylene and excess ethylene is passed through it under 25–30 atmospheres pressure [16]. The products are separated by distillation, and the isobutylene is recycled to the diisobutylene synthesis unit.

In the "Shell Higher Olefins Process," in which ethylene is oligomerized to higher alpha-olefins (Section 4.3), the desired products are C_8–C_{20} terminal olefins. However, since the distribution of products is largely statistical, substantial amounts of higher and lower molecular weight olefins are also produced. These undesired α-olefin fractions are isomerized to their internal isomers [17]. These, in turn, are subjected to metathesis to produce mixtures of olefins in the C_{10}–C_{20} range [18]. These olefins are then hydroformylated (Section 5.5) to produce long-chain alcohols for use in plasticizer and surfactant manufacture.

Another potential application of metathesis that has received much attention is the synthesis of polymers from cycloolefins. The first recognized example of metathesis was ring-opening polymerization of cyclopentene by an MoO_3/Al_2O_3 catalyst [19]. Although cyclohexene resists metathesis for thermodynamic reasons, the 5-membered ring opens readily:

$$n\ \text{(cyclopentene)} \longrightarrow \left[CH_2CH_2CH_2CH{=}CH \right]_n$$

The same ring-opening polymerization of cyclopentene is also catalyzed by a soluble, well-characterized carbene complex, $Ph_2C{=}W(CO)_5$. In this instance the polypentenamer product retains the *cis* stereochemistry of the cycloolefin almost quantitatively [20]. The diphenylcarbene complex also converts cyclobutene to a polymer equivalent to all-*cis* polybutadiene. Cycloheptene, cyclooctene, and norbornene are also converted to all-*cis* polymers with greater than 90% stereoselectivity.

Most medium-sized cycloalkenes (C_7–C_{20}) undergo ring opening with conventional metathesis catalysts such as WCl_6 and an alkylaluminum compound [21]. The polymers obtained with these catalysts often have high percentages of *trans* double bonds in the chain, even when a *cis* cycloalkene is the starting material. The polymers are usually high molecular weight crystalline thermoplastic materials.

The applications mentioned to this point have concerned simple olefins without polar substituents. Extensive research has been done on metathesis of functionally substituted olefins, but few applications have been reported. The major problem appears to be that groups such as C=O and C≡N react with alkylmetal catalyst precursors or nucleophilic carbene intermediates. Greatest success has been achieved with olefinic esters in which the C=C and C=O groups are not conjugated. Esters of oleic and linoleic acids have been metathesized with catalysts prepared from WCl_4 or WCl_6 and tetramethyltin [22]:

$$2RCH{=}CH(CH_2)_nCOOMe \rightleftharpoons RCH{=}CHR + MeOOC(CH_2)_nCH{=}CH(CH_2)_nCOOMe$$

One promising application of metathesis that is beginning to be exploited is synthesis of biologically active organic molecules. *Cis*-9-tricosene, the housefly pheromone, has been synthesized by metathesis with soluble Mo- and W-based catalysts [23,24]. For example, 1-decene and 1-pentadecene gave a mixture of *cis* and *trans* 9-tricosene in 26% yield.

$$H_2C{=}CHC_8H_{17} + H_2C{=}CHC_{13}H_{27} \xrightarrow[EtAlCl_2]{MoCl_2(NO)_2(PPh_3)_2} H_2C{=}CH_2 + C_8H_{17}CH{=}CHC_{13}H_{27}.$$

The synthesis itself is simple, but separation of the desired product can be difficult. The yield is limited by subsequent metathesis reactions of the product. If the separation problems can be solved, commercial production of synthetic insect lures by metathesis may be feasible. Many other insect sex attractants are also olefin derivatives that might be made by metathesis.

9.3 OLEFIN METATHESIS MECHANISMS

Olefin metathesis is so unlike other catalytic reactions of olefins (Chapters 3–6) that its mechanism has attracted intense study. Initial thoughts centered on concerted reaction paths. These generally involved exchange of alkylidene fragments between two olefins via cyclobutane-like intermediates such as **5**. Despite the intellectual appeal of a concerted process a large body of evidence now supports the stepwise mechanism discussed next.

$$M + H_2C{=}CH_2 + PhCH{=}CHPh \rightleftharpoons \underset{\mathbf{5}}{[M(H_2C\cdots CH_2)(PhHC\cdots CHPh)]} \rightleftharpoons M + H_2C{=}CHPh + CH_2{=}CHPh$$

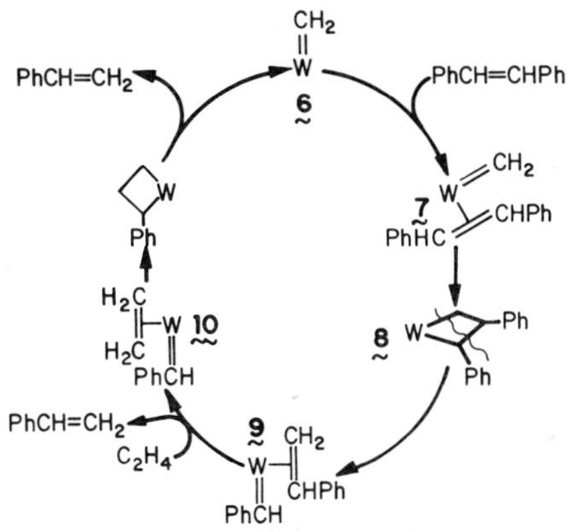

Figure 9.1 A metallocyclic mechanism for ethylene/stilbene metathesis. All the reactions are reversible.

The currently accepted mechanism is illustrated by the stilbene-ethylene metathesis in Figure 9.1. This scheme, based on a proposal by Chauvin [6], invokes a metallocyclic intermediate. The catalyst cycle begins with a carbene complex (**6**) that may be formed by processes like those described in Section 9.1. Coordination of stilbene to **6** yields the olefin complex (**7**). Intramolecular cycloaddition of C=C to W=C produces the metallocycle (**8**). The cycloaddition process seems to be readily reversible. Fission of the ring can either regenerate **7** or can produce the new olefin complex (**9**). Displacement of styrene from **9** by ethylene gives **10**, which can again undergo cycloaddition and ring fission to complete the catalytic cycle.

This mechanism helps explain the apparent inactivity of some catalysts in the metathesis of terminal olefins. The combination of WCl_6, $C_2H_5AlCl_2$ and ethanol catalyzes one of the most rapid homogeneous catalytic reactions. With internal olefins such as 2-butene, rates of 100 catalyst cycles per second may be attained. However, little or no metathesis is detected with terminal olefins. This anomaly is explained by labeling studies. Terminal olefins rapidly exchange CH_2 groups with one another, but CHR groups do not exchange. This result may be rationalized by assuming that a methylene complex such as **6** coordinates a terminal olefin in such a way that the metallocycle

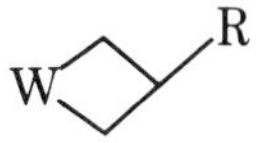

is formed selectively. Fragmentation of this ring can produce CH_2 exchange but does not accomplish metathesis in the usual sense.

Sophisticated product analyses and labeling studies support a mechanism like that of Figure 9.1. However, even more convincing evidence has started to emerge from the chemistry of organometallic compounds that mimic the metathesis catalysts. Complexes analogous to **9** and **10** have been detected by nmr in solutions that contain $PhCH{=}W(CO)_5$ and olefins [25]. In another model system stepwise transfer of $^{13}CH_2$ from olefin to metal to a second olefin has been followed by nmr [12]. The Lewis acid-stabilized methylene complex (**3**) reacts with ^{13}C-labeled isobutylene to form ^{13}C-labeled **3**:

$$\underset{\mathbf{3}}{Cp_2Ti(\mu\text{-}CH_2)(\mu\text{-}Cl)AlMe_2} + H_2C^*{=}CMe_2 \rightleftharpoons \underset{\mathbf{3^*}}{Cp_2Ti(\mu\text{-}{}^*CH_2)(\mu\text{-}Cl)AlMe_2} + H_2C{=}CMe_2$$

This process and the subsequent transfer of $^{13}CH_2$ from **3*** to methylenecyclohexane can be followed by nmr. Chemical evidence points to a metallocycle as an intermediate in this reaction.

The origin of the "carbene" complex (**6**) (Figure 9.1) that initiates metathesis seems quite varied. In organometallic systems like $WCl_6/C_2H_5AlCl_2$, α-hydrogen abstraction from an alkyl group is almost certainly the source. In this instance the cocatalyst also serves as a Lewis acid to stabilize the alkylidene ligand as in **3**. In the metal carbonyl and metal oxide catalysts a different mechanism must be responsible. A likely sequence of events is illustrated for methylene ligand synthesis from a propylene complex:

$$M(\eta^2\text{-}H_2C{=}CHCH_3) \rightleftharpoons H{-}M(\eta^3\text{-}H_2C{\cdots}CH{\cdots}CH_2) \rightleftharpoons M(CH_2)_3 \text{ (metallacyclobutane)} \rightleftharpoons M({=}CH_2)(\eta^2\text{-}H_2C{=}CH_2)$$

All the individual steps in this sequence are known for molybdenum or tungsten complexes. The equilibrium between a propylene complex and its allyl hydride isomer has been observed by nmr [26]:

$$\pi\text{-}C_3H_6Mo(DPPE)_2 \rightleftharpoons \pi\text{-}C_3H_5MoH(DPPE)_2$$

Reaction of cationic allyl complexes with $NaBH_4$ generates metallocyclobutanes analogous to those present in the catalyst cycle for metathesis [27]:

$$[\pi\text{-}C_3H_5M(C_5H_5)_2]^+ \xrightarrow{[H^-]} (CH_2)_3M(C_5H_5)_2 \qquad M = Mo, W$$

The CH_2 or CHR complexes that take part in the metathesis cycle are relatively labile. A major decomposition pathway is dimerization of the ligands to form olefin [3]:

$$M{=}CH_2 + H_2C{=}M \rightarrow M(\mu\text{-}CH_2)_2M \rightarrow M(\eta^2\text{-}H_2C{=}CH_2) + M$$

In the heterogeneous metathesis catalysts this process is inhibited by the isolation of the alkylidene ligands at well-separated sites on the catalyst surface. As mentioned previously, many of the soluble catalysts are stabilized by coordination of the ligand to a Lewis acid.

9.4 ALKANE REACTIONS

Saturated hydrocarbons are generally inert to the action of soluble transition metal complexes [28–30]. The major exception is the activity of $[PtCl_4]^{2-}$ salts in catalysis of H/D exchange between D_2O and the lower alkanes. In contrast to the soluble catalysts some of the precious metals catalyze drastic reactions of alkanes. Hydrogenolysis and isomerization of paraffins by oxide-supported platinum provide the basis for catalytic reforming of petroleum, one of the largest scale processes in modern industry.

Carbenoid ligands may be intermediates in both the homogeneous and heterogeneous catalytic reactions of alkanes. The evidence for their involvement is indirect, but the simple explanations of these processes can be based on such intermediates.

H/D Exchange

Both homogeneous and heterogeneous platinum catalysts effect exchange between D_2O and simple alkanes such as methane and ethane. The best studied soluble catalyst system is a solution of K_2PtCl_4 and $DClO_4$ in aqueous acetic acid, in which all acidic hydrogen is present as D^+. When methane is incubated with such a solution at 100–120°, slow exchange occurs to produce a mixture of CH_3D, CH_2D_2, CHD_3, and CD_4 [28,30]. Two aspects of the reaction especially merit discussion.

- Platinum metal precipitates during the reaction.
- The incorporation of deuterium is not the simple random process one might expect. Much more CH_2D_2, CH_3D, and CD_4 form early in the process than would be predicted on the basis of a stepwise exchange mechanism.

The deposition of metallic platinum during the reaction has raised questions about whether or not the exchange is catalyzed by the metal. Platinum powders are quite active for exchange between methane and D_2 at 100–200° [31,32]. The possibility of metal catalysis is usually discounted on the basis that the recovered metal from a $[PtCl_4]^{2-}$ experiment shows low activity for H/D exchange.

The "multiple exchange" phenomenon responsible for the skewed distribution of polydeuterated alkanes tells something about the nature of the exchange process. The usual interpretation of multiple exchange is that several HD exchanges occur

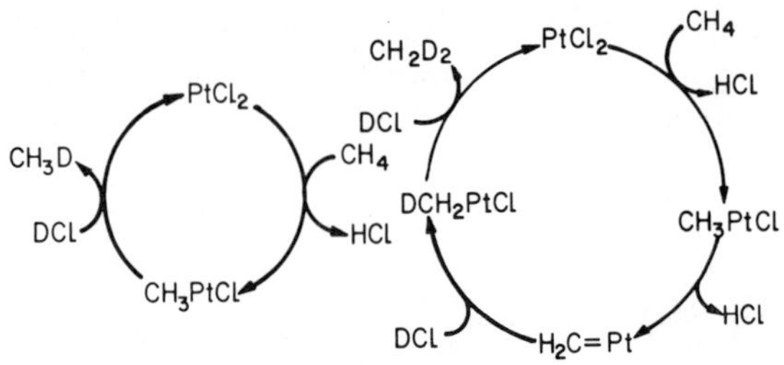

Figure 9.2 Catalytic cycles for single and multiple exchange between D^+ and methane. All reactions in the cycles are reversible.

each time an alkane molecule binds to a platinum atom. Shilov [28] has invoked carbenoid intermediates to explain polydeuteration.

A commonly proposed scheme for H/D exchange is shown in Figure 9.2. The "$PtCl_2$" species shown as the starting point in either single or multiple exchange is probably solvated by water or acetic acid. The initial reaction of methane with this species can be regarded as electrophilic attack or oxidative addition or both. At any event one C—H bond is cleaved to produce a CH_3—Pt intermediate. Reversal of this initial step with participation of D^+ from solvent forms CH_3D, a single act of exchange. The simplest explanation of multiple exchange is illustrated in the right-hand cycle of the figure. Loss of two protons would give a solvated methylene complex, "$H_2C{=}Pt$." Addition of 2 moles of DCl to this intermediate forms CH_2D_2 and regenerates the catalytic species. With higher alkanes such as ethane it is unnecessary to invoke α-hydrogen abstraction. The easier β-hydrogen elimination would give an olefinic intermediate:

$$C_2H_6 + PtCl_2 \xrightarrow{-2HCl} Pt{-}\!\!\left\|\begin{matrix}CH_2\\ CH_2\end{matrix}\right. \xrightarrow{2\ DCl} C_2H_4D_2 + PtCl_2$$

Hydrogenolysis and Isomerization

Although these alkane reactions do not occur with soluble catalysts except when the hydrocarbon is sterically strained, their technical importance merits some discussion. Catalytic hydrogenolysis of C—C bonds is an efficient way to reduce the average molecular weight of a petroleum fraction to increase volatility. Isomerization of linear alkanes to their branched isomers increases the "octane number" of gasoline fractions by increasing the number of *tertiary* C—H bonds. Typically these reactions are carried out by passing the hydrocarbon over a silica- or alumina-supported platinum catalyst at 200–400°.

Although both processes are based on C—C bond cleavage, it is generally agreed that C—H bond cleavage precedes C—C bond breaking [33]. For example, in hydrogenolysis of ethane [34], the first interaction of the alkane with platinum surface is C—H bond rupture, as depicted in Figure 9.3. In this sketch hypothetical

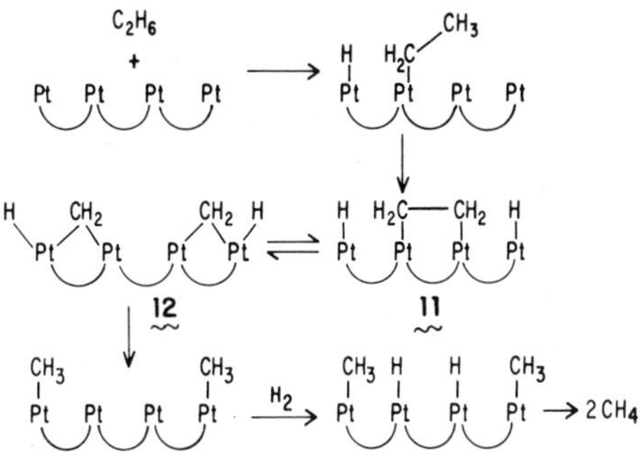

Figure 9.3 The interaction of ethane with a platinum surface and subsequent hydrogenolysis.

organometallic surface species are drawn in their simplest forms. (Practitioners of heterogeneous catalysis usually represent these species in less definite fashion, for example,* -CH_2CH_3. The lack of spectroscopic characterization of surface intermediates fully justifies this conservative approach. However, extrapolation from soluble organometallic complexes can provide useful working hypotheses about the mechanisms of heterogeneous processes.)

The cleavage of the first C—H bond in the figure can be viewed as an oxidative addition in which the H binds to one platinum and the C attaches to a neighbor. Quite possibly, both the H and the C_2H_5 ligands take bridging positions among two or three platinum atoms. Multicenter bonding of this sort may eliminate thermodynamic difficulties that would be associated with oxidative addition to a single metal atom [35].

A second oxidative addition of a C—H bond to an adjacent pair of platinum atoms forms the Pt—CH_2—CH_2—Pt species (**11**). This di-σ-bonded intermediate is isomeric with a simple ethylene complex. It can generate a pair of methylene ligands by C—C bond cleavage. This cleavage is the reverse of the intermolecular CH_2 ligand dimerization discussed at the end of Section 9.3. The methylene ligands in **12** are assigned bridging positions by analogy to a recently isolated methylene-diplatinum complex. Reaction of CH_2N_2 with a Pt—Pt bonded complex forms a stable methylene complex [36]:

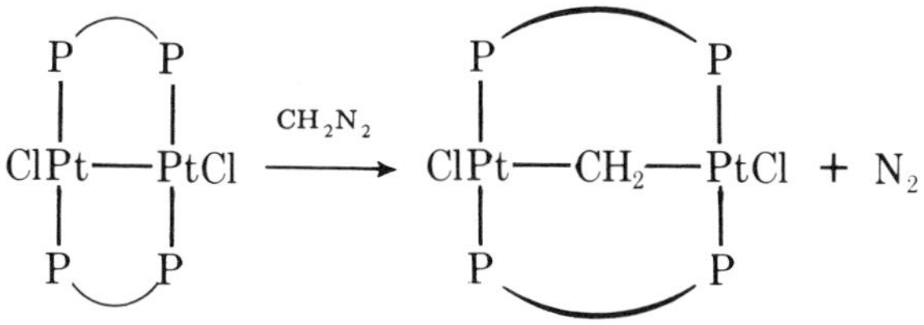

$$\mathrm{ClPt{-}PtCl}(P\frown P)_2 \xrightarrow{CH_2N_2} \mathrm{ClPt{-}CH_2{-}PtCl}(P\frown P)_2 + N_2$$

P⌒P = $Ph_2PCH_2PPh_2$

Once the C—C bond is broken the two CH_2 functions can be hydrogenated by hydrogen transfer from Pt atoms that bear hydride ligands. Hydrogen atoms seem

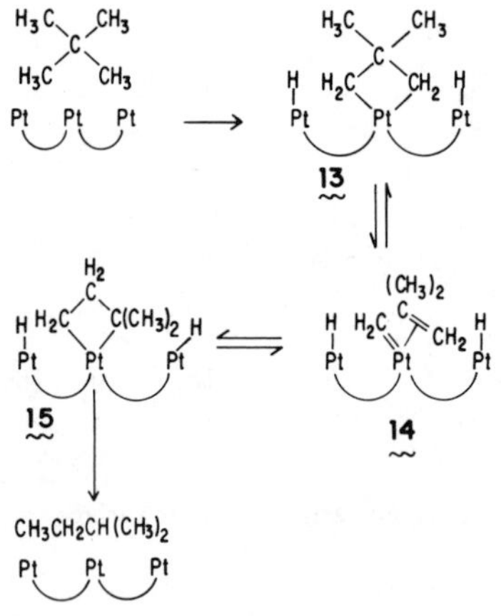

Figure 9.4 Isomerization of neopentane on a platinum surface.

to migrate freely across the metal surface and, to some extent, through the bulk metal. The overall result is hydrogenolysis of ethane to form methane.

Similar reactions can account for alkane isomerization in organometallic terms. The proposed mechanism in Figure 9.4 strongly resembles that currently accepted for olefin metathesis. Good organometallic precedents support the steps in this sequence. As in hydrogenolysis the first step is reaction of C—H bonds with the platinum surface to form Pt—H and Pt—C bonds. In this instance both Pt—C bonds are directed to the same Pt atom because the metallocycle (**13**) thus formed is remarkably stable. Pyrolysis of a bis(neopentyl)platinum complex generates a well-characterized soluble analog of the proposed surface species [37]:

$$L_2Pt(CH_2CMe_3)_2 \longrightarrow L_2Pt\!\!\left\langle\begin{matrix}CH_2\\ \\ CH_2\end{matrix}\right\rangle\!\!CMe_2 + CMe_4$$

16

$$L = R_3P$$

The soluble platinacyclobutane (**16**) is relatively stable, but its nickel analog ($L = Ph_3P$) decomposes to a mixture of isobutylene, ethylene, dimethylcyclopropane, and 2-methyl-2-butene [38]. The formation of isobutylene and ethylene indicates a fragmentation like **13** → **14** in the figure. (The ethylene arises by dimerization of CH_2 ligands.) More significantly, the 2-methylbutene probably arises by an isomerization sequence like that proposed for the surface reaction.

The mechanisms of Figures 9.3 and 9.4 are representations in organometallic terminology of conventional heterogeneous catalytic mechanisms. Although the mechanisms of the figures are oversimplified and speculative, such analogies from solution chemistry are probably valid to some extent. Until spectroscopic characterization of surface intermediates is possible, these analogies should be useful to stimulate thought about detailed surface processes.

REVIEWS ON OLEFIN METATHESIS

Scope and Mechanism

Rooney, J. J., and A. Stewart, *Catalysis,* **1,** 277 (1977).
Katz, T. J., *Adv. Organomet. Chem.,* **16,** 283 (1977).
Grubbs, R. H., *Prog. Inorg. Chem.,* **24,** 1 (1978).
Calderon, N., J. P. Lawrence, and E. A. Ofstead, *Adv. Organomet. Chem.,* **17,** 449 (1979).

Applications

Hughes, W. B., *Chem. Tech.,* 486 (1975).
Streck, R., *Chem. Ztg.,* **99,** 397 (1975).
Mol, J. C., *Chem. Weekbl.,* 467 (1978).

SPECIFIC REFERENCES

1. E. O. Fischer, *Angew. Chem.,* **86,** 651 (1974).
2. R. R. Schrock, *Accounts Chem. Res.,* **12,** 98 (1979).
3. R. R. Schrock and P. R. Sharp, *J. Am. Chem. Soc.,* **100,** 2389 (1978).
4. F. N. Tebbe, G. W. Parshall, and G. S. Reddy, *J. Am. Chem. Soc.,* **100,** 3611 (1978).
5. R. R. Schrock and G. W. Parshall, *Chem. Rev.,* **76,** 243 (1976).
6. J. L. Herisson and Y. Chauvin, *Makromol. Chem.,* **141,** 161 (1970).
7. N. J. Cooper and M. L. H. Green, *J.C.S. Chem. Comm.,* 761 (1974).
8. R. R. Schrock, *J. Organomet. Chem.,* **122,** 209 (1976).
9. V. Malatesta, K. U. Ingold, and R. R. Schrock, *J. Organomet. Chem.,* **152,** C53 (1978).
10. N. Calderon, E. A. Ofstead, J. P. Ward, W. A. Judy, and K. W. Scott, *J. Am. Chem. Soc.,* **90,** 4133 (1968).
11. E. L. Muetterties, *Inorg. Chem.,* **14,** 951 (1975).
12. F. N. Tebbe, G. W. Parshall, and D. W. Ovenall, *J. Am. Chem. Soc.,* **101,** 5074 (1979).
13. R. R. Schrock, *J. Am. Chem. Soc.,* **98,** 5399 (1976).
14. *Hydrocarbon Proc.,* **46,** 232 (Nov. 1967).
15. W. R. Knox, P. D. Montgomery, and R. N. Moore, Ger. Offenleg. 2,500,023 (1975).
16. D. L. Crain and R. E. Reusser, U.S. Patent 3,660,516 (1972).
17. P. A. Verbrugge and G. J. Heiszwolf, British Patent 1,416,317 (1975).
18. A. J. Berger, U.S. Patent 3,726,938 (1973).
19. H. S. Eleuterio, U.S. Patent 3,074,918 (1963).
20. T. J. Katz, S. J. Lee, and N. Acton, *Tetrahedron Lett.,* 4247 (1976).
21. N. Calderon, E. A. Ofstead, and W. A. Judy, *J. Polym. Sci.,* Pt. A-1, **5,** 2209 (1967); G. Natta, G. Dall-Asta, I. W. Bassi, and G. Carrella, *Makromol. Chem.,* **91,** 87 (1966).
22. P. B. van Dam, M. C. Mittelmeijer, and C. Boelhouwer, *J. Am. Oil Chem. Soc.,* **51,** 389 (1974).
23. R. Rossi, *Chim. Ind.* (*Milan*), **57,** 242 (1975).
24. F. W. Kupper and R. Streck, *Chem. Ztg.,* **99,** 464 (1975).
25. C. P. Casey, S. W. Polichnowski, H. E. Tuinstra, and A. J. Schusterman, National Meeting,

American Chemical Society, Miami Beach, FL (1978); *Chem. Eng. News,* 49 (Sept. 25, 1978).

26. J. W. Byrne, H. U. Blaser, and J. A. Osborn, *J. Am. Chem. Soc.,* **97,** 3871 (1975).
27. M. Ephritikhine, B. R. Francis, M. L. H. Green, R. E. Mackenzie, and M. J. Smith, *J. Chem. Soc. Dalton,* 1131 (1977).
28. A. E. Shilov and A. A. Shteinman, *Coord. Chem. Rev.,* **24,** 97 (1977); *Kinet. Katal.,* **18,** 1129 (1977).
29. G. W. Parshall, *Catalysis,* **1,** 335 (1977).
30. D. E. Webster, *Adv. Organomet. Chem.,* **15,** 147 (1977).
31. C. Kemball, *Adv. Catal.,* **11,** 223 (1959).
32. R. L. Moss, *Catalysis,* **1,** 37 (1977).
33. J. R. Anderson, *Adv. Catal.,* **23,** 1 (1973).
34. J. H. Sinfelt, *Catal. Rev.,* **3,** 175 (1969); *J. Catal.,* **27,** 468 (1972).
35. L. Abis, A. Sen, and J. Halpern, *J. Am. Chem. Soc.,* **100,** 2915 (1978).
36. M. P. Brown, J. R. Fisher, S. J. Franklin, R. J. Puddephatt, and K. R. Seddon, *J.C.S. Chem. Comm.,* 749 (1978).
37. P. Foley and G. M. Whitesides, *J. Am. Chem. Soc.,* **101,** 2732 (1979).
38. R. H. Grubbs, Proceedings of the 1st International Symposium on Homogeneous Catalysis, Corpus Christi, TX, Nov. 29–Dec. 1, 1978.

10 OXIDATION OF HYDROCARBONS BY OXYGEN

The largest scale application of homogeneous catalysis is the oxidation of hydrocarbons by molecular oxygen. Oxidation processes and their uses are listed in Table 10.1 in the order in which they are discussed in this chapter. The remarkable scale of processes like the oxidations of *p*-xylene and cyclohexane has been indicated in the summary table in Chapter 1.

The processes discussed in this chapter differ sharply from those of earlier chapters. Although soluble metal complexes enhance the yields and selectivities of these oxidations, the catalytic complex often does not interact with either O_2 or the hydrocarbon. Dozens of metal complexes of molecular oxygen have been isolated and characterized [1], but such complexes do not seem to be involved in the *major* oxidation pathways. Instead, the reaction between the hydrocarbon and oxygen is a radical chain process that operates even in the absence of metal catalysts. The usual product of the radical chain process is an alkyl hydroperoxide, ROOH. These hydroperoxides are unstable under reaction conditions and decompose to give alcohols, ketones, or carboxylic acids.

The major function of the metal complex in many oxidation processes is catalytic decomposition of hydroperoxides. In this way the metal enhances the formation of desirable products and stimulates the production of free radical species that initiate the radical chain reaction between the hydrocarbon and oxygen. These two effects give the chemist substantial control over the yield and the rate of the overall oxidation process.

In the following sections the general reactions of oxygen with metal complexes and with hydrocarbons are discussed first. The interaction between these two kinds of chemistry is illustrated in the oxidation of cyclohexane, an important hydrocarbon substrate. Finally, the oxidations of the major industrial hydrocarbons are considered individually.

TABLE 10.1 Industrial Oxidations of Hydrocarbons

Hydrocarbon	Oxidation Product	Applications
Cyclohexane	Cyclohexanol and cyclohexanone	Converted to adipic acid and caprolactam (polyamide precursors)
Cyclododecane	$C_{12}H_{23}OH$ and $C_{12}H_{22}O$	Oxidized to dodecanedioic acid and lauryl lactam (polyamide precursors)
Butane	Acetic acid	Solvent, vinyl acetate polymers
Toluene	Benzoic acid	Phenol synthesis (Section 7.3)
m-Xylene	Isophthalic acid	Polymers and plasticizers
p-Xylene	Terephthalic acid Terephthalate esters	Polyester fibers, films and plastics (Section 11.1)

10.1 REACTIONS OF O_2 WITH METAL COMPLEXES

In contrast to many other simple diatomic molecules such as N_2 and F_2, dioxygen (O_2) is paramagnetic. It has two unpaired electrons in the ground state. The highest occupied molecular orbitals are a pair of π^* orbitals of equal energy so that the two highest energy electrons have no reason to spin-pair [1]. Consequently, dioxygen may be regarded as a diradical.

The formulation as a diradical is very useful in understanding the chemistry of O_2. Most of its reactions proceed in one-electron steps:

$$O_2 + e^- \rightleftharpoons O_2^{\dot{-}} \xrightarrow{e^-} O_2^{2-}$$

In fact, almost all the reactions discussed in this chapter proceed by one-electron transfer processes. The most effective catalytic metal ions have two stable oxidation states related by transfer of one electron. The most important are cobalt, manganese, and copper, which undergo the following redox reactions easily [2]:

$$Co^{3+} + e^- \rightleftharpoons Co^{2+} \qquad E_0 = 1.8$$

$$Mn^{3+} + e^- \rightleftharpoons Mn^{2+} \qquad E_0 = 1.5$$

$$Cu^{2+} + e^- \rightleftharpoons Cu^+ \qquad E_0 = 0.17$$

These three metals vary considerably in their redox properties. As indicated by the oxidation-reduction potentials, copper(I) ion is easily oxidized by O_2($E_0 \approx 1.2$ at pH = O) in aqueous solution. Cobalt(II) and manganese(II) salts are not oxidized by O_2 under these conditions. However, the redox potentials change substantially with changes in solvent and changes in the ligands bound to the metal ion. When six NH_3 or CN^- ligands are attached to cobalt, the 3+ oxidation state becomes more stable than the 2+ state.

Cobalt(II) salts are the most widely used soluble catalysts for hydrocarbon oxidation reactions. However, the literature reveals that they are often used in

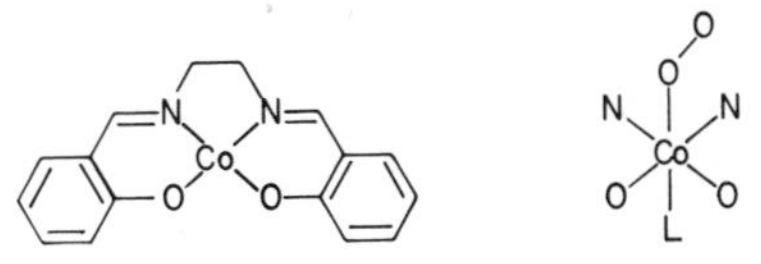

Figure 10.1 The chelate structure of cobalt salen (*left*) and the ligand arrangement about the metal in an O_2 complex. L is an N-donor ligand such as an amine.

combination with Mn, Cu, or Cr salts in the industrial processes discussed in this chapter [2]. Typically, the cobalt salt is a carboxylate. Cobalt(II) acetate is often used for oxidations in acetic acid solution. For reactions in hydrocarbon solvents long-chain carboxylic acid salts are used to gain solubility. Two common varieties are salts of naphthenic acids and those of 2-ethylhexanoic acid ("octoates"). The cobalt(II) carboxylates have complex structures with four to six oxygen atoms arrayed about the metal atom.

Cobalt(II) salts react with O_2 to form labile O_2 complexes when certain N-containing chelate ligands are present [3]. The so-called "cobalt salen" (Figure 10.1) gives a relatively stable O_2 complex in the solid state, but the formation of the O_2 complex is reversible on heating. Complexes of the salen family were used in oxygen storage devices during World War II [4]. The ligand array about cobalt in the complex is essentially square planar. When O_2 coordinates, it occupies an axial position in an octahedral structure. Complexation of O_2 is assisted by coordination of an amine to the other axial position, just as in heme complexes.

The formation of the O_2 complex usually involves some transfer of electron density from the metal to the O_2 ligand. In other words the cobalt is partially oxidized:

$$Co^{2+} + O_2 \rightleftharpoons Co^{3+} + O_2^{\cdot -}$$

This partial oxidation is facilitated by N-donor ligands that compensate for electron withdrawal by O_2. The cobalt carboxylate complexes used as hydrocarbon oxidation catalysts do not react detectably with dioxygen in water, but O_2 complexes form in primary amine solutions [5]. However, O_2 complex formation may not be necessary for catalysis of hydrocarbon oxdiation. The electronic factors that favor interaction with the ·O-O· molecule also favor reaction with the R-O-O· radicals present in hydrocarbon oxidation mixtures.

Two reactions of alkyl hydroperoxides with cobalt ions seem especially important in catalysis [6]. One involves oxidation of cobalt(II) via formation of a complex in which the hydroperoxide becomes a ligand on the cobalt ion:

$$Co^{II} + ROOH \rightarrow [Co(HOOR)] \rightarrow Co^{III}OH + RO\cdot$$

Electron transfer from cobalt to oxygen occurs after complexation. Weakening the O—O bond facilitates breakdown of the complex and formation of the energetic alkoxy radical. The other reaction reduces cobalt(III) with formation of the more stable alkylperoxy radical.

$$Co^{3+} + ROOH \rightarrow Co^{2+} + ROO\cdot + H^+$$

The oxidation potentials of these reactions seem to be balanced so that the two processes occur simultaneously in solution. The rapid shuttling of cobalt between the two oxidation states catalyzes breakdown of ROOH into radicals that initiate hydrocarbon oxidation. The radicals themselves give oxygenated products.

The weakening of the O—O bond in ROO· and ROOH on coordination to cobalt(II) parallels an effect seen with molecular oxygen. Electrons added to the O_2 molecule populate antibonding (π^*) orbitals and weaken the O—O bond. This effect is evident in both the O—O bond length and its dissociation energy [7]:

$$O_2 \xrightarrow{e^-} O_2^{\dot{-}} \xrightarrow{e^-} O_2^{2-}$$

	O_2	$O_2^{\dot{-}}$	O_2^{2-}
Bond energy	118	88	46 kcal/mole
Length	1.21	1.26	1.49 Å

One might expect the bond length and energy of ROO· to correspond to $O_2^{\dot{-}}$. Complexation to cobalt(II) with transfer of an electron from Co to an antibonding molecular orbital largely centered in the O-O region should weaken the bond as observed.

The coordination of ROO· to cobalt probably resembles the "end-on" coordination of O_2 shown in Figure 10.1. In this respect it also resembles O_2 coordination to iron in heme derivatives. In contrast to cobalt and iron, nickel and the platinum metals form "side-on" O_2 complexes [1,8]. A typical example is "Vaska's complex," which is a reversible oxygen carrier. In solution at room temperature it absorbs O_2, but the reaction is reversed when the solution is heated:

$$trans\text{-}IrCl(CO)(PPh_3)_2 + O_2 \rightleftharpoons IrCl(CO)(PPh_3)_2(\eta^2\text{-}O_2)$$

Some of these O_2-binding complexes, especially $RhCl(PPh_3)_3$ and $Pt(PPh_3)_3$, catalyze oxidation of phosphines and also oxidize CO, SO_2, and isonitriles. However, despite much study, only modest catalytic activity in hydrocarbon oxidation is observed.

10.2 REACTION OF O_2 WITH HYDROCARBONS

Most aliphatic hydrocarbons react with oxygen, but the reactions are extremely slow unless a free radical initiator is present. The primary product of alkane oxidation is the hydroperoxide, ROOH, which can be a source of radicals by O—O fission. Hence, alkane oxidation by O_2 is autocatalytic, but induction periods are very long. To achieve oxidation in a practical time period, initiators such as benzoyl peroxide

$$(PhC(=O)\text{—}O\text{—}O\text{—}C(=O)Ph)$$

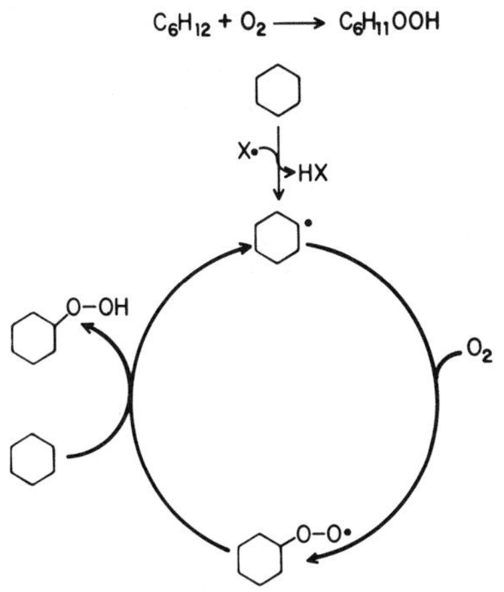

$$C_6H_{12} + O_2 \longrightarrow C_6H_{11}OOH$$

Figure 10.2 Cycle for the oxidation of cyclohexane to cyclohexyl hydroperoxide.

may be added. Alternatively, metal ions such as cobalt(II) may be used to catalyze decomposition of ROOH and accelerate the normal autoxidation process. The latter approach is preferred in practice and is the basis for most industrial processes.

Free radical attack on an aliphatic C—H bond precedes interaction with oxygen in the oxidation process. The attacking reagent X· abstracts hydrogen to form a new radical:

$$\text{R—H} + \text{X·} \rightleftharpoons \text{R·} + \text{HX}$$

This hydrogen abstraction process can be fairly selective with radical reagents such as $Cl_3C·$ and ROO·, which are moderately stable and long lived in solution. With these low-energy radicals, tertiary C—H bonds are attacked in preference to secondary, which, in turn, are more susceptible to attack than primary. Allylic and benzylic C—H bonds are especially vulnerable. Energetic radicals such as RO· are rather indiscriminate in abstraction of hydrogen from carbon, especially at high temperatures.

An alkyl radical, once formed, combines readily with oxygen to give an alkylperoxy radical:

$$\text{R·} + \text{·O—O·} \rightarrow \text{R—O—O·}$$

Since the process generates a species that can abstract aliphatic hydrogen, the radical initiator has started a cyclic oxidation of the hydrocarbon. The cyclic effect is illustrated in Figure 10.2 by the oxidation of cyclohexane. This hydrocarbon is a convenient example because all the C—H bonds are equivalent. In oxidation of linear alkanes (Section 10.5) discrimination among primary and secondary C—H bonds becomes important.

The attack of a radical X· on cyclohexane initiates the oxidation process by abstracting a hydrogen atom. The cyclohexyl radical formed in this way combines with O_2 to form a cyclohexylperoxy radical. When the latter encounters a cyclohexane molecule, it abstracts hydrogen from a C—H bond. The transfer of H from

C to O produces the primary product, cyclohexyl hydroperoxide, and regenerates a cyclohexyl radical to start another reaction cycle.

Cyclohexyl hydroperoxide, the primary product of this oxidation, has only moderate stability. Solutions can be stored for long periods at low temperatures, but decomposition is rapid when the solutions are heated above 100°. Concentrated solutions of hydroperoxides can be treacherously explosive. As a result the hydroperoxide is seldom isolated. Most commonly, the oxidation is done in the presence of a metal ion that decomposes the peroxide almost as rapidly as it is formed. This method of operation eliminates the hazard of uncontrolled peroxide decomposition and provides a source of radicals to maintain the reaction cycle.

The decomposition of an alkyl hydroperoxide can follow many different pathways that include cleavage of O—O, O—H, C—C, and C—H bonds. Some are catalyzed by metal ions, others are simple radical chain processes, and a third group are radical chain sequences *initiated* by a metal ion [2]. The relative rates of these reactions are controlled by the nature of the metal ion present. Because of the great variety of ROOH reactions those of major interest are discussed separately in the following sections. The oxidation of cyclohexane is considered first because it is relatively simple and reasonably well understood.

10.3 ADIPIC ACID SYNTHESIS

Adipic acid, a major intermediate in the production of nylon (Chapter 11), is made by oxidation of cyclohexanone and cyclohexanol [9]. These intermediates, in turn, come from air oxidation of cyclohexane or hydrogenation of phenol (Figure 10.3). They are also intermediates in the production of caprolactam, which can be polymerized to form 6-nylon directly. The two stages in oxidation of cyclohexane to adipic acid are discussed separately here.

Cyclohexane Oxidation

The commercial practice of this reaction is strongly influenced by its chemical peculiarities [9–12]. Grossly, cyclohexane is converted to cyclohexyl hydroperoxide (**1**) by the mechanism of Figure 10.2, and the hydroperoxide is decomposed to the observed products:

$$C_6H_{12} \longrightarrow \underset{\mathbf{1}}{C_6H_{11}OOH} \longrightarrow C_6H_{11}OH + C_6H_{10}{=}O$$

Oxidative attack on the C—H bonds of cyclohexane is slow and requires vigorous reaction conditions. In contrast the hydroperoxide, alcohol, and ketone are easily oxidized. As a result the reaction is generally run with low conversions of cyclohexane to avoid degradation of the desired products.

A typical industrial oxidation may be performed by reacting air with a cyclo-

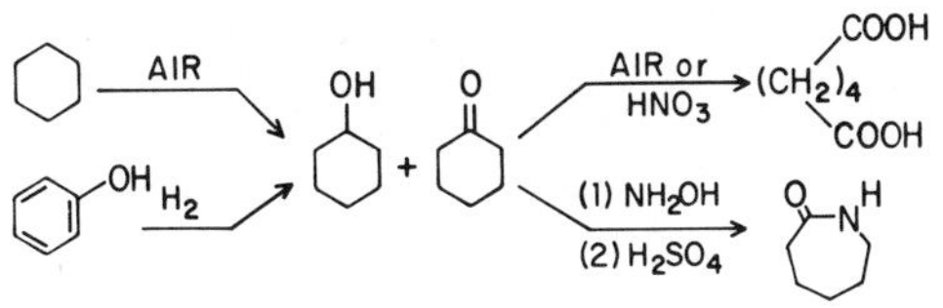

Figure 10.3 Production of nylon intermediates via cyclohexanol and cyclohexanone.

hexane solution of a soluble cobalt(II) salt at about 140–165° and 10 atmospheres' pressure [11]. The reaction takes place in continuous fashion. The residence time in the reactor is limited to produce about 10% conversion of the cyclohexane. Liquid reaction mixture is withdrawn continuously and is distilled. Cyclohexane is recycled to the oxidation reactor. Cyclohexanol and cyclohexanone are sent to another oxidation unit for conversion to adipic acid as described later or to caprolactam (Figure 10.3; ref. 11). Combined yields of alcohol and ketone are 60–70% if conversion is limited to 6–9% [9]. Higher yields can be attained at 10–12% conversion if boric acid is added to stabilize the cyclohexanol as it is formed. However, this mode of operation requires added investment and operating costs for boric acid recycle (cf. Section 10.4).

The catalyst in this process is a hydrocarbon-soluble cobalt(II) carboxylate such as the naphthenate or 2-ethylhexanoate. Other metal ions such as manganese(II) or chromium(III) are frequently used in addition to cobalt to control product distribution. The metal ions probably have no *direct* part in the conversion of cyclohexane to cyclohexyl hydroperoxide because this oxidation is a simple radical chain process (Figure 10.2). However, the ions have a controlling role in the conversion of the hydroperoxide to cyclohexanol and cyclohexanone. In addition, since the metal-catalyzed hydroperoxide reactions supply the free radicals necessary to initiate and maintain the cycle of Figure 10.2, the metal ion concentration provides some control over the overall reaction rate. The kinetic effect is not simple, however, since the metal ions become inhibitors of the radical chain processes when present in high concentrations [13].

Cobalt and manganese ion catalysts produce both cyclohexanol and cyclohexanone and seem to function by similar mechanisms [6]. One of the major catalytic cycles for hydroperoxide decomposition appears in Figure 10.4. Reaction of Co^{2+} or Mn^{2+} with the hydroperoxide produces a complex of M^{3+} and OH^-, along with the alkoxy radical, as discussed in Section 10.2. The highly energetic cyclohexyloxy radical can abstract a hydrogen atom from any organic compound in the system to form cyclohexanol. The Co^{3+} complex reacts with CyOOH to form CyOO· and complete the catalyst cycle. The peroxy radical can give cyclohexanol and cyclohexanone in a bimolecular reaction:

$$2\,\mathrm{CyOO\cdot} \longrightarrow \mathrm{O_2} + \mathrm{CyOH} + \text{(cyclohexanone)}{=}\mathrm{O}$$

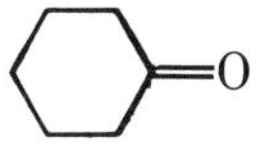

The probability of this bimolecular reaction is enhanced by the relative stability of the CyOO· radical and by coordination to manganese or cobalt [6].

Chromium(III) ions, in contrast to Co and Mn, catalyze decomposition of

$$2CyOOH \longrightarrow CyO\cdot + CyOO\cdot + H_2O$$

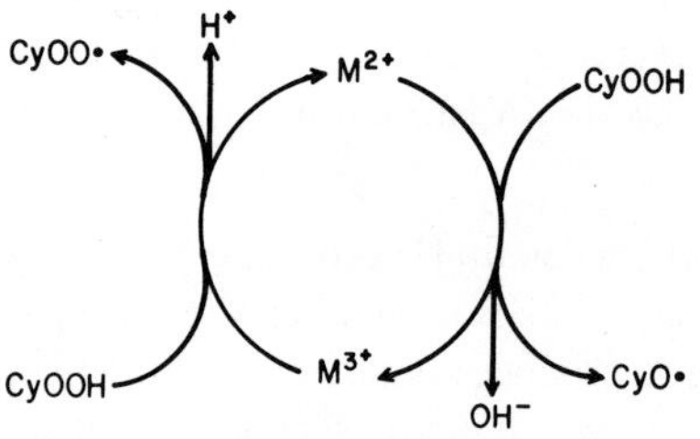

Figure 10.4 Catalytic cycle for decomposition of cyclohexyl hydroperoxide (CyOOH) by Co or Mn ions.

CyOOH to form cyclohexanone as the major product [14]. This reaction is superficially just a dehydration of the hydroperoxide:

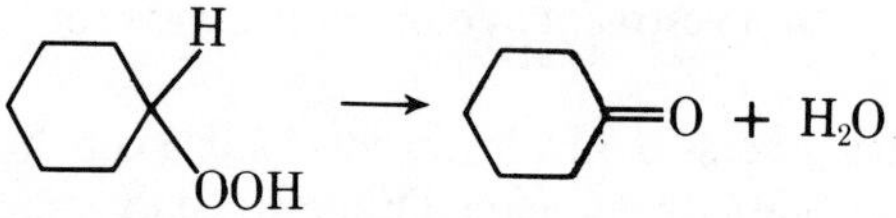

The reaction probably occurs by a nonradical mechanism in which the metal ion assists cleavage of the O—O bond.

In addition to these reactions, which lie on the major pathway from cyclohexane to cyclohexanol and cyclohexanone, literally dozens of other reactions occur in the reaction mixture. Some involve C—C bond cleavage to form lactones and C_4–C_6 dicarboxylic acids. The α-CH_2 groups in cyclohexanone are especially susceptible to radical attack. Recent labeling studies implicate an α-oxy radical (**2** in Figure 10.5) as a precursor to the dicarboxylic acids [15].

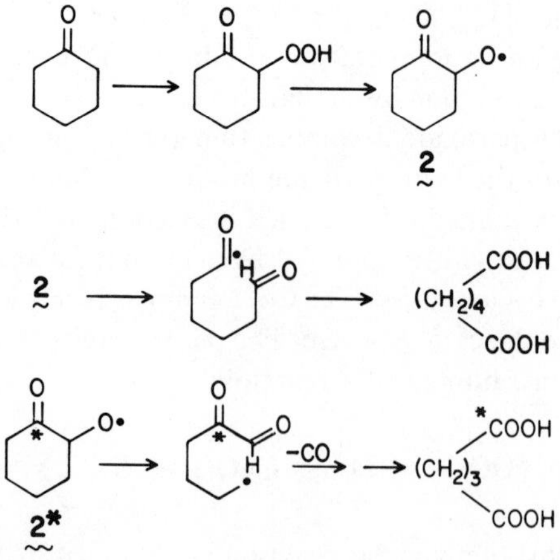

Figure 10.5 Formation of adipic and glutaric acids from cyclohexanone.

A direct oxidation of cyclohexane to adipic acid would be attractive economically [16], but yields in such processes are usually low. Present commercial operation is largely based on a separate oxidation of the alcohol-ketone mixture to adipic acid.

Oxidation of Cyclohexanol and Cyclohexanone

Although the cyclohexanol-cyclohexanone mixture from cyclohexane oxidation can be converted to adipic acid by a cobalt-catalyzed air oxidation, present commercial operations generally use nitric acid as the oxidant [9]. Even with nitric acid, however, air is the ultimate oxidant because the nitrogen oxide byproducts are recycled:

$$\text{cyclohexanone} + \text{cyclohexanol} + HNO_3 \longrightarrow NO + NO_2 + HOOC(CH_2)_4COOH$$

(NO + NO_2 → air → HNO_3)

The mixture of cyclohexanol and cyclohexanone is fed continuously to a solution of $Cu(NO_3)_2$ and NH_4VO_3 in 45–50% nitric acid at 70–80° [17]. The oxidation is complete in a few minutes. The gaseous products, mainly nitrogen oxides, are recycled to a nitric acid synthesis unit. (Some nitric acid is lost to products such as N_2 and N_2O, which are not reoxidized to HNO_3.) The hot acid solution, which contains the organic products, is cooled to crystallize the desired adipic acid. Yields of pure adipic acid typically exceed 90%.

The chemistry in the nitric acid oxidation is very complex. Nearly all the cyclohexanol is oxidized to cyclohexanone in a noncatalytic reaction that is initiated by traces of HNO_2 [18]. The cyclohexanone forms adipic acid by two major pathways (Figure 10.6). Both begin with conversion of the ketone to its 2-nitroso derivative (**3**). The mechanism of this transformation is not well established, but analogy to other oxidations of cyclohexanone suggests that the ketone enolizes in an acid-catalyzed reaction. Loss of a hydrogen atom (or an electron and a proton sequentially) gives a delocalized radical that combines with NO to form 2-nitrosocyclohexanone (**3**).

The nitroso ketone (**3**) forms adipic acid by a noncatalytic path and by a vanadium-catalyzed sequence [19]. Reaction of **3** with HNO_2 gives a nitro derivative (**4**), which hydrolyzes to the "nitrolic acid" (**5**) and to adipic acid. In the catalytic sequence tautomerization of **3** gives an oxime (**6**) that hydrolyzes to 1,2-cyclohexanedione (**7**). This diketone is stoichiometrically oxidized by two VO_2^+ ions to form adipic acid. The vanadium(IV) coproduct is reoxidized easily by nitric acid. Hence, this pathway is catalytic in vanadium usage.

The role of $Cu(NO_3)_2$ in the oxidation is less clear. This salt seems to increase the overall yield of adipic acid by suppressing a side reaction that produces glutaric acid [19].

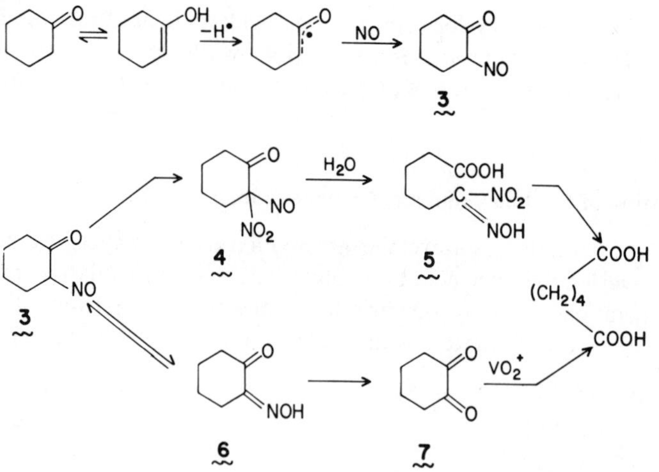

Figure 10.6 Oxidation of cyclohexanone to adipic acid in nitric acid solution.

10.4 OXIDATION OF CYCLODODECANE

The major industrial application of the butadiene oligomerization discussed in Section 4.5 is synthesis of 1,5,9-cyclododecatriene. This compound is converted to dodecanedioic acid [20] and lauryl lactam [21] by the chemistry of Figure 10.7. These two products are intermediates in production of polyamides for several specialty applications. The syntheses of both the lactam and the dicarboxylic acid begin with hydrogenation of the triene (**8**) to cyclododecane (**9**) over a heterogeneous catalyst. The cyclododecane is oxidized with air to a mixture of cyclododecanol and cyclododecanone as described here. This mixture is oxidized to dodecanedioic acid by a nitric acid oxidation like that used to make adipic acid. As in adipic acid synthesis the nitric acid oxidation can be catalyzed by $Cu(NO_3)_2$ and NH_4VO_3 [22].

The air oxidation of cyclododecane to a mixture of alcohol and ketone is even more complex than the comparable oxidation of cyclohexane [20]. Almost certainly, the initial oxidation of the hydrocarbon (**9**) forms cyclododecyl hydroperoxide (**10**) by a mechanism like that of Figure 10.2. The cobalt-catalyzed decomposition of

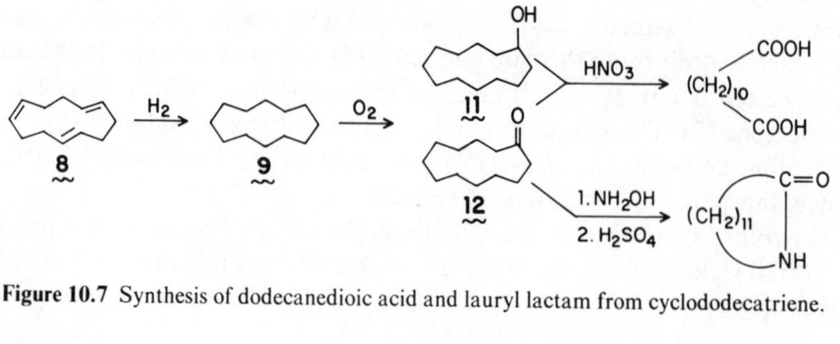

Figure 10.7 Synthesis of dodecanedioic acid and lauryl lactam from cyclododecatriene.

10 yields cyclododecanol (**11**) by a cyclic process analogous to that of Figure 10.4.

$$\underset{\mathbf{9}}{C_{12}H_{24}} \xrightarrow{O_2} \underset{\mathbf{10}}{C_{12}H_{23}OOH} \xrightarrow{Co^{II}} \underset{\mathbf{11}}{C_{12}H_{23}OH}$$

As in the cyclohexane oxidation the alcohol **11** is oxidized by air to give cyclododecanone (**12**) and many undesirable byproducts. Formation of the byproducts can be minimized by removing **11** from the reaction mixture. This removal can be achieved either by low-conversion operation or by complexation with boric acid. If the reaction liquids are separated and distilled after about 23% conversion of **9**, a 90% yield of useful products is obtained. This low-conversion operation is complicated by the high boiling points of the products and the fact that **9** is a solid (m.p. 60°). Because of these problems the oxidation is often done commercially with addition of boric acid.

In a process described by Chemische Werke Hüls [23], moist *meta*-boric acid is added continuously to the reaction mixture to esterify cyclododecanol as it forms. Cyclododecane that contains a trace of a cobalt(II) carboxylate is oxidized with air at 160–180° and 1–3 atmospheres pressure. As the hydrocarbon is converted to the alcohol, the latter reacts with the boric acid to form a borate ester:

$$C_{12}H_{23}OH + H_3BO_3 \rightarrow [C_{12}H_{23}OBO]_3 + H_2O$$

This product is fairly resistant to further oxidation and is easily separated from the reaction mixture. Hydrolysis of the borate ester gives free cyclododecanol. The aqueous boric acid solution is concentrated and recycled to the oxidation reactor. Cyclododecanol and cyclododecanone (5:1) are formed in 80–82% yield at 33% conversion.

10.5 ACETIC ACID SYNTHESIS

Acetic acid is a major intermediate in the organic chemical industry [24,25]. It is used extensively as a solvent in xylene oxidation (Section 10.6) and as a starting material for synthesis of vinyl acetate, which is used in making polyvinyl acetate and polyvinyl alcohol. Acetic anhydride is an acetylating agent for production of cellulose acetate and aspirin.

Whereas future plants for manufacture of acetic acid may be based on carbonylation of methanol (Section 5.3), most synthetic acetic acid is now made by oxidation processes. Oxidation of butane and other aliphatics is the largest scale process. Another major process is the oxidation of acetaldehyde produced from ethylene by the Wacker process (Section 6.1).

The butane and acetaldehyde oxidations are discussed separately here. Both processes employ cobalt(II) and other transition metal salts as catalysts, but the detailed chemistry is substantially different.

Butane Oxidation

In the United States butane from natural gas is relatively cheap and abundant. It has become a major feedstock for acetic acid production by oxidation according to the theoretical equation:

$$C_4H_{10} + 5/2\ O_2 \rightarrow 2CH_3COOH + H_2O$$

In Europe, where natural gas is less abundant, light naphtha fractions from petroleum refining are used. With both feedstocks, yields of acetic acid are only moderate (30–60%), although the yield from butane is higher than that from naphtha. The yield is strongly dependent on operating conditions, especially on conversion of starting material.

The oxidation of alkanes to acetic acid is more complex than the cyclohexane oxidation discussed earlier because it involves C—C bond cleavage in addition to C—H oxygenation. This complexity is especially apparent in the oxidation of petroleum naphtha, in which a mixture of aliphatic hydrocarbons is degraded to C_2 fragments. The success of these oxidations reflects the relative stability of acetic acid toward radical processes: It is the most stable product apart from CO_2 and tends to accumulate under a variety of reaction conditions. This same resistance to oxidation also commends acetic acid as a solvent for oxidation reactions.

Typically, the oxidation is accomplished by sparging air through an acetic acid solution of butane and the catalyst at 160–200° and 60–80 atmospheres pressure [26]. A portion of the reaction mixture is continually withdrawn and distilled to recover unconverted hydrocarbon. The products—acetic, propionic, and butyric acids and some 2-butanone—are separated, and the residual catalyst solution is recycled to the reactor. An acetic acid yield of 45% at 30% conversion may be typical. Economics depend heavily on sale of the other products.

Recent patents emphasize a trend to less severe reaction conditions that give higher yields and conversions. The high reaction temperatures used ordinarily are necessary to sustain a radical chain reaction because butane is relatively inert. One approach to milder conditions involves continuous recycle of 2-butanone byproduct to the oxidation reactor [27]. The ketone is very susceptible to radical attack at the α-methylene group and sustains the radical chain sequence under mild conditions (110–130°). The 2-butanone also contributes to the acetic acid yield [28].

$$C_2H_5COCH_3 + 3/2\ O_2 \rightarrow 2CH_3COOH$$

The overall yield of acetic acid from butane and 2-butanone is about 75% at 85% conversion.

At least part of the beneficial effect of 2-butanone addition is assignable to an increase in the level of cobalt(III) ions in the reaction mixture [28]. (Oxidation of cobalt(II) to cobalt(III) appears to be the slow step in the process.) The oxidation of the ketone proceeds via an α-hydroperoxide,

$$CH_3C(=O)—CH(OOH)CH_3 \qquad [29]$$

which can oxidize cobalt(II) by the mechanism like that of Figure 10.4. Another approach to operation under mild conditions is to preoxidize part of the cobalt(II) salt with ozone [30]. This treatment eliminates the induction period in batch oxidations of butane and permits continuous operation at 110–130° and 35 atmospheres pressure. The oxidation converts cobalt(II) acetate to a μ_3-oxo bridged trimer $Co_3O(OAc)_6(AcOH)_3$, which is a good initiator for alkane oxidations [31].

The use of "promoters" such as ketones, cobalt(III), or bromide ion point to the importance of the initiation step in the overall mechanism of oxidation. In the batch oxidation of butane or in the startup of a continuous process it is necessary to supply a free radical initiator that is sufficiently energetic to abstract a hydrogen atom from secondary C—H bonds. The RO· and ROO· radicals fulfill this requirement and are generally present in substantial quantities in oxidation mixtures at 150°. There is some question about whether or not cobalt(III) ions attack alkanes or cycloalkanes directly [2].

Once initiation has occurred a radical chain reaction like that of Figure 10.2 converts butane to 2-butyl hydroperoxide [29]:

$$n\text{-}C_4H_{10} + O_2 \longrightarrow CH_3\underset{\displaystyle OOH}{\underset{|}{C}}HC_2H_5$$

The selectivity for attack at the secondary carbons is moderate, but substantial amounts of propionic and butyric acids arise from attack at the terminal carbons of butane.

The 2-butyl hydroperoxide that forms initially yields the ultimate oxidation products in many different ways. Simple decomposition in the absence of catalyst is reported to give largely C_4 products [6]:

$$C_2H_5\underset{\displaystyle OOH}{\underset{|}{C}}HCH_3 \longrightarrow C_2H_5\underset{\displaystyle O}{\underset{\|}{C}}CH_3 + C_2H_5\underset{\displaystyle OH}{\underset{|}{C}}HCH_3 + O_2 + H_2O$$

In practice, however, catalytic decomposition by cobalt salts gives $C_4H_9O\cdot$ and $C_4H_9OO\cdot$ radicals that sustain the cyclic oxidation. They replace radicals lost by chain termination processes such as radical dimerization and disproportionation. In addition they provide mechanisms for fragmentation of the C_4 chain into the desired C_2 products, as well as the C_1 and C_3 byproducts always observed. A β-cleavage of the 2-butoxy radical provides one pathway for C—C fission:

$$CH_3\underset{\displaystyle OOH}{\underset{|}{C}}HC_2H_5 \xrightarrow{Co^{2+}} CH_3\underset{\displaystyle O\cdot}{\underset{|}{C}}HC_2H_5 \longrightarrow CH_3CHO + C_2H_5\cdot$$

The CH_3CHO intermediate is easily oxidized to acetic acid, as discussed in the next section.

Another C—C cleavage mechanism probably involves intramolecular reactions of the hydroperoxy radical [6]:

$$\underset{\displaystyle \mathrm{OOH}}{\mathrm{CH_3CHCH_2CH_3}} \xrightarrow{Co^{3+}} \mathrm{CH_3\underset{|}{C}H{-}\underset{\vdots}{C}H_2CH_3} \ (\text{O}{-}\text{O}\cdot) \longrightarrow \mathrm{CH_3CHO + C_2H_5O\cdot}$$

An even more important path probably involves 2-butanone, a product of both thermal and catalytic decomposition of 2-butyl hydroperoxide. The α-methylene C—H bonds of the ketone are especially susceptible to oxidation and form a peroxy radical that produces acetic acid precursors by C—C fission:

$$\mathrm{CH_3\overset{}{\underset{\|}{C}}(O){-}\underset{|}{C}H(OO\cdot)CH_3} \longrightarrow \mathrm{CH_3C(=O)\cdot + CH_3CO_2H}$$

The overall process gives moderate yields of acetic acid. These yields contrast sharply with the high yields of acetic acid obtained by acetaldehyde oxidation, as discussed next.

Acetaldehyde Oxidation

The invention of the Wacker process for production of acetaldehyde from ethylene (Section 6.1) provided an attractive industrial route to acetic acid. An especially desirable aspect was that the technology for oxidation of acetaldehyde was well developed. Grain-based ethanol and acetylene-based acetaldehyde had been major feedstocks for acetic acid production. Both starting materials were oxidized by air with cobalt(II) acetate as the catalyst [32].

Acetaldehyde is much easier to oxidize than butane is and gives much higher yields. The oxidation is typically run at 66° and 1 atmosphere pressure, in contrast to the 150–225° and 60 atmospheres used in butane oxidation [25]. Acetaldehyde diluted with about 20% acetic acid is fed to the reactor along with N_2-diluted oxygen. Cobalt(II) and manganese(II) acetates are used as catalysts while salts of nickel [33], chromium [32], or copper [34] are often added to control the product distribution. With nickel addition the yield of acetic acid exceeds 90% at 92–97% conversion of acetaldehyde. Acetic anhydride can be made the major product when a mixture of cobalt and copper salts is used as the catalyst [35]. Rapid separation of the products by flash evaporation avoids hydrolysis of the anhydride.

The oxidation of acetaldehyde resembles alkane oxidation in that a hydroperoxide is a major intermediate. In this instance the hydroperoxide is peracetic acid:

$$\mathrm{CH_3C(=O)H + O_2 \longrightarrow CH_3C(=O)OOH}$$

The oxidation can be done in such a way as to make peracetic acid a major product [35]. The acetaldehyde is partially oxidized at low temperatures to form peracetic acid as an acetaldehyde addition product:

$$CH_3C(=O)OOH + O{=}C(H)CH_3 \rightleftharpoons CH_3C(=O)\text{—}O\text{—}O\text{—}CH(OH)CH_3$$

The adduct is decomposed thermally to form a 30% solution of peracetic acid in acetic acid. Such solutions are used extensively as epoxidation reagents in pharmaceutical synthesis. They are also used in production of glycerol from allyl alcohol.

The initial oxidation of acetaldehyde to peracetic acid occurs by a radical chain mechanism (Figure 10.8) similar to that of alkane oxidation. The aldehyde C—H is readily attacked by RO·, ROO· or Co(III) to form an acetyl radical that can initiate the cyclic oxidation process. A dioxygen adduct of a cobalt porphyrin complex can also abstract the aldehyde hydrogen even at 10° [36]. The acetyl radical combines with O_2 to give the acetylperoxy radical. This species selectively abstracts aldehyde hydrogen to form peracetic acid and an acetyl radical that repeats the cycle. The yield of peracetic acid is almost quantitative at low temperature and low conversion. As noted earlier, the peracetic acid is present largely as its acetaldehyde adduct.

The metal ion catalysts affect the ultimate product distribution by selective decomposition of peracetic acid and its aldehyde adduct [35,36]. Cobalt and manganese ions appear to form acetic acid by a redox cycle somewhat analogous to that of Figure 10.4. One major segment is:

$$\text{AcOOH} + M^{2+} \rightarrow \text{AcO}\cdot \xrightarrow{\text{AcH}} \text{AcOH} + \text{Ac}\cdot$$
$$+ M^{3+} + OH^-$$

The reduction of cobalt(III) or manganese(III) ions in the second part of the cycle probably occurs by reaction with the aldehyde rather than with peracetic acid. When copper(II) ion is present, it can divert some of the acetyl radical into anhy-

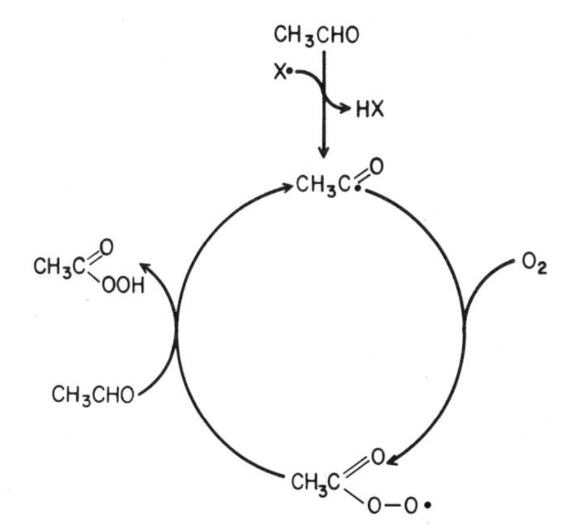

Figure 10.8 Cycle for oxidation of acetaldehyde to peracetic acid.

dride formation:

$$Ac\cdot + Cu^{2+} \rightarrow Ac^+ \xrightarrow{AcOH} Ac{-}O{-}Ac + H^+ + Cu^+$$

It does not seem to bring about decomposition of peracetic acid solutions at 30° in the absence of cobalt or manganese [35]. Hence, a combination of the metal ions is synergistic for acetic anhydride production.

Technology similar to that for acetaldehyde oxidation is used to produce butyric acid, 2-ethylhexanoic acid, and some fatty acids. In each case an aldehyde obtained by hydroformylation (Chapter 5) is oxidized to a carboxylic acid. The mechanisms of these oxidations closely resemble that described earlier [37].

10.6 OXIDATION OF METHYLBENZENES

The largest scale homogeneous catalytic oxidations are those of *p*-xylene to terephthalic acid and its esters. On a smaller but still substantial scale *m*-xylene is oxidized to isophthalic acid, and toluene is oxidized to benzoic acid. Benzoic acid is primarily an intermediate in production of other compounds such as phenol (Section 7.3) and terephthalic acid. Some of the latter is produced by a disproportionation of potassium benzoate catalyzed by carbonate salts [38]:

$$2\, C_6H_5COOK \xrightarrow[K_2CO_3]{CO_2} C_6H_6 + KOOC{-}C_6H_4{-}COOK$$

The oxidation of methylbenzenes resembles the oxidation of alkanes and cycloalkanes in some respects but is generally much easier [2]. Benzylic C—H bonds are more susceptible to free radical attack than alkyl C—H bonds are. Even mildly energetic radicals such as bromine atoms can attack benzylic C—H bonds directly. These reactions probably initiate the complex sequence by which a methyl group is converted to a carboxyl function:

$$ArCH_3 \rightarrow ArCH_2OOH \rightarrow ArCH_2OH + ArCHO \rightarrow ArC(O)OOH \rightarrow ArCOOH$$

Even though methylbenzenes as a class are relatively easy to oxidize, individual differences strongly influence the conditions required. Generally, the ease of oxidation parallels the electron density of the aromatic ring. Methylnaphthalene and *p*-methoxytoluene are more easily oxidized than toluene and can be oxidized by a different mechanism [2]. Toluene and the xylenes are comparable in their initial susceptibility to oxidation. However, once one methyl group of a xylene is oxidized, the remaining methyl group is deactivated by the electron-withdrawing effect of the carboxyl group. As a result the oxidation of *p*-xylene sometimes takes place stepwise:

$$H_3C{-}C_6H_4{-}CH_3 \rightarrow H_3C{-}C_6H_4{-}COOH \rightarrow HOOC{-}C_6H_4{-}COOH$$

More vigorous conditions are required for the second step than for the first.

Toluene Oxidation

The oxidation of toluene differs from many xylene oxidation processes in that it is accomplished in the absence of a solvent [39–41]. Typically, a toluene solution of cobalt(II) 2-ethylhexanoate is reacted with air at 190° and 10 atmospheres pressure [40]. Liquid reaction mixture is withdrawn at a rate such that toluene conversion is 40–65%. Distillation or crystallization separates toluene for recycle from the crude benzoic acid. Yields of benzoic acid are about 80% after purification by redistillation or recrystallization. Superior results are said to be obtained with a mixture of nickel and manganese salts [41].

Xylene Oxidation

Many processes have been developed for oxidation of *p*-xylene to terephthalic acid or dimethyl terephthalate. Generally, these processes use air as the oxidant, cobalt and manganese salts as the catalyst, and give high yields. Two major processes are discussed here. The Mid-Century process, which was commercialized by Amoco Chemicals [42], produces terephthalic acid by a one-step oxidation of *p*-xylene. The Dynamit Nobel process yields dimethyl terephthalate in multistep process that avoids the use of a solvent. Each process has particular advantages. As noted in Chapter 11 most polymeric terephthalate esters are made from dimethyl terephthalate by ester exchange, but terephthalic acid is used increasingly as a starting material for polymer production.

The Mid-Century/Amoco process may be used for oxidation of both *m*- and *p*-xylenes. Typically, the oxidation is performed in acetic acid with a mixture of cobalt(II) and manganese(II) acetates and bromides as the catalyst. Bromide is said to be unique among the halogens in its catalytic activity [43]. The acetic acid solvent keeps the intermediates and byproducts in solution. The terephthalic acid is virtually insoluble in the acetic acid, as well as in other organic solvents.

A recent patent describes a process in which *p*-xylene is oxidized in acetic acid at about 225° and 15 atmospheres pressure of air with a mixture of cobalt acetate, manganese acetate, and sodium bromide as the catalyst [44]. During a 90-minute residence time in the reactor most of the xylene is converted to terephthalic acid, which crystallizes in about 99.95% purity. The slurry of product in acetic acid is withdrawn continuously, the crude terephthalic acid is separated, and the acetic acid is recycled. The terephthalic acid is recrystallized from aqueous acetic acid under pressure to attain a temperature at which the terephthalic acid has significant solubility. A product of about 99.96% purity is reported [44].

The Dynamit Nobel process [45] is based on a series of oxidations and esterifications:

$$\underset{CH_3}{\overset{CH_3}{\bigcirc}} \xrightarrow{O_2} \underset{CH_3}{\overset{COOH}{\bigcirc}} \xrightarrow{MeOH} \underset{CH_3}{\overset{COOMe}{\bigcirc}} \xrightarrow{O_2} \underset{COOH}{\overset{COOMe}{\bigcirc}} \xrightarrow{MeOH} \underset{COOMe}{\overset{COOMe}{\bigcirc}}$$

Although it is more complex than the Mid-Century/Amoco process in a chemical sense, the engineering aspects may be simpler because the reaction mixtures are less corrosive. The interesting aspect of this oxidation/esterification process is that it uses the easy xylene oxidation to promote the more difficult oxidation of a toluic acid derivative. Neat *p*-xylene can be oxidized under the conditions described for toluene oxidation, but the major product is toluic acid. When, however, the toluic acid is esterified and cooxidized with *p*-xylene, the major product is monomethyl terephthalate. Evidently, the radical intermediates formed in xylene oxidation assist attack on the *p*-toluate methyl group in a manner similar to bromine atoms in the Mid-Century/Amoco process.

In a semiworks scale experiment [45] that appears to simulate the Dynamit Nobel process described earlier the oxidation of *p*-xylene is started by blowing air at 6 atmospheres pressure through a xylene solution of cobalt(II) 2-ethylhexanoate at 150°. Once the oxidation is started, methyl *p*-toluate is added, sometimes accompanied by manganese(II) 2-ethylhexanoate as a cocatalyst. When the oxidation is done continuously [46], part of the liquid reaction mixture is withdrawn, and the *p*-toluic acid and monomethyl terephthalate products are esterified with methanol at 250–280° and 20–25 atmospheres. The esters are separated by distillation. Methyl *p*-toluate is recycled to the oxidation reactor. The dimethyl terephthalate is crystallized to obtain the purity required for polymer production.

Many variations of these processes are practiced commercially. One process is conceptually related to the Dynamit Nobel process in that it uses the oxidation of 2-butanone as a source of radicals to promote the relatively sluggish oxidation of *p*-toluic acid.

Other Substrates

In addition to toluene and the xylenes many other oxidations of methylbenzenes have been proposed or operated on a modest scale. Pyromellitic dianhydride, obtained by oxidation of 1,2,4,5-tetramethylbenzene (durene) is an intermediate for synthesis of polyimide resins and films. The 2,6-naphthalene dicarboxylic and 4,4′-biphenyldicarboxylic acids, obtained by oxidation of the corresponding dimethyl arenes, have received serious consideration as polyester intermediates.

Mechanism

The oxidation of methylbenzenes is much like that of butane and cyclohexane in that it is necessary to generate benzylic radicals that can couple with O_2 in a radical chain process like that of Figure 10.2. A major difference is that the initiation step is much easier than for aliphatic hydrocarbons. Two initiation mechanisms can be discerned [2]:

1. Electron transfer from the arene to a cobalt(III) ion gives an arene radical cation, which, in turn, forms a benzyl radical by proton loss. This mechanism is naturally not available with alkanes.

2. Abstraction of benzylic hydrogen by bromine atoms, R·, RO·, and ROO· radicals and possibly even dioxygen complexes. This abstraction reaction is much easier for methylbenzenes than for alkanes because the benzyl radical is relatively stable.

The toluene oxidation process and the first step in the Dynamit Nobel process probably involve mechanism **1** to a considerable extent. Mechanism **2** operates in all oxidations of methylbenzenes but seems especially important in the Mid-Century/Amoco process and the second step in the Dynamit Nobel process.

Catalysis of toluene and xylene oxidation by cobalt salts is characterized by an induction period in which cobalt(II) ions are oxidized to cobalt(III) [47]. Monomeric cobalt(III) ion is a powerful oxidizing agent ($E_0 = 1.82$) when it is surrounded only by O-donor ligands such as water or OH^- ions or $RCOO^-$ ions. Cobalt(III) oxidizes toluene and the xylenes to radical cations by electron transfer [47–49]. The cations yield benzyl radicals by proton loss. In the oxidation of toluene a major pathway is the following:

$$C_6H_5CH_3 \xrightarrow{Co^{3+}} [C_6H_5CH_3]^{+\cdot} \xrightarrow{-H^+} C_6H_5\dot{C}H_2 \xrightarrow{O_2} C_6H_5CH_2OO\cdot$$

This sequence opens the way to oxygenation of the methyl group to form benzyl alcohol and benzaldehyde. These compounds are relatively easy to oxidize to benzoic acid by mechanisms like those discussed for acetaldehyde oxidation.

The cobalt(III) initiation pathway is effective in many oxidations, but it has severe limitations. It is strongly inhibited by cobalt(II) ions, which seem to form dimers with cobalt(III) [47]. The dimers are not sufficiently potent oxidizing agents to oxidize toluene and the xylenes directly. As a result the rate of oxidation is inversely dependent on cobalt(II) concentration. This phenomenon puts a practical limit on total catalyst concentration and the rate that can be attained by cobalt(III) initiation. Even more seriously, the cobalt(III) ion does not seem very effective in oxidation of *p*-toluic acid, the intermediate in *p*-xylene oxidation. Evidently, the electron-withdrawing carboxyl group raises the oxidation potential of *p*-toluic acid to the point that it is inert to cobalt(III) ion.

The hydrogen abstraction mechanism **2** is less sensitive to arene *pi* electron density than the electron transfer mechanism is. Odd electron species such as bromine atoms and R·, RO·, or ROO· radicals abstract hydrogen from methylbenzenes. Although some of these reagents can be quite discriminating in hydrogen abstraction, they are capable of hydrogen removal from the methyl group of *p*-toluic acid [50]. As noted earlier both the Dynamit Nobel and Mid-Century/Amoco processes apparently use this approach to initiation of the difficult second step in xylene oxidation.

The Dynamit Nobel process uses a combination of mechanisms **1** and **2**, as illustrated in Figure 10.9. The electron transfer mechanism **1** provides a supply of

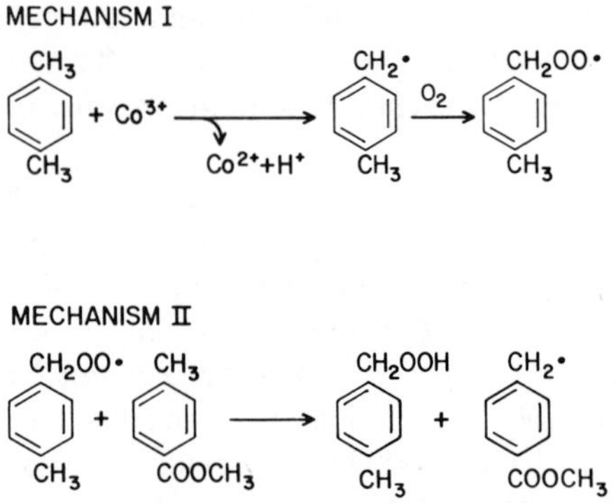

MECHANISM I

MECHANISM II

13 14

Figure 10.9 Combination of electron transfer and hydrogen abstraction mechanisms in co-oxidation of *p*-xylene and methyl *p*-toluate.

p-methylbenzyl radicals which react with O_2 in a now familiar pattern to form alkylperoxy radicals. These radicals can attack the *p*-methyl group of methyl *p*-toluate to abstract hydrogen and generate a new radical. The *p*-methylbenzyl hydroperoxide (**13**) can form *p*-methylbenzyl alcohol and *p*-methylbenzaldehyde by standard mechanisms. Similarly, the new benzylic radical (**14**) can oxygenate and can initiate new reaction cycles in its own right. In the course of doing so, it is converted to terephthalate precursors.

The Mid-Century/Amoco process catalyst, reportedly a mixture of manganese, cobalt, and bromide salts [38], functions largely by hydrogen abstraction (mechanism **2**). The manganese(III) ion is not a sufficiently potent oxidizing agent to abstract an electron from *p*-xylene [51]. The cobalt and manganese ions perform the usual function of hydroperoxide decomposition to produce RO· and ROO· radicals. However, these ions also play another major role in oxidation of bromide ions to bromine atoms [50,52]. This electron transfer is rapid and provides a constant supply of bromine atoms. The latter are extremely effective in abstraction of hydrogen atoms from methyl groups. At low temperatures [52] the major initiation process for methyl group oxidation is hydrogen abstraction by bromine atoms. Under commercial conditions, R·, RO·, and ROO· may also be significant initiators.

The bromide-bromine cycle probably involves both free and metal-complexed species, as indicated in Figure 10.10 [50,52]. Coordination of bromide ion to cobalt(II) may facilitate electron transfer to oxygen or peroxy species to form a cobalt(III) complex. Cobalt(III) ion is a powerful oxidizing agent that can abstract an electron from a bromide ligand to yield a bromine atom. The free or complexed bromine atom can then abstract a hydrogen atom from a methyl group to complete the cycle [50].

Another significant species in oxidations in acetic acid is the $\cdot CH_2COOH$ radical. Manganese(III) acetate reportedly decomposes to give this radical and $Mn(OAc)_2$ [51]. This process is significant in several ways. The carboxylmethyl

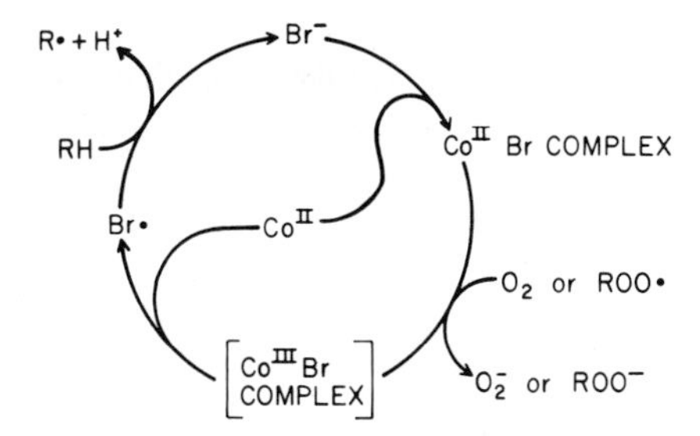

Figure 10.10 The bromine cycle in a bromide-promoted oxidation of a hydrocarbon.

radical can abstract hydrogen from methylbenzenes to initiate the desired oxidation process. However, there are two undesirable effects. The $\cdot CH_2COOH$ radical can add to the aromatic ring or couple with benzylic radicals to form unwanted by-products. Even more seriously, it provides a pathway for the oxidation of acetic acid to carbon dioxide and water. This destruction of the acetic acid solvent is an economic handicap.

The oxidation of methylbenzenes, like the oxidation of alkanes and cycloalkanes, is obviously an extremely complex process. Even the nature of the major chain-carrying steps is not totally understood, let alone a good assessment of the relative rates of the dozens of reactions that occur simultaneously. A complete kinetic description of the processes seems beyond our grasp at present.

GENERAL REFERENCES

Dumas, T., and W. Bulani, *Oxidation of Petrochemicals: Chemistry and Technology,* Halstead Press, 1974.

Emanuel, N. M., *The Oxidation of Hydrocarbons in the Liquid Phase,* Macmillan, 1965.

Emanuel, N. M., E. T. Denisov, and Z. K. Maizus, *Liquid Phase Oxidation of Hydrocarbons,* Plenum Press, 1967.

Denisov, E. T., N. I. Mitskevich, and V. E. Agabekov, *Liquid Phase Oxidation of Oxygen-Containing Compounds,* Consultants Bureau, 1977.

Kochi, J. K., *Organometallic Mechanisms and Catalysis,* Academic Press, 1978, provides an up-to-date treatment of the mechanism of metal-catalyzed oxidation processes (see also ref. 2).

SPECIFIC REFERENCES

1. J. S. Valentine, *Chem. Rev.,* **73,** 235 (1973).
2. R. A. Sheldon and J. K. Kochi, *Adv. Catal.,* **25,** 272 (1976).
3. R. G. Wilkins, "Uptake of Oxygen by Cobalt(II) Complexes in Solution," in R. Dessy, J. Dillard, and L. Taylor, Eds, *Bioinorganic Chemistry, Advances in Chemistry Series* **100,** American Chemical Society, 1971, pp. 111–134.
4. A. E. Martell and M. Calvin, *Chemistry of the Metal Chelate Compounds,* Prentice-Hall, 1952, pp. 336–350.

5. G. Henrici-Olive and S. Olive, *J. Organomet. Chem.*, **52**, C49 (1973).
6. N. M. Emanuel, Z. K. Maizus, and I. P. Skibida, *Angew. Chem. Int. Ed.*, **8**, 97 (1969).
7. F. A. Cotton and G. Wilkinson, *Advanced Inorganic Chemistry,* Wiley-Interscience, 1966 p. 91.
8. L. Vaska, *Accounts Chem. Res.*, **9**, 175 (1976).
9. V. D. Luedecke, "Adipic Acid," in J. J. McKetta and W. A. Cunningham, Eds., *Encyclopedia of Chemical Processing and Design,* Vol. 2, Marcel Dekker, 1977, pp. 128–146.
10. I. V. Berezin, E. T. Denisov, and N. M. Emanuel, *The Oxidation of Cyclohexane,* Pergamon, 1966.
11. H. J. Boonstra and P. Zwietering, *Chem. Ind.*, 2039 (1966).
12. S. A. Miller, *Chem. Proc. Eng.*, **50**, 63 (June 1969).
13. J. F. Black, *J. Am. Chem. Soc.*, **100**, 527 (1978).
14. G. F. Pustarnakova, V. M. Solianikov, and E. T. Denisov, *Izvest. Akad. Nauk SSSR, Ser. Khim,* 547 (1975).
15. J. D. Druliner, *J. Org. Chem.*, **43**, 2069 (1978).
16. K. Tanaka, *Chem. Tech.*, 555 (1974).
17. A. F. Lindsay, *Chem. Eng. Sci.*, **3** (Special Supplement) 78–93 (1954).
18. W. J. van Asselt and D. W. van Krevelen, *Chem. Eng., Sci.*, **18**, 471 (1963).
19. W. J. van Asselt and D. W. van Krevelen, *Rec. Trav. Chim.*, **82**, 51, 429 (1963).
20. O. S. Shchatinskaya, S. M. Sedova, T. L. Vesel'chakova, and A. M. Gol'dman, *Soviet Chemical Industry,* **8**, 502 (1976).
21. W. Griehl and D. Ruestem, *Ind. Eng. Chem.*, **62**, 16 (Mar. 1970).
22. J. O. White and D. D. Davis, U.S. Patent 3,637,832 (1972).
23. F. Broich and H. Grasemann, *Erdöl Kohle-Erdgas-Petrochemie,* **18**, 360 (1965); British Patent 1,110,396 (1968).
24. K. S. McMahon, "Acetic Acid," in J. J. McKetta and W. A. Cunningham, Eds., *Encyclopedia of Chemical Processing and Design,* Vol. 1, Marcel Dekker, 1976, pp. 216, 240, 258.
25. R. P. Lowry and A. Aguilo, *Hydrocarbon Proc.*, 103 (Nov. 1974).
26. F. Broich, H. Höfermann, W. Hunsmann, and H. Simmrock, *Erdöl Kohle-Erdgas Petrochemie,* **16**, 284 (1963).
27. J. G. D. Schulz and R. Seekircher, U.S. Patent 4,032,570 (1977).
28. A. Onopchenko and J. G. D. Schulz, *J. Org. Chem.*, **38**, 909 (1973).
29. F. Broich, *Chem. Ing. Tech.*, **36**, 417 (1964).
30. J. S. Bartlett, B. Hudson, and J. Pennington, U.S. Patent 4,086,267 (1978).
31. T. Szymanska-Buzar and J. J. Ziolkowski, *Koord. Khim.*, **2**, 1172 (1976).
32. D. C. Hull, U.S. Patent 2,578, 306 (1951).
33. British Patent 1,483,724 (1977).
34. G. C. Allen and A. Aguilo, "Metal-Ion Catalyzed Oxidation of Acetaldehyde," in F. R. Mayo, Ed., *Oxidation of Organic Compounds* Vol. 2, *Advances in Chemistry Series* **76**, American Chemical Society, 1968, pp. 363–381.
35. G. H. Twigg, *Chem. Ind.*, 476 (1966).
36. M. Tezuka, O. Sekiguchi, Y. Ohkatsu, and T. Osa, *Bull. Chem. Soc. Japan,* **49**, 2765 (1976).
37. A. P. Litovka, V. V. Baluev, and S. A. Rakhimzhanova, *Kinet. Katal.*, **19**, 567 (1978).
38. Y. Ichikawa and Y. Takeuchi, *Hydrocarbon Proc.*, **51**, 103 (Nov. 1972).
39. R. W. Ingwalson and G. D. Kyker, "Benzoic Acid," in J. J. McKetta and W. A. Cunningham, Eds., *Encyclopedia of Chemical Processing and Design,* Vol. 4, Marcel Dekker, 1977, pp. 296–308.

40. C. H. Bell, U.S. Patent 3,631,204 (1971).
41. K. Namie, T. Harada, and T. Fuji, U.S. Patent 3,903,148 (1975).
42. R. Landau and A. Saffer, *Chem. Eng. Prog.*, **64,** 20 (Oct. 1968); P. H. Towle and R. H. Baldwin, *Hydrocarbon Proc.*, **43,** 149 (Nov. 1964).
43. D. E. Burney, G. H. Weisemann, and N. Fragen, *Petrol. Refiner,* **38,** 186 (1959); A. Saffer and R. S. Barker, U.S. Patent 2,833,816 (1958).
44. C. M. Park and D. G. Micklewright, U.S. Patent 4,053,506 (1977).
45. G. Hoffman, K. Irlweck, and R. Cordes, British Patent 1,344,383 (1974).
46. *Hydrocarbon Proc.*, **56,** 147 (Nov. 1977).
47. E. J. Y. Scott and A. W. Chester, *J. Phys. Chem.*, **76,** 1520 (1972).
48. E. I. Heiba, R. M. Dessau, and W. J. Koehl, *J. Am. Chem. Soc.*, **91,** 6830 (1969).
49. K. Sakota, Y. Kamiya, and N. Ohta, *Can. J. Chem.*, **47,** 387 (1969).
50. D. A. S. Ravens, *Trans. Faraday Soc.*, **55,** 1768 (1959).
51. E. I. Heiba, R. M. Dessau, and W. J. Koehl, *J. Am. Chem. Soc.*, **91,** 138 (1969).
52. Y. Kamiya, *J. Catal.*, **33,** 480 (1974).

11 CONDENSATION POLYMERIZATION AND RELATED PROCESSES

Many of the industrial homogeneous catalytic processes discussed earlier (especially in Chapter 10) produce intermediates for synthesis of condensation polymers such as polyesters and polyamides. Indeed, some of the largest applications of homogeneous catalysis are in production of polymer intermediates. The selectivity of a homogeneous catalyst is a major advantage when the final product must be better than 99.9% pure.

In addition to its use in synthesis of polymer intermediates, homogeneous catalysis is also used in some polymerization processes. Soluble metal compounds are used as catalysts in most processes for manufacture of polyesters. Similar catalysts are often used for production of polyurethanes. Polyamide synthesis is usually not catalytic, but the topic is considered in this chapter to present an orderly description of the role of homogeneous catalysis in the production of nylon intermediates.

11.1 POLYESTER SYNTHESIS

The most widely used synthetic fiber is poly(ethylene terephthalate) (**3**) which is sold under trade names such as Dacron®, Terylene®, or Kodel®. Since production of polyesters began about 30 years ago the major processes have been based on dimethyl terephthalate (**1**) [1–3]. As shown in Figure 11.1, this approach involves two steps, both of which are catalyzed by soluble metal compounds. Transesterification of the dimethyl ester with ethylene glycol gives methanol and bis(hydroxyethyl) terephthalate (**2**). The hydroxyethyl ester is then heated under vacuum to drive out ethylene glycol and form the polyester (**3**).

Recently, the availability of high-purity terephthalic acid from new processes such as the Mid-Century/Amoco process discussed in Section 10.6 has made the acid attractive as a starting material for polyester synthesis. The acid is generally

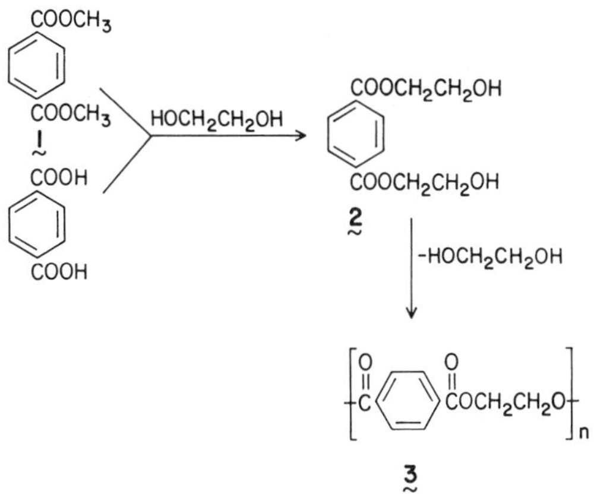

$COOCH_3$ / $COOCH_3$ (1) or $COOH$ / $COOH$ + $HOCH_2CH_2OH$ → $COOCH_2CH_2OH$ / $COOCH_2CH_2OH$ (2) → $-HOCH_2CH_2OH$ → $[C(=O)C_6H_4COCH_2CH_2O]_n$ (3)

Figure 11.1 Steps in the synthesis of poly(ethylene terephthalate) from dimethyl terephthalate or terephthalic acid.

the same price by weight as dimethyl terephthalate but is cheaper on a molar basis. In addition there is no need to recycle methanol when the acid is the starting material. Direct esterification of terephthalic acid often employs a different sort of catalyst from that used in transesterification. However, the conversion of **2** to **3** ("polycondensation") may use the same catalyst whether ester or acid is the ultimate starting material.

Transesterification and direct esterification processes are discussed here with the synthesis of polyethylene terephthalate as an example. Similar reactions occur in the production of terephthalate polyesters from 1,4-butanediol (**4**) or cyclohexanedimethanol (**5**).

$HO(CH_2)_4OH$ (4)

$HOCH_2-C_6H_{10}-CH_2OH$ (5)

Analogous chemistry takes place in the synthesis of esters of adipic and phthalic acids for use as plasticizers and synthetic lubricants.

Transesterification

The usual preparation of poly(ethylene terephthalate) uses two different catalyst systems [4]. The first step in the process of Figure 11.1, the *transesterification* of dimethyl terephthalate with ethylene glycol, is catalyzed by a divalent metal ion. Acetates of Zn, Co, and Mn are used commercially [1]. Although these salts are very effective in the first step, which is performed at 150–200°, they cause undesirable side reactions at the higher temperatures necessary for the second step. The *polycondensation* of bis(hydroxyethyl) terephthalate is accomplished at 250–300° under vacuum to remove ethylene glycol as it is produced. Antimony compounds almost always are used to catalyze this reaction, which is also a transesterification of sorts. Even though antimony does not catalyze transesterification at low temperatures, it cleanly polymerizes the hydroxyethyl ester at about 275°. Presumably antimony catalyzes transesterification by a different mechanism from that of the simple metal ions. It works poorly in the presence of free alcohols [4].

Polyester synthesis is simple from a chemist's viewpoint and can be done in the laboratory in a reaction sequence that simulates commercial practice. In the laboratory preparation [5] the transesterification catalyst, $Ca(OAc)_2{\cdot}2H_2O$, and the polycondensation catalyst, Sb_2O_3, are mixed with dimethyl terephthalate and a slight excess of ethylene glycol. The mixture is heated at 197° and is bubbled with N_2 to expel methanol as it is produced by transesterification. After 3 hours at 197°, the solution of bis(hydroxyethyl) terephthalate and low oligomers is heated to 222° to distill excess ethylene glycol.

The polycondensation step is effected by heating the mixture at 283° under vacuum. Ethylene glycol from the transesterification of two molecules of the hydroxyethyl ester distills. Three hours' heating produces high-molecular-weight polymer that can be melt-spun to form fibers or hot-pressed to produce films.

Industrial polymerization is effected very similarly in either a batch or a continuous process [1]. The transesterification catalyst is an acetate of cobalt, zinc, or manganese, all of which produce rapid reaction of ethylene glycol with the terephthalate ester. Unfortunately, they also catalyze ether formation from the glycol at the high temperatures used in polycondensation:

$$2HOCH_2CH_2OH \rightleftharpoons (HOCH_2CH_2)_2O + H_2O$$

The diethylene glycol formed in this way is incorporated into the final polymer and changes the chemical and physical properties. Consequently, it is common practice to add an "inhibitor" such as phosphoric acid or triphenyl phosphate [6] before the polycondensation step. The inhibitor converts the metal salt to a complex phosphate that has little catalytic activity and does not interfere with the antimony-based polycondensation catalyst.

In a commercial polymerization [1], ethylene glycol, dimethyl terephthalate, and the metal acetate catalyst may be added continuously to a transesterification reactor that is heated to 150–200°. Methanol and some glycol distill. The molten bis(hydroxyethyl) terephthalate that results is added to a polycondensation reactor, along with the inhibitor and antimony(III) oxide or acetate. As in the laboratory experiment the mixture is gradually heated to 270–280° under vacuum to distill glycol as it forms. The molten polymer may be fabricated to fiber or film directly, or it may be converted to solid chips or granules for processing later.

The divalent metal ions are said to act as Lewis acids in the catalysis of transesterification [7]. Coordination of the carbonyl group of the ester to the cation is believed to activate it to nucleophilic attack by the oxygen of an alcohol. The complexation and activation of dimethyl terephthalate by calcium acetate are depicted in Figure 11.2. When the carbonyl group coordinates to a calcium (2+) ion, electron density is transferred to the metal ion. The electron depletion of the carbonyl is most pronounced at the carbon atom. This effect makes the carbon especially susceptible to nucleophilic attack.

The attack by a hydroxyl group is often written with a four-center transition

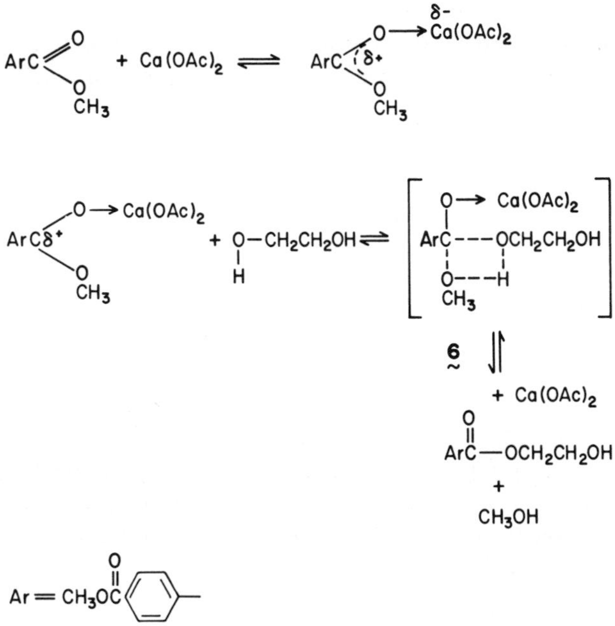

$$ArC(=O)OCH_3 + Ca(OAc)_2 \rightleftharpoons ArC^{\delta+}(\text{-}O^{\delta-}\rightarrow Ca(OAc)_2)OCH_3$$

$$ArC^{\delta+}(\text{-}O\rightarrow Ca(OAc)_2)OCH_3 + HO\text{—}CH_2CH_2OH \rightleftharpoons \left[ArC(\text{-}O\rightarrow Ca(OAc)_2)\text{---}OCH_2CH_2OH \;\; | \;\; CH_3O\text{---}H \right] \; \mathbf{6}$$

$$\mathbf{6} \rightleftharpoons ArC(=O)\text{—}OCH_2CH_2OH + CH_3OH + Ca(OAc)_2$$

$$Ar = CH_3OC(=O)\text{—}C_6H_4\text{—}$$

Figure 11.2 The initial steps in the calcium acetate-catalyzed transesterification of dimethyl terephthalate with ethylene glycol.

state (**6**), as shown in the figure [7]. Regardless of the timing of events the overall effect is breaking and making of C—O and O—H bonds to produce a glycol ester and free methanol. Dissociation of the metal ion from the carbonyl group completes the catalytic cycle.

The same kinds of reactions occur during the first stages in growth of the polymer chain. The Ar group in the figure can be a terephthalate unit at the carboxyl end of a growing polymer chain. Similarly, the ethylene glycol in this scheme can be replaced by a glycol-terminated chain:

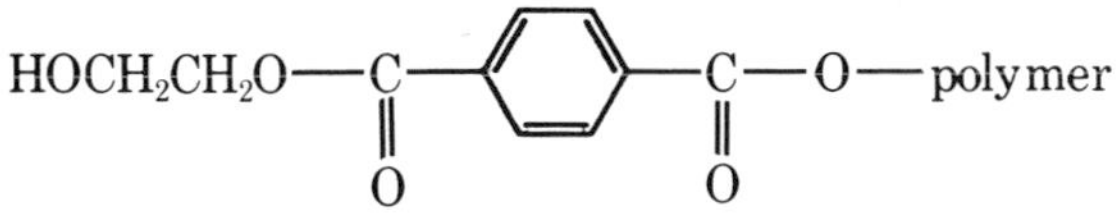

$$HOCH_2CH_2O\text{—}\underset{\underset{O}{\|}}{C}\text{—}C_6H_4\text{—}\underset{\underset{O}{\|}}{C}\text{—}O\text{—}\text{polymer}$$

Under these conditions a substantial increase in molecular weight can occur. Since, however, all the reactions are reversible, residual methanol or glycol in the system can lead to chain cleavage. For this reason the transesterification and polycondensation reactions are conducted with continuous removal of the alcohols.

The mechanism of Figure 11.2 seems quite plausible for the transesterification catalyzed by divalent metal ions but is not universally accepted. It was proposed earlier [8] that the metal ion reacts with ethylene glycol to form an alkoxide, for example, $[Ca\text{-}OCH_2CH_2OH]^+$. The nucleophilic alkoxide ion then attacks the carbonyl group of a coordinated ester. This proposal was supported by the kinetics of the transesterification catalyzed by M^{2+} ions. The reported kinetics (first order each in glycol, ester, and catalyst) would also be consistent with the Lewis acid mechanism if the three reactants were assembled in a complex such as **7** [8,9].

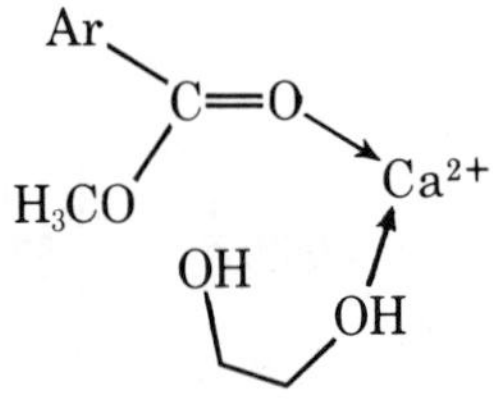

Ar
C=O → Ca^{2+}
H_3CO
OH
OH

7

The mechanism of the antimony-catalyzed polycondensation reaction is also unresolved. It has been suggested [10] that hydroxyethyl endgroups of two growing polymer chains are joined to an antimony center as alkoxides in an intermediate such as **8** in Figure 11.3. Ligand exchange within the coordination sphere of antimony forms a new ester link that couples the two growing polymer chains. The other product is an antimony glycolate (**9**), which can release ethylene glycol by reaction with other hydroxyethyl groups:

$$\mathbf{9} + 2HOCH_2CH_2O\overset{\overset{O}{\|}}{C}Ar \longrightarrow HOCH_2CH_2OH + ROSb(OCH_2CH_2O\overset{\overset{O}{\|}}{C}Ar)_2$$

This mechanism is speculative, but it seems consistent with the kinetic data [11] now available.

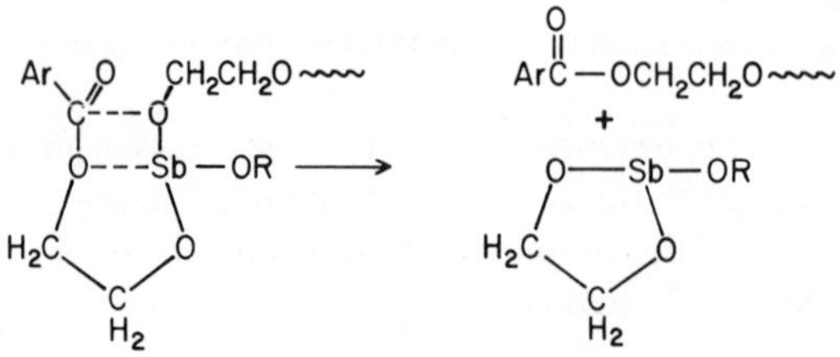

Ar, O, CH_2CH_2O~~~, C---O, O---Sb—OR, H_2C, O, C, H_2 ⟶ Ar$\overset{\overset{O}{\|}}{C}$—$OCH_2CH_2O$~~~ + O—Sb—OR, H_2C, O, C, H_2

8 **9**

Figure 11.3 A concerted mechanism for antimony-catalyzed polycondensation.

Direct Esterification

The change from dimethyl terephthalate to terephthalic acid as a starting material for polyester production requires a change in catalyst. The divalent metal salts usually used as *transesterification* catalysts have little activity in the *esterification* of a carboxylic acid with an alcohol [12]. Extensive screening of potential catalysts for esterification [12,13] has focused attention on compounds of tin and titanium. The tin compounds reported to have good activity in direct esterification include tin(II) oxide and oxalate and organotin(IV) compounds such as dialkyltin(IV) oxides and carboxylates. The titanium catalysts are typically titanium(IV) alkoxides.

The industrial application of titanium alkoxide catalysts is typified by a laboratory or semiworks experiment that may simulate plant operation [14]. A slurry of terephthalic acid in excess ethylene glycol is fed continuously to a reactor along with titanium(IV) tetraisopropoxide. The hot (250–300°) reaction mixture, which contains bis(hydroxyethyl) terephthalate and low-molecular-weight polymer, dissolves the terephthalic acid and permits esterification to proceed in a relatively homogeneous medium. Continuous distillation removes water as it is formed, as well as some of the excess glycol.

The hot esterified material is fed to a polycondensation reactor in which it is heated at 260–300° under reduced pressure to remove ethylene glycol. The titanium compound serves as a polycondensation catalyst, but it may be supplemented with a conventional catalyst such as Sb_2O_3 [14]. As in the transesterification process, formation of ether linkages by condensation of hydroxyethyl groups is an unwanted side reaction.

The insolubility of terephthalic acid has impeded mechanistic studies on polyester synthesis by direct esterification. Careful kinetic studies have, however, been done on the reactions of benzoic acid, isophthalic acid, and substituted isophthalic acids with ethylene glycol [12,13,15]. Tin(II) and titanium(IV) show similar kinetic dependencies, although the order of the reaction varies with the acid being esterified. Since the rate depends on the concentrations of acid, glycol, and catalyst, it appears that all three components must be assembled in a reactive complex such as $(ArCOO)_xM(OCH_2CH_2OH)_y$ [12].

The role of such a complex in esterification of benzoic acids is illustrated in Figure 11.4. The illustrative catalyst, dimethyltin oxide, is normally an oligomer, but much of its chemistry is what would be expected from the hypothetical monomer $Me_2Sn{=}O$. The initial reaction within the catalytic cycle is the addition of the

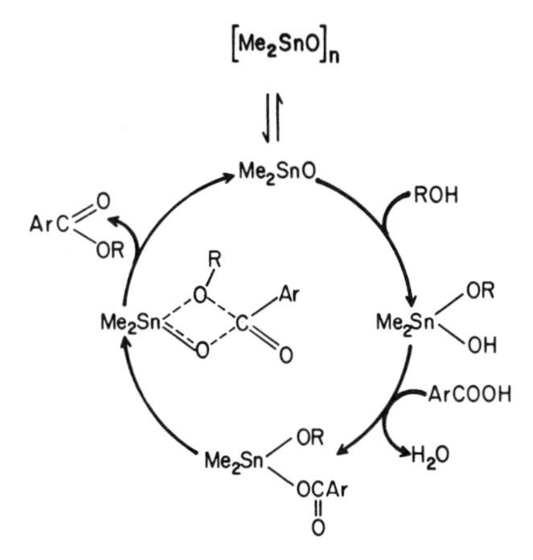

Figure 11.4 A possible cycle for catalysis of esterification by dimethyltin oxide (ROH = ethylene glycol).

alcohol O—H to form a hydroxy alkoxy derivative. Esterification of this compound with the arenecarboxylic acid generates the proposed $(ArCOO)M(OCH_2CH_2OH)$ intermediate. Intramolecular nucleophilic attack on the carbonyl group by the alkoxide ligand produces ester and Me_2SnO to complete the cycle.

This mechanism seems quite plausible for the direct esterification, but much additional work is required to establish it conclusively. It is ironic that such a large industrial application of homogeneous catalysis has received so little scientific examination.

11.2 POLYURETHANE SYNTHESIS

Polyurethane fibers, foams, and adhesives are usually prepared by reaction of diisocyanates with dihydroxy compounds. A simple example is the synthesis of Perlon U, the first commercial polyurethane resin [16]:

$$HO(CH_2)_4OH + OCN(CH_2)_6NCO \longrightarrow -\!\!\left[O(CH_2)_4O-\overset{\overset{O}{\|}}{C}-\overset{H}{N}(CH_2)_6\overset{H}{N}-\overset{\overset{O}{\|}}{C}\right]_n\!\!-$$

(A urethane is an N- and O-substituted derivative of carbamic acid, $H_2NC(=O)OH$.) Most modern polyurethanes are prepared from aromatic diisocyanates such as MDI (Figure 11.5). The glycol component is often a polyether such as PTMEG or PPG. As indicated in the figure, homogeneous catalysis is used in many steps in the preparation of these intermediates [17].

Polyurethanes can also be prepared from simple carbamic esters by catalytic tranesterification [18]:

$$2ArNHC(=O)OR + [HO-CH_2CH_2]_2 \longrightarrow \left[ArNHC(=O)OCH_2CH_2-\right]_2 + 2ROH$$

This polymerization process is much less common than the isocyanate-based process, but it finds some use in the synthesis of plasticizers. It is a true condensation polymerization like polyester production because a mole of alcohol is eliminated in the "condensation" reaction that forms the urethane group. The carbamate ester process may be used more frequently when these starting materials become available from the reaction of nitrobenzenes with CO and alcohols.

The reaction of a diisocyanate with a glycol occurs readily, and, for laboratory-scale operations, no catalyst is required [19]. Catalysts are, however, used to accelerate polyurethane formation in most commercial processes. The most common catalysts are tertiary amines, but metal complexes are used in situations

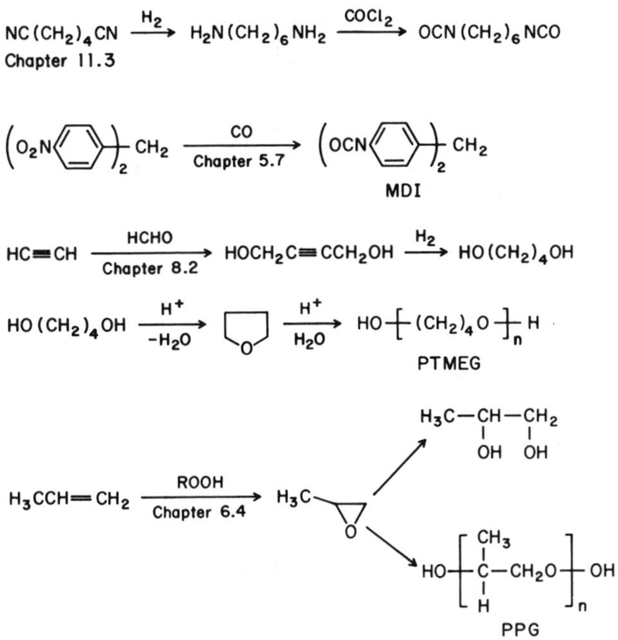

$$NC(CH_2)_4CN \xrightarrow{H_2} H_2N(CH_2)_6NH_2 \xrightarrow{COCl_2} OCN(CH_2)_6NCO$$

Chapter 11.3

$$(O_2N\text{-}C_6H_4)_2CH_2 \xrightarrow[\text{Chapter 5.7}]{CO} (OCN\text{-}C_6H_4)_2CH_2$$

MDI

$$HC{\equiv}CH \xrightarrow[\text{Chapter 8.2}]{HCHO} HOCH_2C{\equiv}CCH_2OH \xrightarrow{H_2} HO(CH_2)_4OH$$

$$HO(CH_2)_4OH \xrightarrow[-H_2O]{H^+} \text{THF} \xrightarrow[H_2O]{H^+} HO{-}[(CH_2)_4O]_n{-}H$$

PTMEG

$$H_3CCH{=}CH_2 \xrightarrow[\text{Chapter 6.4}]{ROOH} \text{propylene oxide} \rightarrow H_3C{-}CH(OH){-}CH_2(OH)$$

$$\text{propylene oxide} \rightarrow HO{-}[C(CH_3)(H){-}CH_2O]_n{-}OH$$

PPG

Figure 11.5 Some applications of homogeneous catalysis in synthesis of polyurethane intermediates.

in which an amine would contribute an unpleasant odor to the product. The most widely used metal catalysts are tin derivatives, although titanium tetraalkoxides are very effective with both isocyanates and carbamic esters. Polyurethane formation is also catalyzed by many of the other metal compounds that are used for polyester production [20]. This similarity in catalysts is not surprising, because the urethane group is similar to an ester in structure and chemistry.

Polyurethanes are used in many different ways—fibers, coatings, adhesives—but one of the largest uses is in the preparation of foams. The polymerization is done in the presence of a "blowing agent" that releases a gas such as N_2, CO_2, or a fluorocarbon. The gas forms bubbles in the viscous prepolymer. The foam structure is retained after the polymer attains high molecular weight and becomes solid. Trifunctional reactants such as triols provide crosslinks for rigidity. The process is illustrated by a recent patent [21] in which a nonvolatile amine and tin(II) octoate are used as cocatalysts to produce both flexible and rigid foams. A mixture of the polyalcohol (a propylene oxide adduct of glycerin), the catalysts, the blowing agent, and a trace of water are vigorously stirred while 2,4-toluenediisocyanate is added. The foaming mixture is poured into a heated box and cured at 100–105° for 1 hour to produce a cube of polyurethane foam. The rigidity of the foam depends on the number of polymer crosslinks provided by the alcohol component.

The mechanism of catalytic polyurethane production may be analogous to that of polyester production. Lewis acidic M^{2+} salts can activate the isocyanate function

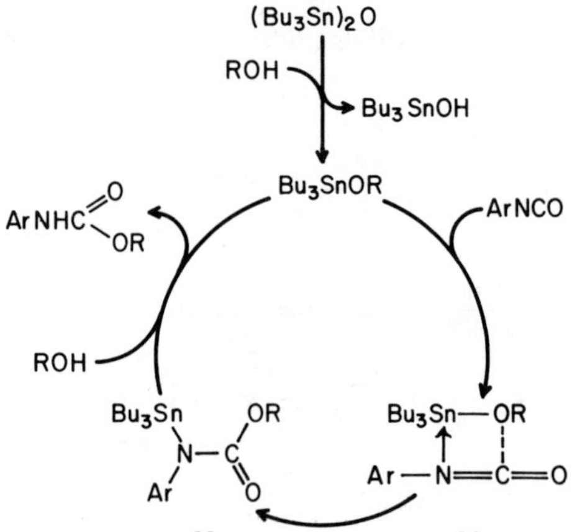

Figure 11.6 A possible mechanism for polyurethane formation with bis(tri-*n*-butyltin) oxide as the catalyst.

by coordination to oxygen [20] as in transesterification. Organotin compounds reportedly [22] act via alkoxide and carbamate intermediates, as shown in Figure 11.6. Coordination of the isocyanate to the tin in **10** activates the C—N group to intramolecular nucleophilic attack by alkoxide to form a tin carbamate (**11**). Alcoholysis of the Sn—N bond in **11** liberates the carbamate-containing polymer and regenerates the tin alkoxide for another cycle of reaction.

11.3 POLYAMIDES AND INTERMEDIATES

The most widely used polyamide in the United States is poly(hexamethyleneadipamide), usually known as 6,6-nylon [23]. (The numerical designation indicates that it is prepared from a six-carbon diamine and a six-carbon dicarboxylic acid.) The usual preparation begins with the reaction of the two components to form a salt. In the polymerization step the salt is heated under vacuum to expel water and form amide bonds [24]:

$$H_2N(CH_2)_6NH_2 + HOOC(CH_2)_4COOH \longrightarrow$$

$$[H_3N(CH_2)_6NH_3][OOC(CH_2)_4COO] \xrightarrow{-H_2O} -\!\!\!\!+\!NH(CH_2)_6NHCO(CH_2)_4CO\!+_{n}\!\!\!\!-$$

Although the amidation reaction ordinarily does not require a catalyst, the commercial manufacture of nylon intermediates is heavily dependent on homogeneous catalysis [25]. The oxidation of cyclohexane to adipic acid (Section 10.3) employs soluble catalysts in two steps. In a similar process cyclododecane is oxidized to dodecanedioic acid, a component of 6,12-nylon and other specialty polyamides (Section 10.4).

In contrast to adipic acid, which is made by similar processes by most producers,

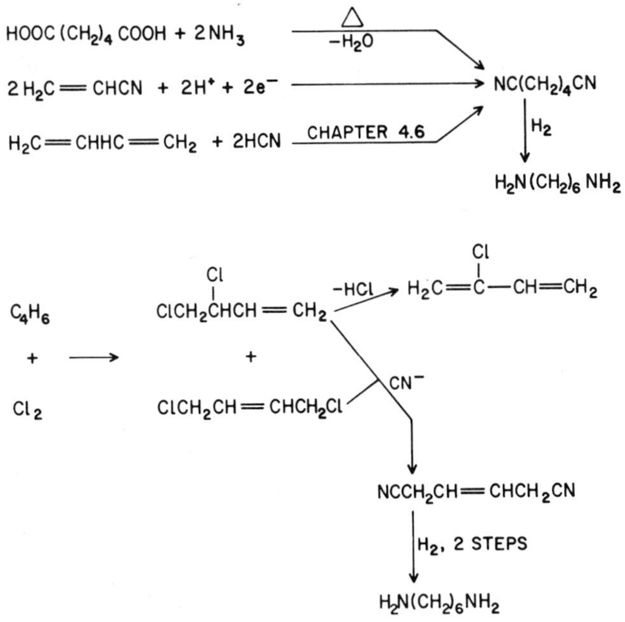

$$HOOC(CH_2)_4COOH + 2NH_3 \xrightarrow[-H_2O]{\Delta} NC(CH_2)_4CN$$

$$2H_2C{=}CHCN + 2H^+ + 2e^- \longrightarrow NC(CH_2)_4CN$$

$$H_2C{=}CHHC{=}CH_2 + 2HCN \xrightarrow{\text{CHAPTER 4.6}} NC(CH_2)_4CN$$

$$NC(CH_2)_4CN \xrightarrow{H_2} H_2N(CH_2)_6NH_2$$

$$C_4H_6 + Cl_2 \longrightarrow ClCH_2CHClCH{=}CH_2 + ClCH_2CH{=}CHCH_2Cl$$

$$ClCH_2CHClCH{=}CH_2 \xrightarrow{-HCl} H_2C{=}CCl{-}CH{=}CH_2$$

$$ClCH_2CHClCH{=}CH_2 + ClCH_2CH{=}CHCH_2Cl \xrightarrow{CN^-} NCCH_2CH{=}CHCH_2CN \xrightarrow{H_2,\ 2\ \text{STEPS}} H_2N(CH_2)_6NH_2$$

Figure 11.7 Commercial synthesis of 1,6-hexanediamine and chloroprene.

1,6-hexanediamine is manufactured by many different processes. These processes are largely based on homogeneous catalysis, as illustrated in Figure 11.7. All the major processes give adiponitrile as an intermediate. The dinitrile is hydrogenated smoothly to the diamine with a heterogeneous catalyst. The oldest route is reaction of adipic acid with ammonia to give ammonium adipate that dehydrates to form adiponitrile when it is heated strongly. Electrolytic reductive dimerization of acrylonitrile to adiponitrile is used commercially in both the United States and Japan. Attempts to dimerize acrylonitrile catalytically have had only moderate success. The addition of two moles of HCN to butadiene regioselectively to form adiponitrile is one of the most sophisticated commercial applications of homogeneous catalysis.

The lower part of Figure 11.7 shows another butadiene-based route. This chemistry, which is discussed in detail later, is obsolescent for diamine synthesis but is widely used for production of chloroprene, a monomer for oil-resistant synthetic rubbers [26]. The uncatalyzed chlorination of butadiene produces a mixture of about 33% 3,4-dichloro-1-butene and 66% *cis*- and *trans*-1,4-dichloro-2-butene [27]. For production of chloroprene the 3,4-dichloro isomer is preferred, and so the unwanted 1,4-dichloro-2-butene is catalytically isomerized to the desired 3,4-dichloro compound. Both the isomerization and the reaction of dichlorobutenes with sodium cyanide are catalyzed by soluble copper(I) salts. The butadiene-based chloroprene synthesis replaces the expensive and hazardous acetylene-based process described in Section 8.2.

Dichlorobutene Isomerization

The vapor-phase chlorination of butadiene produces a mixture of dichlorobutenes, as mentioned earlier. The 3,4-dichloro-1-butene desired for chloroprene synthesis boils at 123° and is easily separated from the 1,4-dichlorobutenes (b.p. 155°) by distillation [28]. To avoid waste of the 1,4-isomers, however, they are isomerized to the 3,4-dichloro derivative. This isomerization is catalyzed by copper(I) complexes:

$ClCH_2(H)C{=}C(H)CH_2Cl$ (trans) $\rightleftharpoons$ $CH_2Cl{-}CHCl{-}CH{=}CH_2$

$\updownarrow$

$ClCH_2(H)C{=}C(H)CH_2Cl$ (cis) $\rightleftharpoons$ $CH_2Cl{-}CHCl{-}CH{=}CH_2$

In one mode of operation [29] the crude dichlorobutenes are fed to a distillation column that contains the catalyst in the stillpot. The 3,4-dichloro-2-butene fed to the column distills continuously. The higher boiling isomers contact the catalyst and are converted to an equilibrium mixture from which the 3,4-dichloro compound distills. Eventually most of the chlorination product is converted to the desired 3,4 isomer. The same chemistry can be used in a different engineering context to produce the optimum mixture of dichlorobutenes for reaction with sodium cyanide to make 1,4-dicyanobutene.

The isomerization is catalyzed by many copper(I) compounds, but the recent patent literature cites advantages for nitrile complexes such as PhCN·CuCl [30] and chloride complexes like $R_4N^+ \ CuCl_2^-$ [31]. The mechanism of the copper-catalyzed isomerization has not been reported, but a kinetic study of dichlorobutene isomerization with a soluble iron catalyst led to a proposal based on π-allyl intermediates [32]. An analogous mechanism based on π-allylcopper intermediates is shown in Figure 11.8.

Presumably, the first step is coordination of a dichlorobutene molecule to copper(I) chloride as a simple olefin complex. Oxidative addition of a C—Cl bond to the Cu(I) ion generates a π-allyl complex in which the metal is formally trivalent. This sequence of events can occur with any of the dichlorobutene isomers. The thermodynamically disfavored *cis*-1,4-dichloro complex (**12**) forms the *anti*-π-allyl complex (**13**) in which the chloromethyl group is *trans* to the hydrogen on C-2. If the chloride migration from carbon to copper is reversed, but the chloride returns to C-3 rather than C-1 from which it came, the 3,4-dichloro-1-butene complex (**14**) forms. The *anti*-π-allyl complex (**13**) can isomerize to its *syn* isomer (**15**). In this instance return of a chloride ion from copper to C-1 generates the complex of *trans*-1,4-dichloro-2-butene.

The isomerization of dichlorobutenes via allylcopper intermediates seems

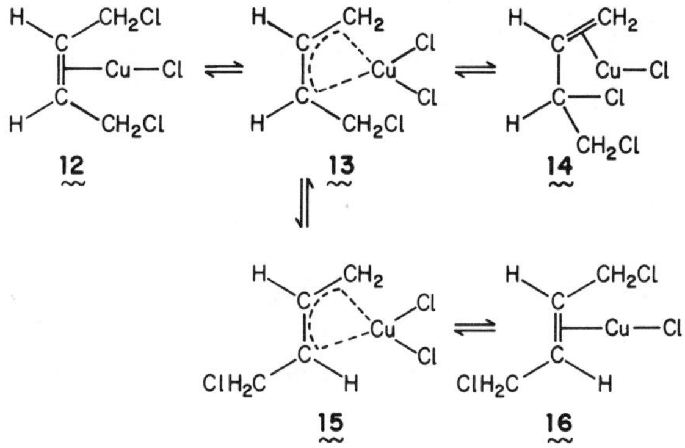

Figure 11.8 A possible mechanism for copper-catalyzed isomerization of dichlorobutenes.

plausible, but the evidence for such a mechanism is scant. Few organocopper(III) complexes have been characterized, even though they seem to be involved in many reactions catalyzed by copper(I) salts. π-Allylcopper intermediates have been suggested in the CuCl-catalyzed hydrolysis of allyl chloride [33] and the reaction of allylic acetates with methylcuprate salts [34]. π-Allylcopper intermediates probably also occur in the copper-catalyzed cyanation of dichlorobutenes discussed next.

Dicyanobutene Synthesis

The cyanation of either 1,4-dichloro-2-butene or 3,4-dichloro-1-butene gives 1,4-dicyano-2-butene as a major product [27,35,36]:

$$\begin{matrix} ClCH_2CH{=}CHCH_2Cl \\ \text{or} \\ H_2C{=}CHCHClCH_2Cl \end{matrix} \xrightarrow{CN^-} NCCH_2CH{=}CHCH_2CN$$

The reaction is catalyzed by copper salts. Detailed descriptions of the commercial process have not been published, but a large-scale laboratory experiment [35] illustrates operable conditions. A 40% solution of sodium cyanide in water was added to an acidic aqueous solution of 3,4-dichlorobutene-1 and $CuCl_2$ at 85–95°. A rapid, exothermic reaction consumed all the cyanide. 1,4-Dicyano-2-butene and 5-chloropentenenitrile were formed in 95% yield. It seems likely that the copper(II) chloride was reduced to Cu(I) under these conditions. Most patent examples use copper(I) chloride or cyanide as the catalyst. In the presence of high concentrations of CN^-, the metal probably forms $[Cu(CN)_2]^-$ and $[Cu(CN)_3]^{2-}$ complexes.

Mechanistic information about this reaction is not available, but analogies suggest allylcopper intermediates like those proposed in Figure 11.8. Exchange of cyanide for chloride within the coordination sphere of copper in a complex such as **13**-CN provides a simple mechanism for cyanation:

13-CN $\xrightarrow{CN^-}$ 17 $\longrightarrow$ 18

This exchange converts **13** to its cyanide analog (**17**). Migration of cyanide from the metal to C-1 produces **18,** the CuCN complex of 5-chloropentenenitrile. Since **13** is accessible from any of the dichlorobutenes, it seems quite reasonable that all can form **18** and, by further cyanation, the desired 1,4-dicyanobutene.

SPECIFIC REFERENCES

1. H. Ludewig, *Polyester Fibres,* Wiley-Interscience, 1971.
2. M. Katz, "Preparations of Linear Saturated Polyesters," in C. E. Schildknecht and I. Skeist, Eds., *Polymerization Processes,* Wiley-Interscience, 1977, pp. 468–496.
3. R. E. Wilfong, *J. Poly. Sci.,* **54,** 385 (1961).
4. S. G. Hovenkamp, *J. Polym. Sci.,* Pt. A-1, **9,** 3617 (1971).
5. F. B. Cramer, *Macromol. Syn.,* **1,** 17 (1963).
6. H. Zimmerman, *Faserforsch. Textiltech,* **19,** 372 (1968).
7. K-H. Wolf, B. Küster, H. Herlinger, C-J. Tschang, and E. Schrollmeyer, *Angew. Makromol. Chem.,* **68,** 23 (1978).
8. C. M. Fontana, *J. Poly. Sci.,* **6,** 2343 (1968).
9. K. Yoda, K. Kimoto, and T. Toda, *J. Chem. Soc. Japan, Ind. Chem. Sect.,* **67,** 909 (1964).
10. S. B. Maerov, *J. Polym. Sci.,* Poly. Chem. Ed., *17,* 4033 (1979).
11. R. Lasarova and K. Dimov, *Angew. Makromol. Chem.,* **55,** 1 (1976).
12. O. M. O. Habib and J. Malek, *Coll. Czech. Chem. Comm.,* **41,** 2724 (1976).
13. L. Nondek and J. Malek, *Makromol. Chem.,* **178,** 2211 (1977).
14. Research Disclosure 16710 (Mar. 1978).
15. J. Vejrosta, E. Zelena, and J. Malek, *Coll. Czech. Chem. Comm.,* **43,** 424 (1978).
16. J. H. Saunders and K. C. Frisch, "Polyurethanes: Chemistry and Technology," Pt. 1, Wiley-Interscience, 1962.
17. J. K. Backus, "Polyurethanes," in C. E. Schildknecht and I. Skeist, Eds., *Polymerization Processes,* Wiley-Interscience, 1977, pp. 642–680.
18. L. Heiss, Ger. Offenleg. 2,655,741 (1978).
19. T. W. Brooks, C. Bledsoe, and J. Rodriguez, *Macromol. Syn. Coll.,* **1,** 381 (1977).
20. J. W. Britain and P. G. Gemeinhardt, *J. Appl. Poly. Sci.,* **4,** 207 (1960).
21. M. Cuscurida and G. P. Speranza, U.S. Patent 4,101,462 (1978).
22. A. J. Bloodworth and A. G. Davies, *J. Chem. Soc.,* 5238 (1965).
23. D. B. Jacobs and J. Zimmerman, "Preparation of 6,6 Nylon and Related Polyamides," in C. E. Schildknecht and I. Skeist, Eds., *Polymerization Processes,* Wiley-Interscience, 1977, pp. 424–467.
24. P. E. Beck and E. E. Magat, *Macromol. Syn., Coll.,* **1,** 317 (1977).
25. J. K. Stille and T. W. Campbell, Eds., *Condensation Monomers,* Wiley-Interscience, 1972; G. W. Parshall, *J. Mol. Catal.,* **4,** 243 (1978).

26. C. E. Hollis, "Chloroprene and Polychloroprene Rubbers," *Chem. Ind.*, 1030 (1969).
27. V. D. Luedecke, "Adiponitrile," in J. J. McKetta and W. A. Cunningham, Eds., *Encyclopedia of Chemical Processing and Design,* Vol. 2, Marcel Dekker, 1977, pp. 146–162.
28. L. A. Smith, U.S. Patent 4,089,751 (1978).
29. F. J. Bellringer and C. E. Hollis, *Hydrocarbon Proc.,* **47** (11) 127 (1968).
30. D. D. Wild, U.S. Patent 3,515,760 (1970).
31. B. T. Nakata and E. D. Wilhoit, Ger. Offenleg. 2,248,668 (1973).
32. G. Henrici-Olivé and S. Olivé, *J. Organomet. Chem.,* **29,** 307 (1971).
33. L. F. Hatsch and R. R. Estes, *J. Am. Chem. Soc.,* **67,** 1730 (1945).
34. J. Levisalles, M. Rudler-Chauvin, and H. Rudler, *J. Organomet. Chem.,* **136,** 103 (1977).
35. I. D. Webb and G. Tabet, U.S. Patent 2,477,672 (1949).
36. M. W. Farlow, U.S. Patent 2,518,608 (1950).

12 TRENDS IN HOMOGENEOUS CATALYSIS

Several significant trends in homogeneous catalytic research have emerged in the 1970s and seem likely to persist [1]. Some are a response to the needs of the chemical industry. These include a shift to new types of feedstocks and the development of "hybrid catalysts" that combine the practical advantages of homogeneous and heterogeneous catalysts. Other trends reflect the increasing sophistication of catalytic science as it matures. One example is a trend to ever-increasing selectivity in homogeneous catalytic processes. Another is the development of new types of catalysts that are activated by light or those that contain more than one metal center.

Four of these trends are outlined in this chapter. The detailed chemistry has largely appeared in earlier chapters, but it may be profitable to generalize on the trends as perceived by an industrial chemist.

12.1 NEW FEEDSTOCKS

The "oil crises" of 1973 and 1979 have produced some realization that global hydrocarbon reserves are limited. Although the response to this realization differs among nations and industries, a need for decreased dependence on petroleum and natural gas is generally accepted. One response in the chemical industry has been increased interest in synthesis gas (CO/H_2 mixtures) as a basic feedstock. Synthesis gas is available from coal and from organic wastes, as well as from hydrocarbon feedstocks. Hence, chemistry based on it is relatively independent of the ultimate carbon source. The synthesis gas-based chemistry is versatile in its applications and uses homogeneous catalysis to good advantage. It seems most desirable as a feedstock for oxygen-containing products such as alcohols and acids. The Fischer-Tropsch synthesis of hydrocarbons from CO and H_2 will remain expensive for years to come.

According to some the latter part of the 20th century could be a "synthesis gas era" in the history of homogeneous catalysis. This viewpoint may be exaggerated, but four feedstock "eras" can be postulated:

1910–1950	Acetylene
1950–1980	Olefins
1980–2000	Synthesis gas
2000.	Biomass

As pointed out earlier the first industrial application of homogeneous catalysis was in the chlorination of acetylene. Until the 1940s, the use of soluble transition metal catalysts was largely confined to acetylene reactions. Research in Germany in the 1940s and 1950s led to use of ethylene in catalytic reactions for production of α-olefins, linear polyethylene, and acetaldehyde.

The use of synthesis gas as a feedstock in homogeneous catalysis also dates to German research in the 1940s. The olefin hydroformylation processes that were developed in this period prospered in the "olefins era." Acetic acid processes based on synthesis gas were commercialized in the 1960s, but the greatest economic incentive for this chemistry was the oil crisis of 1973. The Monsanto acetic acid process, which was commercialized in 1970, has benefited from good timing.

The inclusion of a fourth era in which biological materials such as garbage, manure, wood chips, and cornstalks are major feedstocks is speculative. The use of these materials as feedstocks is desirable both to conserve petroleum and to dispose of these wastes economically. The role of homogeneous catalysis in waste-based processes is unclear but may be a profitable area for research.

The first three eras are nicely delineated in the history of commercial production of acetic acid and of vinyl acetate (Figure 12.1). From 1930 to 1960 a large part of the production of these chemicals was based on the addition of water and acetic acid to the C≡C bond of acetylene. A soluble metal catalyst was used in both processes (Section 8.3), as well as in the oxidation of acetaldehyde to acetic acid (Section 10.5). The Wacker process, which was commercialized in 1960, permitted the use of ethylene as a feedstock for acetaldehyde synthesis (Section 6.1). The economic success of this process led to the development of closely related chemistry for acetoxylation of ethylene with both soluble and heterogeneous catalysts. The commercial vinyl acetate synthesis catalyst may be regarded as an early example of a "hybrid catalyst" (Section 12.3) in which solution-type chemistry occurs on the surface of a solid.

Synthesis gas is the starting material for production of methanol by an efficient heterogeneous catalytic process. In turn, methanol can be reacted with CO from synthesis gas to produce acetic acid. In the BASF and Monsanto processes (Section 5.3) soluble metal complexes are used, but "hybrid catalysts" have also been demonstrated for this chemistry and may have industrial advantages. The synthesis gas-based acetic acid processes were advantageous from an economic viewpoint even before the "oil crisis." The cost advantage has become even greater recently.

ACETYLENE

$$HC{\equiv}CH \xrightarrow{H_2O} CH_3CHO \xrightarrow{O_2} CH_3COOH$$

$$HC{\equiv}CH \xrightarrow{AcOH} H_2C{=}CHOAc$$

ETHYLENE

$$H_2C{=}CH_2 \xrightarrow[O_2]{H_2O} CH_3CHO \xrightarrow{O_2} CH_3COOH$$

$$H_2C{=}CH_2 \xrightarrow[O_2]{AcOH} H_2C{=}CHOAc$$

SYNTHESIS GAS

$$CO + H_2 \longrightarrow CH_3OH \xrightarrow{CO} CH_3COOH$$

$$2\,CH_3OAc + 2\,CO + H_2 \longrightarrow CH_3CH(OAc)_2 + HOAc \xrightarrow{\Delta} H_2C{=}CHOAc + HOAc$$

Figure 12.1. A historical progression in feedstocks and potential feedstocks for production of acetic acid and vinyl acetate.

This economic incentive and the raw material flexibility of synthesis gas make it the feedstock of choice for the foreseeable future.

Synthesis gas can also be used for production of vinyl acetate according to a recent patent [2]. Methyl acetate, made from methanol and acetic acid, can be converted to 1,1-diacetoxyethane by treatment with synthesis gas in the presence of soluble rhodium or palladium complexes. The diacetoxyethane pyrolyzes cleanly to vinyl acetate in well-established technology. Although this vinyl acetate process is not yet used commercially, it seems to have a potential economic advantage like that of acetic acid processes based on synthesis gas.

Acetic acid technology also illustrates other points not shown in the figure. These are that *alkanes* and *biological materials* are attractive feedstocks for the future. At present and in the near future alkanes are much cheaper than olefins and compete well with synthesis gas in economic terms. The manufacture of acetic acid by oxidation of butane or light naphtha (Section 10.5) has been attractive ever since its introduction in the early 1950s in spite of relatively low yields. Uses for these feedstocks other than as fuels are somewhat limited. As a consequence light alkanes should be considered as feedstocks for a number of important industrial chemicals. Homogeneous catalysts for alkane reactions other than radical chain oxidations are rare, but fundamental knowledge about alkane C—H and C—C activation is being developed [3]. This basic knowledge should assist in development of practical processes.

Fermentation and other enzymatic processes have always been a significant source of acetic acid. Food-grade acetic acid (vinegar) is made by all-biological processes, but "synthetic" acetic acid was also produced up to about 1950 by homogeneous catalytic oxidation of grain alcohol. Now grain alcohol and ethylene-derived ethanol are again nearly equal in price. A few more years may see biomass-based ethanol as an attractive feedstock. The same advantage may also exist with furfural derived from cornstalks and bagasse (sugar cane refuse). It was a major starting material for production of tetrahydrofuran and 1,6-hexanediamine before natural gas and petroleum became the dominant feedstocks for organic chemicals. A reversion to feedstock economics like those before 1950 could present many new challenges for homogeneous catalysis.

12.2 ENHANCED SELECTIVITY

One of the greatest virtues of homogeneous catalysis is its selectivity in complex organic reactions. One conspicuous example is the synthesis of adiponitrile by regioselective addition of 2 moles of HCN to the double bonds of butadiene (Section 4.6). In each of the two additions there is little difference in the activation energies for internal versus terminal CN placement. Nevertheless, the mild conditions accessible through homogeneous catalysis permit both additions to occur with very high selectivity to form the desired product isomer.

Such selectivity may be most useful in the pharmaceutical industry because biological activity is so often maximized in one particular isomer of a complex substance. For example, the *l* isomer of 3,4-dihydroxyphenylalanine is active in treatment of Parkinson's disease. It has been possible to produce this optical isomer selectively by hydrogenation of a prochiral olefin with a soluble rhodium catalyst that bears a chiral ligand (Section 3.6). Although olefin hydrogenation has been studied intensively as a method of asymmetric synthesis, many other homogeneous catalytic processes can also be used to induce asymmetry (Figure 12.2) [8–12]. Several articles review these reactions and their applications [4–7].

One nearly unique characteristic of the pharmaceutical industry is the high value and small volume of its product. For this reason expensive catalysts and reagents can be used if they permit a complex reaction sequence to be carried out quickly and easily. A nice example of this is the selective isomerization of an ethynyl alcohol to an α,β-unsaturated aldehyde in one step with a silyl vanadate catalyst [13]:

$$\mathrm{RR'C(OH)C{\equiv}CH} \xrightarrow{(Ph_3SiO)_3VO} \mathrm{RR'C{=}CH{-}CHO}$$

Many applications in synthesis of steroids and prostaglandins seem likely. When organic chemists become more comfortable with the use of soluble metal complexes as reagents and catalysts, these applications should expand rapidly.

The beneficial effects of the selectivity of organometallic catalysts are not necessarily limited to small-volume applications. The polyolefins field demonstrates

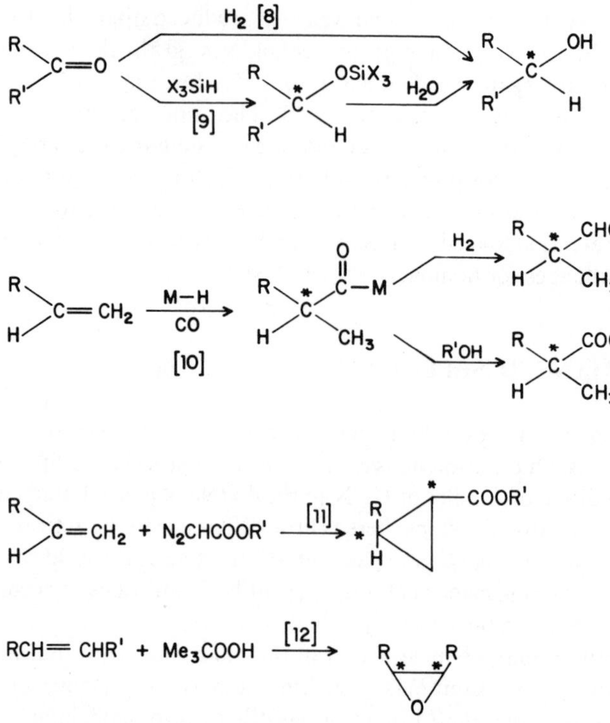

Figure 12.2 Some synthetically useful reactions in which optical asymmetry has been induced by use of soluble catalysts with chiral ligands.

the utility of stereoselectivity in large-volume products. Commercial isotactic polypropylene, which contains hundreds of ordered asymmetric centers in each polymer chain, has greatly different properties from nonordered (atactic) polypropylene. Similarly commercial *cis*-1,4-polybutadiene represents a very high degree of selectivity in formation of one particular isomer.

One of the most elegant applications of stereoselective polymerization is synthesis of block copolymers by catalyst modification during polymerization. A polybutadiene that contains "blocks" of *cis*-1,4 and *trans*-1,4 units in the polymer chain has been prepared in this way [14]. This effect is achieved by changing the reaction conditions while the polymer is still "alive," that is, while it contains a reactive organometallic end group. This technique of modifying a "living polymer" can also be used to make a polymer in which one half or "block" is distinctly different from the other. Addition of a lactone or an epoxide to a growing polybutadiene chain will produce an extended segment of polylactone or polyether in the final polymer molecule. Such copolymers with blocks of different polarity appear useful as blending agents to promote formation of homogeneous polymer composites from substances such as polyvinyl chloride and polystyrene that are partially immiscible otherwise [15].

12.3 HYBRID CATALYSTS

The selectivity of soluble transition metal catalysts sometimes makes them preferable to conventional heterogeneous catalysts for a particular reaction, but there are disadvantages as well. A major disadvantage from an engineering viewpoint is that, at the end of the reaction, it may be difficult to separate the product from the catalyst [16]. With volatile products such as acetaldehyde (b.p. 21°), distillation is an easy answer. However, separation of adiponitrile (b.p. 295°) from a thermally sensitive hydrocyanation catalyst is obviously much more difficult.

One common solution to this problem is to embed a catalyst species that is ordinarily soluble in an insoluble catalyst support. Such hybrid catalysts that combine the virtues of homogeneous and heterogeneous catalysts are popular subjects for research at present. However, several commercial catalysts have used this principle for many years. One of the oldest is used in production of sulfuric acid. In the catalytic oxidation of SO_2 to SO_3, a mixture of SO_2 and O_2 is passed over a hybrid catalyst. The catalytic species is a vanadate salt dissolved in a molten alkali sulfate eutectic that is dispersed in a porous silica support [17]. Thus the catalyst is in solution but is in a different phase from most of the reactants and products, a true hybrid catalyst. A similar situation probably exists in the commercial synthesis of vinyl acetate from ethylene, oxygen, and acetic acid (Section 6.2). The catalyst is a potassium acetate-impregnated palladium/gold alloy supported on silica. It seems likely that the catalytic reaction occurs in an acetic acid film on the catalyst surface. This technique for carrying out a homogeneous catalytic reaction on a solid support may have considerable application [18]. Its main limitation is that the reactants and products must be volatile at the reaction temperature.

A more general approach is synthesis of a heterogeneous catalyst by chemically bonding a transition metal complex to a solid support. The most widely studied supports are metal oxides such as silica and alumina [19] and polymers that have pendant phosphine groups [20,21]. The reactions used to bind a metal complex to such supports are illustrated in Figure 12.3.

The use of a metal oxide support is illustrated by the preparation of the Union Carbide polyethylene catalyst (Section 4.1). Chromocene reacts with hydroxyl groups on a silica surface. The initial step is cleavage of a cyclopentadienyl group from chromium by protonolysis [22]. Subsequently oxidative addition of another OH group to the chromium(II) ion may form a metal hydride species as shown in the figure [23]. The Cr—H bond would be an attractive starting point for growth of a polymer chain by repetitive insertion of ethylene molecules. In practice this catalyst is highly efficient, and the catalyst residues are so innocuous that they can be left in the finished polymer.

More attention in academic laboratories has been directed to use of phosphinated polystyrene as a support for conventional soluble catalysts. Several phosphine-containing polymers have been prepared by substitution reactions on preformed polystyrene or by copolymerization of monomers such as *p*-diphenylphosphinostyrene with styrene. These phosphine-containing polymers attach to the catalytic

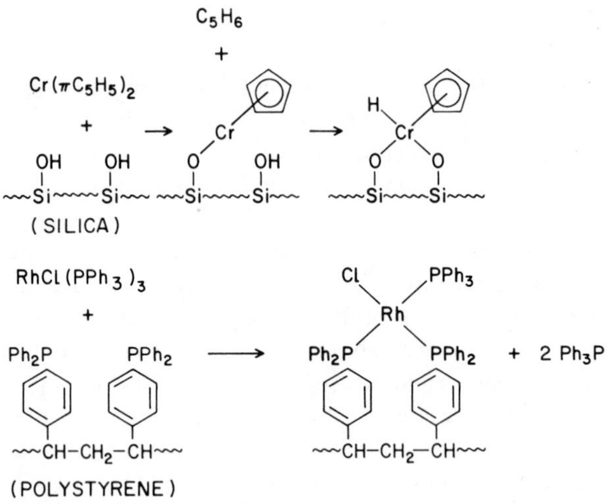

Figure 12.3. Synthesis of hybrid catalysts by reaction of metal complexes with silica and phosphinated polystyrene surfaces.

complex as donor ligands, as shown at the bottom of Figure 12.3. In this reaction two labile triphenylphosphine ligands of Wilkinson's catalyst are displaced by phosphine groups of the polymer. When the displaced triphenylphosphine has been washed from the system, the rhodium complex is firmly anchored to the polymer. This product can be used as a catalyst for many reactions of olefins such as hydrogenation and hydroformylation. Such catalysts are not used commercially but may have potential practical applications.

The hybrid catalysts of Figure 12.3 have both the advantages and disadvantages of conventional heterogeneous catalysts. One major advantage is ease of catalyst separation from product. If the catalyst is used in a "fixed bed" mode in which reactants are passed through a column of the solid catalyst, the separation process is minimal. However, with polymeric phosphine-supported metal catalysts like that of the figure, there is some tendency for rhodium to migrate along the column and even to be leached away in prolonged use [24]. The catalyst can also be used as a slurry in the reaction mixture and be removed by filtration at the end of the reaction. This technique is desirable for highly exothermic reactions because heat removal from a fixed bed of polymeric catalyst is difficult.

As with conventional heterogeneous catalysts, diffusion of reactants and products within a catalyst particle is often a rate-limiting step. This situation is especially complex with polymeric catalysts that are used in organic media. Unless the polymer is highly crosslinked, extensive swelling occurs because solvent diffuses into the polymer lattice. The swelling may be desirable to increase the diffusion of reactants and products but must be controlled to prevent disruption of the catalyst particle.

With metal oxide support materials, swelling does not occur and transport occurs by diffusion through the pore structure of the support. For this reason the silica and alumina supports used for conventional heterogeneous catalysts have received most industrial attention. Catalyst species like Wilkinson's complex can be bound to the metal oxides if the support has been pretreated with a reagent such as $(EtO)_3SiCH_2CH_2PPh_2$ [25]. The silicon end of the reagent binds to silica by Si—O—Si bonds and leaves the phosphine available as a ligand for coordination to a transition metal ion.

The hybrid catalysts seem likely to find a place in industrial practice in the future, although commercial application may be slow. As in the case of the polyethylene catalysts of Section 4.1, the hybrids may be used initially because they have unusual catalytic properties rather than because of the suggested engineering advantages. One fringe benefit of research on hybrid catalysts may be that organometallic chemists become aware of the capabilities and challenges of heterogeneous catalysts.

12.4 NEW CATALYST CHEMISTRY

Another apparent trend is the development of more active soluble catalysts through new concepts in catalyst design. Two approaches are especially noteworthy. Photoactivated catalysts open the way to reactions at even lower temperatures than those used with conventional homogeneous catalysts [26]. Metal cluster catalysts should provide multiple reaction sites for particularly difficult transformations such as alkane isomerization and hydrogenolysis.

Photoactivated Catalysts

The rate-limiting step in many homogeneous catalytic reactions is dissociation of a ligand from the metal ion. This step is necessary to create a vacant coordination site for interaction with a substrate such as an olefin. Sometimes, as in hydrogenations with $Cr(CO)_6$ as a catalyst, temperatures in excess of 100° are needed to obtain a reasonable dissociation rate. The energy necessary for CO dissociation can also be supplied by light. Thus, a solution of $Cr(CO)_6$ catalyzes hydrogenation of dienes at 10° after radiation with 300–380 nm light [27]. The low reaction temperature permits even greater selectivity in the hydrogenation of dienes to monoenes than with the thermally activated catalyst. The catalytic activity of an irradiated $Cr(CO)_6$ solution persists for several hours in the dark.

Selective hydrogenation is the reaction most intensively studied with photoactivated catalysts, but several other homogeneous catalytic processes are also susceptible to this form of activation. For example, hydrosilylation is catalyzed by irradiated solutions of $Fe(CO)_5$ [26]. Most attention has been devoted to metal carbonyl catalysts because photodissociation of CO is facile. It would, however, be surprising if other π-bonding ligands were not susceptible to similar photoexpulsion processes.

Metal Cluster Catalysts

Many multistep reactions such as alkane isomerization and N_2 hydrogenation occur readily with heterogeneous catalysts but not with homogeneous catalysts. The usual explanation is that several catalyst "sites" are necessary for these complex transformations. Such multiple sites are available on the surface of a metal crystallite but not in a conventional soluble catalyst centered on a single metal ion. In principle a metal cluster compound such as $Rh_6(CO)_{16}$ could provide several reaction sites or could act as a reservoir for electrons in a multielectron redox process.

Polynuclear complexes that contain metal-metal bonds are being explored intensively as catalysts for reactions of olefins and of carbon monoxide [28]. Many of the metal carbonyl cluster compounds exhibit considerable activity. For example, hydroformylation of 1-pentene with $Ru_3(CO)_{12}$, $H_4Ru_4(CO)_{12}$, or $[H_3Ru_4(CO)_{12}]^-$ gives *n*-hexanol in high yield [29]. These ruthenium clusters also catalyze the "water gas shift reaction":

$$CO + H_2O \rightleftharpoons CO_2 + H_2$$

Unfortunately, it is difficult to establish that the cluster remains intact during the catalytic reaction. While it seems likely that cluster compounds are catalysts for some reactions, this point has never been proved unequivocally. There is always a nagging doubt that some very active mononuclear species is the catalyst, even though its concentration is undetectably small.

Photoactivation of $Ru_3(CO)_{12}$ and $Ru_3(CO)_9(PPh_3)_3$ seems to produce catalysts that retain the cluster structure, at least in part [30]. Visible light converts these trimers to active catalysts for isomerization of 1-pentene to *cis*- and *trans*-2-pentene at room temperature. The initial product distributions and low energy of the activating light distinguish these catalysts from that generated from $Ru(CO)_4(PPh_3)$. Photoactivation seems to be an especially promising approach to cluster catalysis.

The concern about mononuclear catalysts formed by dissociation of a cluster seems less pressing for alkane fragmentation reactions. Small clusters of nickel atoms formed by condensing nickel vapor in a pentane matrix cleave both C—H and C—C bonds of the pentane to produce C_1-C_4 alkanes [31]. These reactions occur even at −130°. Similarly, Fe_2 clusters prepared by evaporation of iron attack C—H bonds in a methane matrix at −250° [32]. These small metal clusters appear to have reactivity like a metal surface.

Despite the difficulty in characterization of metal cluster catalysts, catalytic activity seems well established. These compounds are promising starting points for exploration of new forms of catalytic activity. The attachment of metal clusters to surfaces may provide heterogeneous catalysts with distinctive selectivity [33].

SPECIFIC REFERENCES

1. F. Basolo and R. L. Burwell, Eds., *Catalysis, Progress in Research*, Plenum Press, 1973.
2. N. Rizkalla and C. N. Winnick, Ger. Offenleg. 2,610,035 (1976).
3. G. W. Parshall, *Catalysis*, **1**, 335 (1977); D. E. Webster, *Adv. Organomet. Chem.*, **15**, 147 (1977); A. E. Shilov and A. A. Shteinman, *Coord. Chem. Rev.*, **24**, 97 (1977).

4. R. Pearce, *Catalysis*, **2,** 176 (1978).

5. J. D. Morrison, W. F. Masler, and M. K. Neuberg, "Asymmetric Homogeneous Hydrogenation," *Adv. Catal.* **25,** 81 (1976).

6. J. W. Scott and D. Valentine, *Science*, **184,** 943 (1974); idem, *Synthesis*, 329 (1978).

7. B. Bogdanovic, *Angew. Chem. Int. Ed.*, **12,** 954 (1973).

8. J. Solodar, *Chem. Tech.*, 421 (1975).

9. I. Ojima, K. Yamamoto, and M. Kumada, *Aspects Homog. Catal.*, **3,** 186 (1977).

10. G. Consiglio, P. Pino, et al., *Angew. Chem. Int. Ed.*, **12,** 669 (1973); *Helv. Chim. Acta*, **59,** 642 (1976).

11. A. Nakamura, A. Konishi, Y. Tatsuno, and S. Otsuka, *J. Am. Chem. Soc.*, **100,** 3443, 3449, 6544 (1978).

12. K. B. Sharpless et al., *J. Am. Chem. Soc.*, **99,** 1990 (1977); S. Yamada et al., ibid, **99,** 1988 (1977); C. Dobler and E. Höft, *Z. Chem.*, **18,** 218 (1978).

13. H. Pauling, D. A. Andrews, and N. C. Hindley, *Helv. Chim. Acta*, **59,** 1233 (1976).

14. Ph. Teyssie, Proceedings, 1st International Symposium on Homogeneous Catalysis, Corpus Christi, TX (1978).

15. Ph. Teyssie et al., in T. Saegusa and E. Goethals, Eds., "Ring Opening Polymerization," American Chemical Society, Washington, 1977, pp. 165–177.

16. G. W. Parshall, *J. Am. Chem. Soc.*, **94,** 8716 (1972); P. R. Rony., *Ann. N.Y. Acad. Sci.*, **172,** 238 (1970).

17. H. F. A. Topsoe and A. Neilsen, *Trans. Danish Acad. Tech. Sci.*, 3, 18 (1948).

18. P. R. Rony and J. F. Roth, *J. Mol. Catal.*, **1,** 13 (1975); U.S. Patent 3,855,307 (1974).

19. L. L. Murrell, in J. J. Burton and R. L. Garten, Eds, *Advanced Materials in Catalysis*, Academic Press, 1977, p. 235.

20. F. R. Hartley and P. N. Vezey, *Adv. Organomet. Chem.*, **15,** 189 (1977).

21. Y. Chauvin, D. Commereuc, and F. Dawans, *Prog. Poly. Sci.*, **5** (3–4) 95 (1977).

22. M. H. Chisholm, F. A. Cotton, M. W. Extine, and D. C. Rideout, *Inorg. Chem.*, **18,** 120 (1979).

23. D. T. Laverty, J. J. Rooney, and A. Stewart, *J. Catal.*, **45,** 110 (1976).

24. W. H. Lang, A. T. Jurewicz, W. O. Haag, D. D. Whitehurst, and L. D. Rollmann, *J. Organomet. Chem.*, **134,** 85 (1977).

25. K. G. Allum, R. D. Hancock, I. V. Howell, T. E. Lester, S. McKenzie, R. C. Pitkethly, and P. J. Robinson, *J. Organomet. Chem.*, **107,** 393 (1976); A. A. Oswald and L. L. Murrell, U.S. Patent 4,083,803 (1978).

26. M. S. Wrighton, D. S. Ginley, M. A. Schroeder, and D. L. Morse, *Pure Appl. Chem.*, **41,** 671 (1975).

27. M. S. Wrighton and M. A. Schroeder, *J. Am. Chem. Soc.*, **95,** 5764 (1973).

28. E. L. Muetterties, Science, **196,** 839 (1977); C. U. Pittman and R. C. Ryan, *Chem. Tech.*, **8,** 170 (1978).

29. R. M. Laine, *J. Am. Chem. Soc.*, **100,** 6451 (1978).

30. J. L. Graff, R. D. Sanner, and M. S. Wrighton, *J. Am. Chem. Soc.*, **101,** 273 (1979).

31. S. C. Davis and K. J. Klabunde, *J. Am. Chem. Soc.*, **100,** 5973 (1978).

32. P. H. Barrett, M. Pasternak, and R. G. Pearson, *J. Am. Chem. Soc.*, **101,** 222 (1979).

33. J. M. Bassett and R. Ugo, *Aspects Homog. Catal.*, **3,** 137 (1977).

INDEX

Acetaldehyde:
from acetylene, 155
from ethylene, 102
oxidation, 198
Acetic acid:
from acetaldehyde, 198
from alkanes, 196
from CO and methanol, 80
trends, 223
Acetic anhydride, 98, 200
Acetoxylation:
benzene, 127
ethylene, 104, 113
propylene, 106, 108
Acetylene:
addition of HCl, 153, 157
addition of HCN, 157
addition of HOAc, 156
addition reactions, 154
bonding characteristics, 148
carbonylation, 159
chlorination, 158
complexes, 148
dimerization, 152
hydration, 155
oligomerization, 163
oxidative coupling, 150
reaction with aldehydes, 151
reaction with HgX_2, 149
reaction with ketones, 151
trimerization, 165
Acetylide ion:
bonding characteristics, 148
Acrylate esters, 159
Acrylic acid, 159
Acrylonitrile:
dimerization, 60, 217
hydrosilylation, 73
synthesis, 111, 157
Acyl chlorides:
decarbonylation, 91
Acyl complexes:
in acetic acid synthesis, 81
in carboxylation, 82, 83, 129
in decarbonylation, 92
in hydroformylation, 88, 90
in ROH homologation, 97, 98
Adipic acid, 190
polyamidation, 216
polyesterification, 209
Adiponitrile, 38, 60, 70, 217
Alcohols, linear, 57, 59, 85
Aldehydes:
decarbonylation, 91
synthesis, 85
Alkanes:
C–H bond cleavage, 179, 230
H/D exchange, 179
hydrogenolysis, 180
isomerization, 182
reaction with oxygen, 188
Alkyl hydroperoxides:
decomposition, 196, 198
formation, 190, 203
as oxidants, 115
reaction with metal ions, 187
Alkylidene ligands:
formation, 171, 178
in heterogeneous catalysis, 180
in metathesis, 177

in olefin dimerization, 62
Allyl acetate, 106, 108
Allyl ligands:
in arene hydrogenation, 141
bonding characteristics, 25
in dichlorobutene reactions, 218
geometry, 26
in isomerization, 33, 35
in olefin hydrogenation, 43
in olefin oxidation, 108
in polymerization, 55
Alpha-Olefins:
hydroformylation, 85
synthesis, 56, 58
Aluminum compounds:
in olefin oligomerization, 56
triethylaluminum, 57
Antimony compounds:
in polyesterification, 210, 212
Arenes, *see specific compound*
Asymmetric synthesis:
applications, 226
epoxidation, 117
hydroformylation, 86
hydrogenation, 43

Benzene:
acetoxylation, 127
bonding characteristics, 121
carboxylation, 129
chlorination, 120
copper complex, 123
coupling to olefins, 124
hydrogenation, 140, 141, 143
oxidative coupling, 126
substitution reactions, 120, 122, 127
synthesis, 165
Benzoic acid:
conversion to terephthalic acid, 200
decarbonylation, 130
synthesis, 129, 201
Benzoquinone, 132
Benzyl chloride:
oxidative addition, 14
reaction with CO, 80
Bifurandione, 160
Biomass as a feedstock, 223, 224
Biphenyl, 126, 136
Biphenyldicarboxylic acids, 126, 202
Block copolymers, 55, 226
Bonding:
diene complexes, 24
dinitrogen complexes, 8
Bromide ion:
in oxidation catalysis, 201, 204
Bromobenzene:
carbonylation, 129
reductive coupling, 136
o-Bromonitrobenzene:
reductive coupling, 136
Butadiene:
chlorination, 217
codimerization with ethylene, 63
complexes of, 24, 55
copolymers, 226
dimerization, 65
hydrocyanation, 70
oxidation, 109
polymerization, 54
trimerization, 65, 68
Butane:
oxidation, 196
1,4-Butanediol:
polyesterification, 209
reaction with isocyanates, 214
synthesis, 109, 110, 151
Butyne-1,4-diol, 151
1-Butene, synthesis, 62
2-Butene, synthesis, 174
Butyraldehyde:
synthesis, 85, 87, 89

Carbamic acid and esters, 93, 214
Carbene complexes, *see* Alkylidene ligands
Carboalkoxy complexes, 84
Carbon monoxide:
bonding, 77
hydrogenation, 95
metal complexes, 6, 13, 77
oxidation, 94
Carbonylation reaction:
benzyl chloride, 80
ethanol, 80
methanol, 80
Carboxylation reaction:
olefins, 82
Carvone, hydrogenation, 38
Catalysis, general, 17
Catalysis, homogeneous:
history, 1, 147, 222
reviews, 3
trends, 222
uses, 2, 186
Catalysts, heterogeneous:
alkane reactions, 179
olefin acetoxylation, 106, 109
olefin polymerization, 50, 227
propylene oxidation, 109

Catalysts, hybrid, 50, 227
Chlorination:
  acetylene, 147, 158
  benzene, 120
  butadiene, 217
2-Chlorobutadiene (chloroprene), 152, 217
Chromium complexes:
  $Cr(CO)_6$:
    bonding, 6
    CO replacement, 7
    photolysis, 8, 229
  $Cr(CO)_3$ (arene):
    bonding, 121
    protonation, 122
    reaction with nucleophiles, 122
  $Cr(C_6H_6)_2$:
    structure, 121
  in hydrogenation, 42, 229
  in oxidation, 191
  in polymerization, 50, 228
Cluster catalysts, 95, 229
Cobalt complexes:
  in acetaldehyde oxidation, 198
  in acetylene trimerization, 166
  in alkane oxidation, 196
  in arene hydrogenation, 139, 140, 141
  in bifurandione synthesis, 160
  in butane oxidation, 196
  in carboxylation of olefins, 82
  in carbonylation of ROH, 80, 96
  $Co_2(CO)_8$:
    ligand replacement, 13
    photolysis, 13
  in hexadiene synthesis, 65
  in hydroformylation, 86, 88
  in hydrogenation of benzene, 139, 140, 141
  in hydrogenation of CO, 95
  in hydrogenation of olefins, 38, 40, 42
  in methanol carbonylation, 80, 96
  in oxidation of cyclohexane, 191
  in oxidation of toluene, 201
  in oxidation of xylenes, 201, 202
  oxygen complexes, 187
  in polymerization of dienes, 54
  in transesterification, 209
Copper (I) acetylides:
  in acetylene dimerization, 153
  in butynediol synthesis, 152
  in oxidative coupling, 150
  properties, 149
  structure, 148
Copper (I) benzoate:
  decarbonylation, 130
Copper (II) benzoate:
  oxidative decarboxylation, 130
Copper (II) bromide:
  in ethylene oxidation, 114
Copper (I) chloride:
  in acetylene dimerization, 152
  in hydrosilylation, 73
  in isomerization, 154, 218
  oxidation product, 133, 150
Copper (II) chloride:
  in ethylene oxidation, 102
Copper (II) phenoxide, 134
Copper (I) trifluoromethanesulfonate, 136
Cumene, 137
Cyanation, 219
Cyclododecane:
  oxidation, 194
Cyclododecanol, 194
1,5,9-Cyclododecatriene:
  hydrogenation, 41
  synthesis, 66, 68
Cyclododecene:
  epoxidation, 116
  synthesis, 41
Cyclohexane:
  from benzene, 140, 141, 143
  oxidation, 190
Cyclohexanol:
  oxidation, 193
  synthesis, 190
Cyclohexanone:
  oxidation, 193
  synthesis, 190
Cyclohexene, 143
Cyclohexyl hydroperoxide, 190
1,5-Cyclooctadiene:
  hydrogenation, 42
  synthesis, 66
Cyclooctatetraene, 164
Cyclooctene, 42

Decarbonylation:
  acyl chlorides, 91
  aldehydes, 91
1-Decene:
  oxidation, 103
1,1-Diacetoxyethane:
  from acetylene, 156
  synthesis from CO, 97, 224
o-Di-*n*-butylbenzene, 137
Dichlorobutenes:
  cyanation, 219
  isomerization, 217, 218
  synthesis, 217

1,4-Dicyanobutene:
  hydrogenation, 38
  synthesis, 219
Dienes:
  hydrogenation, 41
1,4-Dihydroxybenzene, 161
3,3-Dimethyl-1-butene, 175
Dimethylbutenes, 60, 62
Dinitrogen complexes, *see* Nitrogen complexes
DIOP ligand, 44, 86
Dioxygen, *see* Oxygen
Dodecanedioic acid, 194
L-dopa, 43, 225

18-Electron rule, 6
Elimination reaction, 30
  α hydrogen, 172
  β hydrogen, 16
Epoxidation, 115
Ethane:
  hydrogenolysis, 180
Ethanol:
  oxidation, 198
  reaction with CO, 80
  synthesis from MeOH, 96
Ethylene:
  aluminum alkyl reaction, 56
  codimerization with butadiene, 63
  codimerization with norbornadiene, 63
  copolymers, 52
  dimerization, 62
  oligomerization, 56, 58
  oxidation, 102, 105, 113
  polymerization, 49
  reaction with aryl bromides, 126
  reaction with benzene, 124
  synthesis by metathesis, 174
Ethylene glycol:
  acetate esters, 105, 107, 113
  ether formation, 210
  polyesterification, 209
  synthesis from CO, 95
2-Ethylhexanoic acid, 82, 85
2-Ethylhexanol, 85
Ethylidene diacetate, *see* 1,1-Diacetoxyethane
Ethylidenenorbornene, 63

Feedstocks, 222
Formaldehyde:
  reaction with C_2H_2, 151

Grignard reagents:
  coupling, 137

1,4-Hexadiene:
  isomerization, 35
  polymerization, 52
  synthesis, 63
1,6-Hexanediamine:
  polyamidation, 216
  synthesis, 217
Hexaphenylbenzene, 166
Hydride complexes:
  bonding, 5
  in hexadiene synthesis, 64
  in hydrocyanation, 71
  in hydrogenation, 40, 141
  in hydrosilylation, 72
  in isomerization, 33, 35
Hydrocyanation:
  acetylene, 157
  butadiene, 70
Hydroformylation, 85
  by cluster catalysts, 230
  mechanism, 88
Hydrogen:
  isotopic exchange, 179
Hydrogenation:
  arenes, 139
  asymmetric, 43
  carbon monoxide, 95
  catalysts, 37, 42
  olefins, 37
Hydrogen chloride:
  addition to C≡C, 153, 157
Hydrogen cyanide:
  addition to C≡C, 157
  addition to olefins, 70
α-Hydrogen elimination, 172
β-Hydrogen elimination, 16, 30, 33
Hydrogenolysis:
  alkanes, 180
Hydroquinone, 133, 161
Hydrosilylation:
  ketones, 226
  mechanism, 72
  olefins, 71

Inert gas rule, 6
Insertion reactions:
  acetylenes, 17, 153, 167
  carbon monoxide, 16, 79
  olefins, 16, 29
Iron carbonyls:
  acetylene reactions, 162
  in hydroquinone synthesis, 162

in olefin isomerization, 34
photolysis, 229
Iron (III) chloride, 120, 159
Isocyanates:
reaction with ROH, 214
synthesis, 93
Isomerization:
alkanes, 182
1-butene, 17
catalysts, 32
chlorobutadienes, 153, 154
in decarbonylation, 91
dichlorobutenes, 218
ethynylcarbinols, 225
mechanisms, 33
olefins, 17, 31, 59
pentenenitriles, 31
Isophthalic acid, 200
Isoprene:
dimerization, 66
polymerization, 54
Isopropylbenzene, 137

Laurylactam, 66, 194
Ligands:
bonding characteristics, 6
chiral, 44, 86, 117, 225
dissociation, 7, 12, 18
electronic effects, 10
replacement, 7, 12
steric effects, 10
Linolenate esters:
hydrogenation, 41
Lithium compounds:
butyllithium, 54

Manganese complexes:
$Mn(COCH_3)(CO)_5$, 79, 84
in oxidation, 191, 199, 201
in transesterification, 209
Mercury salts:
in acetylene reactions, 149, 155, 157
Metal carbonyls, 6, 13, 77
Metal clusters, 95, 229
Metallocycles, 62, 177
Metathesis of olefins, 58, 173
Methane:
H/D exchange, 179
Methanol:
reaction with CO, 80, 96
Methyl acetate:
carbonylation, 97
Methyl isobutyrate, 83, 84
Methylene-bis(phenylisocyanate), 93, 215
Molybdenum complexes:
as epoxidation catalysts, 116
$Mo(CO)_6$:
CO replacement, 7, 12
in epoxidation, 115
in olefin metathesis, 174
Muconic acid, 132

Naphtha:
oxidation, 196
Naphthalene:
hydrogenation, 139
Neohexene, 175
Neopentane:
isomerization, 182
Nickel complexes:
in acetylene carbonylation, 159
in acetylene oligomerization, 165
in arene hydrogenation, 140
in butadiene dimerization, 66
in butadiene trimerization, 69
in cyclooctatetraene synthesis, 164
in haloarene coupling, 136, 137
in hexadiene synthesis, 64
in hydrocyanation, 70
in hydrogenation, 41
in isomerization, 32
$Ni(PX_3)_4$:
ligand dissociation, 11
protonation, 11, 18, 71
structure, 11
in olefin dimerization, 60
in olefin oligomerization, 58
in polymerization, 54
Nitric acid, 193, 194
Nitriles:
cotrimerization with C_2H_2, 168
Nitro compounds:
conversion to isocyanates, 93
Nitrogen complexes, 8
Nonanal, 87
Nylon, 190, 216

Octadienes, 42
Octanedioic acid, 160
1,8-Octanediol, 160
1,3,7-Octatriene:
hydrogenation, 42
synthesis, 66
1-Octene:
epoxidation, 116
hydroformylation, 87
Olefins:
bonding characteristics, 22, 24

carboxylation, 82
coupling to arenes, 124
dimerization, 59
hydroformylation, 59, 85
hydrogenation, 36
hydrosilylation, 71
insertion reactions, 29
isomerization, 17, 31, 58
metathesis, 58, 173, 183
oligomerization, 56
oxidation, 101
polymerization, 48
ring-opening polymerization, 175
skeletal isomerization, 35
Olefin complexes:
nucleophilic addition, 27, 104, 108
stability, 23
Oligomerization:
acetylenes, 163
ethylene, 56
Oxidation:
acetaldehyde, 198
butane, 196
cyclododecane, 194
cyclohexane, 190
cyclohexanol, 193
general principles, 13, 185
naphtha, 196
toluene, 201
xylenes, 201
Oxidative addition, 14
Oxidative coupling:
olefins, 112
phenols, 133
Oxygen:
metal complexes, 187
reaction with alkanes, 188

Palladium complexes:
in acetylene trimerization, 166
in arene reactions, 123, 139
in CO oxidation, 95
in carboxylation of olefins, 82
in decarbonylation, 91
in dimerization of dienes, 67
in halobenzene coupling, 139
in isocyanate synthesis, 93
in isomerization, 34
in oxidation of C_2H_4, 102, 105
in oxidation of C_3H_6, 106
in oxidation of dienes, 109
Pentenenitriles:
isomerization, 31, 70
synthesis, 70
Peracetic acid, 198
Phenol:
oxidation, 132
synthesis, 127, 130
Phenyl acetate, 127
Phenylacetic acid, 80
p-Phenylene oxide, 132
Pheromones:
from olefin metathesis, 176
Phosphomolybdovanadates:
in arene coupling, 126
in ethylene oxidation, 103
Photolysis:
$Co_2(CO)_8$, 13
$Cr(CO)_6$, 8, 229
$Cr(CO)_3$ (arene), 43
$Ru_3(CO)_{12}$, 230
Platinum complexes:
in hydrogenation, 38, 40, 42
in hydrosilylation, 72
$[Pt(C_2H_4)Cl_3]^-$, 22, 24
$Pt(PEt_3)_3$, 14
Polyamides, 190, 216
Polybutadiene, 54, 226
Polyester, 208
Polyethylene, 49
Polyisoprene:
hydrogenation, 37
structure, 53
synthesis, 54
Polymerization:
applications, 49, 208
see also specific reactants
Polymers:
hydrogenation, 37
Polypropylene, 51
Polyurethanes, 214
Propargyl alcohol:
oligomerization, 164
synthesis, 152
Propionic acid:
synthesis from ethanol, 80
synthesis from ethylene, 82
Propylene:
carboxylation, 83
copolymers, 52
dimerization, 56, 59, 60, 62
epoxidation, 115
hydroformylation, 85, 87, 89
metathesis, 174
oxidation, 106, 114
polymerization, 51
Propylene oxide, 115
Pyridines, 167, 168

Pyromellitic dianhydride, 202

Reductive elimination, 14, 71
Rhodium complexes:
  in acetylene dimerization, 153
  in carbonylation of MeOH, 80
  in carbonylation of MeOAc, 98
  in decarbonylation, 91
  in ethylene dimerization, 60
  in hexadiene synthesis, 64
  in hydroformylation, 89
  in hydrogenation of CO, 95
  in hydrogenation of olefins, 37, 39, 42, 44
  in olefin isomerization, 32
  $Rh(CO)(H)(PPh_3)_3$, 7, 89
  $[Rh(CO)_2I_2]^-$, 80, 94
  $RhCl(PPh_3)_3$:
    catalysis, 32, 37, 39, 91, 153
    ligand dissociation, 12
    oxidative addition, 15
    polymer support, 228
    structure, 8, 10
  in shift reaction, 94
Ruthenium complexes:
  in arene hydrogenation, 143
  in hydroquinone synthesis, 162
  in isomerization, 33
  in olefin hydrogenation, 42
  $Ru(C_6Me_6)_2$, structure, 143
  $Ru_3(CO)_{12}$, photolysis, 230

Shift reaction, 82, 94, 160
Speier's catalyst, 72
Stilbene:
  metathesis, 174, 175, 177
Styrene:
  reaction with benzene, 124
  synthesis, 124, 174, 175
Suberic acid, 160
Sulfur dioxide:
  oxidation, 227
Synthesis gas, 77, 222

Tantalum complexes, 62, 172
Tellurium catalysts:
  ethylene oxidation, 113
Terephthalic acid, 200
  polyesterification, 209, 212
1,1,2,2-Tetrachloroethane, 158
Tin compounds:
  in esterification, 212
  in polyurethane synthesis, 215
Titanium complexes:
  in butadiene trimerization, 68, 70
  in esterification, 212
  in olefin polymerization, 49, 51, 54
  in polyurethane synthesis, 215
  $Ti(C_5H_5)_2CH_2AlClMe_2$, 173, 178
Toluene:
  carboxylation, 129
  oxidation, 175, 201
2,4-Toluenediisocyanate, 93, 215
p-Toluic acid, 129, 201
Transesterification, 106, 156, 209
Transition metals:
  acetylene complexes, 148
  alkylidene complexes, 171
  bonding characteristics, 5
  carbon monoxide complexes, 6, 13, 77
  diene complexes, 24
  dinitrogen complexes, 8
  olefin complexes, 22
  oxygen complexes, 187
  redox potentials, 186
  review articles, 19
Trichloroethylene, 158
9-Tricosene, 176
Tungsten complexes:
  in olefin metathesis, 174

Vanadium complexes:
  as epoxidation catalysts, 116
  in isomerization, 225
  in oxidation by HNO_3, 193, 194
  in polymerization, 53
  in H_2SO_4 synthesis, 227
Vinyl acetate:
  from acetylene, 156
  from ethylene, 105
  from synthesis gas, 97, 224
Vinylacetylene, 152, 154
Vinyl chloride, 111, 157
Vinyl esters, 106, 156

Wacker process, 101, 102
Water gas shift reaction, *see* Shift reaction
Wilkinson's catalyst, *see* $RhCl(PPh_3)_3$

Xylenes:
  hydrogenation, 140
  oxidation, 201
2,6-Xylenol:
  oxidation, 132

Zeise's salt, 22, 24
Ziegler catalysts:
  in arene hydrogenation, 140

in butadiene dimerization, 66
in copolymerization, 53
in hexadiene synthesis, 64
in hydrogenation, 38
in olefin dimerization, 60
in polymerization, 49, 51, 54
Zinc compounds:
in transesterification, 209